T0331236

Introduction to Econophysics

Introduction to Econophysics

Contemporary Approaches with Python Simulations

Carlo Requião da Cunha

CRC Press is an imprint of the
Taylor & Francis Group, an **informa** business

First edition published 2022
by CRC Press
6000 Broken Sound Parkway NW, Suite 300, Boca Raton, FL 33487-2742

and by CRC Press
2 Park Square, Milton Park, Abingdon, Oxon, OX14 4RN

CRC Press is an imprint of Taylor & Francis Group, LLC

Library of Congress Cataloging-in-Publication Data

Names: Requião da Cunha, Carlo, author.
Title: Introduction to econophysics : contemporary approaches with Python simulations / Carlo Requião da Cunha.
Description: 1st edition. | Boca Raton, FL : CRC Press, [2022] | Includes bibliographical references and index.
Identifiers: LCCN 2021020191 | ISBN 9780367648459 (hardback) | ISBN 9780367651282 (paperback) | ISBN 9781003127956 (ebook)
Subjects: LCSH: Econophysics. | Economics--Statistical methods. | Python (Computer program language)
Classification: LCC HB137 .R47 2022 | DDC 330.01/5195--dc23
LC record available at https://lccn.loc.gov/2021020191

ISBN: 978-0-367-64845-9 (hbk)
ISBN: 978-0-367-65128-2 (pbk)
ISBN: 978-1-003-12795-6 (ebk)

DOI: 10.1201/9781003127956

Typeset in Nimbus font
by KnowledgeWorks Global Ltd.

http://www.filosofisica.com/books/econophysics.html

Dedication

---

for Marco, Mônica, Ingrid, Sofia, and Arthur

Contents

Preface

It was 2017 when I was approached by a group of undergraduate students asking me to offer an elective course in econophysics. The course was listed in the university catalog but had never been offered before. As I was progressively being attracted to intersectional areas of physics I took the challenge.

The first obstacle that I had to overcome was the limited number of resources. Econophysics is still a relatively new field and there were not many books available on the topic if compared to books about well-established fields such as quantum mechanics. Moreover, most of the books about econophysics that I had the opportunity to find were aimed at graduate level or above. On the other hand, the public demanding the course had barely completed their first course on statistical mechanics. This made me structure the discipline with adjacent topics to make the subject easier to tackle by undergraduate students. Finding all these topics in the same book, however, was impossible for me at that time.

It was also important to motivate the students with real-life examples before introducing some mathematically intense theories. Therefore, the book often begins a chapter with a real story related to the subject and many applications are discussed throughout the chapters. Furthermore, brief code snippets are included to show how sophisticated theories can be translated and simulated in a current computer language: Python. In many cases, compiled languages such as C and Fortran would render faster simulations, but Python was chosen for its simplicity and also for the opportunity it offers for the students to advance in a language that is highly demanded by today's industry.

The book was written in two languages: English and mathematics. Although many authors prefer to produce more conceptual materials, I advocate that quite often math can speak for itself and say much more than a spoken language. The only constraint is that the reader must know how to read this language. Nonetheless, this book was written for physics students, and, although there is some good level of math, it lacks the deep formal rigor that a mathematician would expect. Also, since the book was written with undergraduate students in mind, many derivations are explicitly shown. I also found it the perfect opportunity to introduce some advanced math that physics students would not typically encounter in an undergraduate program such as measure theory, stochastic calculus, and catastrophe theory. Being firstly aimed at undergraduate students, though, it does not mean that the book cannot be used by graduate students and researchers in general. On the contrary, the material developed here had also been used in a graduate level course in physics at UFRGS.

Econophysics, as it will be discussed, tries to build an interface between physics and economics. The latter, though, unlike physics, can have many different schools of thought. Typically, the school of economic thought an author tries to connect with is not made explicit in most materials about econophysics. Schools of thought that experimentally failed are vehemently discarded in this book, unless they are useful to highlight improved theories from other schools. Consequently, it is not by mere chance that the reader will find many references to the Austrian school of economic thought. Although it is not perfect, this school establishes a good link between physics and economics as it will be discussed along the book.

Finally, I support Paul Feyerabend's view that science, its history and philosophy, make an inseparable triad. When one fails, the other two are also compromised and knowledge eventually halts. Therefore, the reader will notice several endnotes with very short biographies about some of the authors who helped construct the theories we have today.

PLAN OF THE BOOK

This book was primarily designed to be useful for those learning econophysics. Therefore, every major theory necessary to understanding fundamental achievements in econophysics are included. Most of the derivations are present but those that are not are skipped because they are well beyond the scope of the book. All snippets are tested and should reproduce most of the presented figures. Moreover, several appendices are included to help the reader dive more deeply into some topics.

The book begins with a conceptual review of economics drawing some links with physics. Still on the first chapter the efficient market hypothesis is introduced and some strategies with derivatives are used to illustrate the behavior of investors.

Chapter two mixes probability theory with some concepts used in econophysics such as log-returns. The reader will become familiarized with mathematical terminology such as sigma-algebras and probability spaces but will also study how probability distributions can have curious behaviors such as heavy tails. The Pareto distribution is used to discuss some common topics in economics such as the Lorenz curve and the Gini index. The chapter continues by exploring some basic econometrics for financial time series. Particularly, the GARCH model is discussed making a bridge to stylized facts.

Chapter three deals with stochastic calculus for physicists. It begins with the connection between martingales and fair games and moves to Markov chains. The latter is used to model a three-state market that can be bullish, bearish or *crabish*, a concept that appeared in one of the classes and was joyfully adopted. Mean reverting processes are then discussed bringing the Langevin equation to the table. The chapter ends with point processes that are currently being used to model some economic problems such as the Epps effect. This is another mathematically intense chapter that explores the fundamental properties of each of the models. Although it is not strictly about the intersection of economics with physics, it holds the fundamental knowledge to deal with some important concepts in later chapters.

Chapter four uses many concepts derived in the previous chapter to study options pricing. It begins defining Wiener processes and introduces the Fokker-Planck equation. Two important models are then discussed: binomial trees and the Black-Scholes-Merton model. Some additional stochastic calculus is used throughout the chapter.

Chapter five introduces the concept of portfolio. It begins with the well-known Markowitz portfolio and CAPM models and then makes a connection between portfolios and random matrix theory, speculating if some of the tools used in quantum mechanics can be used to study these portfolios.

Chapter six is dedicated to crises, cycles and collapses. It begins with a review of the Austrian business cycle theory and then advances to catastrophe theory, a topic that finds connections with phase transitions in physics, but is uncommon in most undergraduate programs. The chapter closes with self-organized criticality under a cellular automata approach.

The next topic, game theory, is shown in chapter seven mainly through examples without dwelling too much on its mathematical formalities. Some famous examples such as the minority, cooperative and coordination games are discussed. By the end of the chapter, we discuss evolutionary game theory and the prey-predator model.

This book ends with a treatment of agent-based simulations. First, an introduction to complex networks, their properties and models are presented. Some important socioeconomic models that resemble the famous Ising model are then presented. The chapter closes with some brief introduction to interacting multiagent systems in a Boltzmann equation approach.

They say that "an author never finishes a book, he only abandons it," and I couldn't agree more! Non-linear effects and chaos theory, for instance, are currently important topics in econophysics. Nonetheless, these and many other topics are left for future editions of this book.

ACKNOWLEDGMENTS

I am most grateful to friends at Arizona State University where parts of this book were written. I particularly acknowledge with gratitude the hospitality of Ying-Cheng Lai in his group and am indebted to David Ferry who indirectly helped to bring this project up to life. I am also grateful to the AFOSR for financial support under award number FA9550-20-1-0377.

Carlo Requião da Cunha
Porto Alegre, March 2021

1 Review of Economics

Economics is a science that deals with complex systems. These systems are not composed of inanimate particles, but of human beings that have their own feelings, desires, that take action. This book analyzes whether these systems have any particular dynamics for which we can use the tools developed for studying physical problems. The theory that emerges from this endeavor is called **econophysics** and explicitly develops in a *per analogiam* fashion [2].

The idea of tackling problems of economics from the perspective of physics is as old as Copernicus[1]. Copernicus proposed, for instance, the *quantity theory of money* (QTM) that states that prices are functions of the money supply. Copernicus also proposed the *monetae cudendae ratio* that states that people tend to hoard good money (undervalued or stable in value) and use bad money (overvalued) instead[2]. Quetelet[3] was another scientist that went even further coining the term 'social physics' for his work that used statistical methods for social problems.

The same happened in the opposite direction with social scientists trying to use early concepts of physics in their works. For instance, Comte[4] was a philosopher who wanted to model social sciences as an evolution of the physical sciences. Even Adam Smith[5] is known to have been inspired by Newton's[6] works.

It was much more recent that the term *econophysics* was coined and used by physicists such as Stanley[7] and Mantegna[8] when using the tools of statistical mechanics to solve problems related to finances [3,4].

There are some concepts developed in economics that make a good connection with physics. The first one is the concept of *methodological individualism* developed by Weber[9] and further explained by Schumpeter[10]. There the focus of analysis of social problems is put on the action-theoretical level, rather than in the whole. It is not to be confused, though with atomism. This is a different term adopted by some economists, including Menger[11]. Atomism is based on the idea that once we fully understand the individual psychology, it would possible to deduce the behavior of group of individuals. Rather, methodological individualism does not rely on a reduction of the social sciences to psychology. It perceives social phenomena as emerging from the action of individual agents. Moreover, these phenomena can emerge even as *unintended consequences* of purposeful individual actions[12]. This will prove useful, for instance, when we study game theory and agent based models throughout this book.

Dispersed knowledge is another economic concept that we will encounter in our study of econophysics. As the agents individually take action, they produce a knowledge that is dispersed among individuals and no agent has access to all this information at every instant [5]. Although, it may seem obvious, it has huge implications such as the production of genuine uncertainty. On the other hand, this dispersed knowledge can produce regular large-scale consequences constituting *spontaneous order*. Hayek[13] provides the example of a virgin forest [6]. The first person walking

DOI: 10.1201/9781003127956-1

through this forest creates small modifications that helps other individuals cross this forest. Eventually, as a significant number of individuals cross this forest, a path is naturally formed. The same process happens with prices. Through the interaction of buyers and sellers, prices are created without the need of any central planner. This is known by economists as a *price discovery process*. Moreover, as Hayek points out, prices carry information that helps agents to coordinate their separate actions towards the increase and decrease of production and consumption. This spontaneous "order brought about by the mutual adjustment of many individual economies in a market" is called *catallaxy* by Hayek [7].

In order to begin our study of econophysics, let's review three more important concepts in economics that will help us build a solid groundwork: supply and demand, the efficient market hypothesis and the concept of stocks and derivatives.

1.1 SUPPLY AND DEMAND

Imagine that you are starting a business. As an entrepreneur you have to pay employees, you probably have to pay bank loans and also need to make some profit for yourself. Thus, it is likely that you are willing to sell your products at a high price. Just like you, other entrepreneurs want to do the same. The information about the price is not absolute, but dispersed in the society. Therefore, some entrepreneurs may advertise their goods at lower prices and there can even be the cases where the entrepreneurs need to sell a product as soon as possible. Thus, it is expected that many entrepreneurs are willing to sell a product at high prices whereas few entrepreneurs are willing to sell the same product at low prices.

The situation is the opposite for costumers. They have a list of needs and limited money. However, the information is also not certain for them. A product that some consider expensive might be considered not so expensive by others. Also, there can be situations where a product is needed no matter the cost. Therefore, it is expected that many customers are willing to buy a product at low prices whereas not so many customers are willing to buy the same product at high prices.

This competition between agents has captured the attention of many economists from John Lock[14] [8] to Adam Smith [9]. It was Cournot[15] [10] and Marshall[16] [11], however, that formalized supply and demand curves as shown in Fig. 1.1.

Let's see the supply and demand curves in action. Figure 1.1 illustrates a situation where the price is fixed at a high price P_s. In this case, only a few consumers Q_{ds} are willing to pay this price and many entrepreneurs Q_{os} are willing to sell. However, when sales do not materialize, the entrepreneurs are encouraged to lower their prices. On the other hand, not so many entrepreneurs can produce or sell the product at this low price. Therefore, O_{os} tends to move to Q_e and since more consumers are expected to agree with this new price, Q_{ds} is pushed towards Q_e.

Any deviation from this 'equilibrium point' tends to produce stimuli that encourages both the consumers and entrepreneurs move back to it. For instance, if there is a movement towards a lower price P_i, fewer entrepreneurs Q_{oi} are encouraged to produce, but there would probably be many consumers Q_{id} willing to buy. This pushes entrepreneurs to try to satisfy these potential consumers. Thus, chances are that the entrepreneurs observe a rise in demand and increase their productions and prices. Once again the tendency is that Q_{oi} and Q_{di} move towards Q_e, and P_i moves towards P_e.

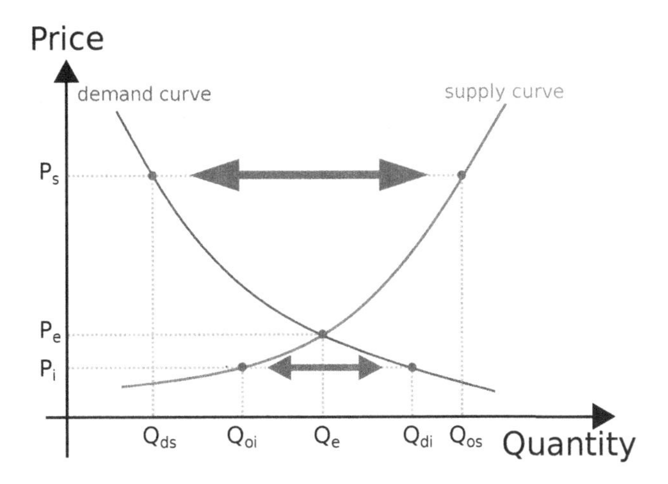

Figure 1.1: Supply and demand curves showing an equilibrium point and the situation where there are a) low and high production, and b) low and high prices

At this point, supply and demand are in equilibrium. We can intuitively notice that this point may not be stable and oscillations can occur due to a set of reasons such as information imperfection as we discussed before. Nonetheless, the incentive is such that a free economic system always tends to move towards the point Q_e, P_e. We also notice that the price works as a *system of information* that allows entrepreneurs to adjust their production and prices according to the demand that exists for their products.

The cobweb (or Verhulst[17]) diagram is a tool borrowed from the field of dynamical systems [12] that can be used to visualize this behavior. The model proposed by Kaldor[18] [13] is used to explain the price oscillation in some markets. We start with a standard supply and demand diagram as shown in Fig. 1.2 and consider a hypothetical scenario where farmers have a small production but the demand for their products is high. This automatically leads to an increase in prices. This raise in prices encourages the farmers to increase their productions to the next period. When this next period comes, the high production leads to a competition among farmers that want to sell their productions resulting in a lowering of the prices. This dynamics continues successively until an equilibrium situation is achieved (intersection of the curves). This is known as a *convergent* case. However, it is also possible to conceive a *divergent* case where the oscillation increases and a *limit cycle* case where the oscillation is constant in amplitude.

In order to have a convergent system, it is necessary that the demand be more *elastic* than the supply in the equilibrium point Q_e, P_e. Elasticity is a measure of percentual sensitivity of one variable with respect to another. The elasticity of demand

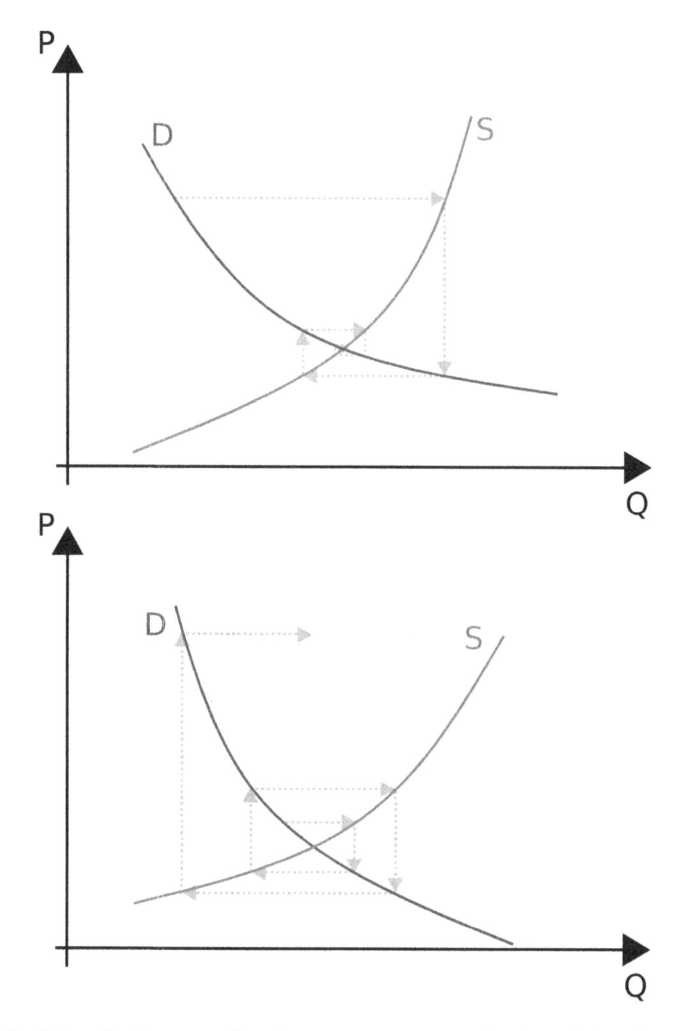

Figure 1.2: Cobweb diagrams for the convergent case (top) and the divergent case (bottom)

(d) or supply (s) is given by:

$$e_{d,s} = \frac{dQ^{d,s}/Q}{dP^{d,s}/P}. \tag{1.1}$$

Economists define $|e| > 1$ as an elastic case. This implies that the change of quantity is higher than the change in price of a certain product. The opposite case where $|e| < 1$ is known as inelastic.

Therefore, in order to have a convergent case, it is necessary that:

$$\left|\frac{dQ^s/Q}{dP^s/P}\right|_{Q_e,P_e} < \left|\frac{dQ^d/Q}{dP^d/P}\right|_{Q_e,P_e}. \tag{1.2}$$

This can be visualized in the diagrams of Fig. 1.2. If the absolute value of the slope of the supply curve is higher than that of the demand curve at the equilibrium point then the system is convergent.

This adaptation of the behavior of the economic agents according to the events is known as *adaptive expectations*. This theory, however, suffers many criticism, specially when one considers the divergent case. Why would the entrepreneurs insist in an action that has made their situations only worse? This is the main objection of Muth[19], for instance. Notwithstanding, many price oscillations have been explained using this theory. Mathematically, an adaptive expectation could be modeled as:

$$p_{n+1}^e = p_n^e + \lambda(p - p_n^e), \tag{1.3}$$

where p_n^e is the price expectation for the the nth period whereas p is the real price. This, for instance, explains the *monetary illusion* pointed by Fisher[20] [14]. This is a phenomenon where the agents confuse the face value of money by its past purchase power. Fiat currencies nowadays have no convertibility to any finite reserve, no intrinsic value or any use value. Therefore the purchase power assigned to them has to be based on past experienced levels.

The adaptive expectations theory is currently understood as a particular case of a wider theory developed by Muth and Lucas[21], known as *theory of rational expectations*. In their theory, all agents shape their expectations of future economic indicators based solely on reason, the information available to them, and their past experiences [15]. This theory asserts that although people may eventually be wrong about those indicators, on the average they tend to convert to the correct values because they learn from past mistakes and adapt. For example, if an inflationary policy (expansion of the monetary base) was practiced and that lowered unemployment, the same result may not happen in the following year if the inflationary policy is practiced again. According to Muth and Lucas, that would be because people adjust their expectations immediately after the first inflationary policy. The relationship between the inflation rate and the unemployment rate is actually known as *Phillips curve*[22] and the rational expectations theory predicts that it is inelastic in the short-term.

More recently, studies by researchers such as Tversky[23], Kahneman[24], Shiller[25] and Thaler[26] have pointed out some flaws in the theory of rational expectations. For instance, agents are not perfectly rational and have a *bounded rationality*[27] [16] also limited by dispersed and incomplete information. Furthermore, as stated before, the agents are humans that have feeling, passions and often biases and discrimination [17]. Sometimes these agents can also be carried out by suggestions (nudge effect [18]) or conform with a herding behavior [19]. The theory that emerges from this program of merging the psychology of agents with economics is known as *behavioral economics*. We will return to this topic in Chapter 8 when we study agent-based models.

1.2 EFFICIENT MARKET HYPOTHESIS

It was perhaps Regnault[28] [20] who first suggested that the stock prices followed a random walk. This hypothesis was further developed by mathematicians such as Bachelier[29] [21] and Mandelbrot[30] [22] but it was with Fama[31] [23] and his adviser Samuelson[32] [24] that it became popularized and received the name efficient market hypothesis (EMH).

According to this hypothesis, a market is considered efficient when the available information is fully incorporated into the prices. In this case, the price should follow a random walk and by doing so past trends do not tell us anything about the current price of a stock. Consequently, predicting or beating the market should be impossible because the market incorporates all available information and that is arbitraged away. Indeed, if a stock price follows a random walk, the price increment should follow a Gaussian distribution and its expected value should always be zero as we will further explore later in this book. If markets were perfectly efficient, a consequence would be that the best investment scenario would be adopting passive portfolios that would only profit by the long term natural growth of the market.

The EMH was actually categorized by Fama in three formulations. In its weak formulation, only public market information is known by the agents, whereas in the semi-strong formulation stock prices quickly adjust to reflect any new information. Finally in the strong formulation of the EMH all public and private information are immediately incorporated into prices such that not even inside trading should have any effect on prices.

If markets were perfectly efficient, however, it would be impossible for bubbles to exist since the information about the occurring bubble would be immediately incorporated into the prices. Also, if markets were efficient, there would be no reason for investors to trade and markets would eventually cease to exist. This is known as the *Grossman[33]-Stiglitz[34] paradox* [25]. Furthermore, it fails to explain the *neglected firm effect* where lesser-known firms can produce higher returns than the average market. This can be explained, though, by the fact that information about these firms is not much available. This can be used by investors to demand a premium on the returns. Moreover, it is simple to argue that agents are plural and have different views about information. Therefore, they value stocks differently. Also, stock prices can be affected by human error. And finally, prices can take some time to respond to new information. Therefore, agents who first use this new information can take some advantage.

Nonetheless, as the quality of information increases with technology, more efficient the market tends to become and less opportunities for arbitrage tend to happen.

1.3 STOCKS AND DERIVATIVES

Let's consider that you have a business. You observe an increase in sales but you need more capital to expand your business. One way of solving this problem is by adopting a strategy used since the Roman republic: you can issue *stocks* or *shares* (units of stocks). This means that you are selling small portions of ownership of your business, that is now public, to raise capital. As your business makes profit, you now have to share these *dividends* with your new partners or *company shareholders*. These shares are also now negotiable in an *equity market* or *stock market* and constitute one type of *security* together with other *financial instrument* such as cash and bonds .

Shares can be *ordinary shares* (common shares) that allows holders to vote or *preference shares* (preferred stocks) that guarantee preference in receiving dividends. These securities are usually negotiated in multiples of a fixed number (typically 100). This constitutes a *round lot*. Fewer shares can also be negotiated and these fractions of a standard lot are known as *odd lot*.

Now that the capital of your business is open, you as a businessman have the fiduciary responsibility to operate the company according with not only your interests but the interests of the shareholders.

The administrative board may now raise more capital to a further expansion. This time, however, the board does not want to issue stocks, rather they want to issue debt payable in the future with some interest. These are called *debentures* and are not backed by any collateral. In order to reduce the risk, some agencies offer credit rankings to help investors access the associated risk.

You want now to purchase a specific quantity of a commodity, say titanium, as raw material for your business. However, for some reason you can only purchase it three months from now. Your income is in one currency, but the price of this commodity is negotiated in another currency. The exchange rate between the two currencies oscillates violently and you are concerned that the price of the commodity is going to be much higher in the delivery date. In order to fix the purchase price of the commodity, you can issue a purchase contract executable in three months with a predetermined price. This is an example of a *forward contract* and it can be used as a *hedge* to protect your business against uncertainty. This kind of financial security whose value is derived from another asset is known as a *derivative*.

Instead of fixing the price of the commodity (spot price), it is possible to daily adjust it in order to obtain a better approximation to the spot price. This is called a *future contract* and it is negotiable in future markets. In order to operate futures, an initial *margin* amount is necessary. If you are long positioned in an operation and the future price increases 1 %, your account receives the respective amount. On the other hand, if you are short positioned, this amount is subtracted from your account. If your company is producing a new product that will only be sold in the future, it may be interesting to sell a future contract and invest the money in some transaction indexed to the inflation rate, for example.

We are considering that your business sells products in one currency but purchase raw materials in another. There may be other businesses whose major concern is not the exchange rate but the inflation rate. One of these businesses may even believe that the exchange rate will not be as bad as the inflation rate in the future. You two can then make a contract exchanging the debt indexes. This kind of operation where two parties exchange the liability of different financial instruments is known as *swap*.

1.3.1 OPTIONS

It is also possible to purchase a contract that gives you the right to buy or sell an asset at a specific price in the future (*exercise price* or *strike*). When you purchase the right to buy a product at a specific price, this is called *call option*, whereas the right to sell a product is called *put option* [26]. If the current price at the exercise date is higher than the strike price, one can execute the right to buy the asset at the lower value. On the other hand, if the current price at the exercise date is lower than the strike price, one can execute the right to sell the asset at the higher value. Both situations correspond to options *in the money* (ITM). If the current price and the strike are the same, we say this option is *at the money* (ATM). Also, if the transaction is not favorable, though, then the option is *out of the money* (OTM). Unlike future contracts, you are not required to execute an option. Nonetheless, you have to pay a premium for the option.

If the option you purchased can only be exercised on the expiration date, this is called *European option*. If, on the other hand, you can execute it any time before the expiration date, then it is called *American option*. There are also options whose payoff is given by the average price of the underlying asset over a period. These are called *Asian options*.

A call option for a hypothetical stock has a strike of $40 and this stock is trading at the exercise day for $42. In this case we say that its *intrinsic value* is $2. If the option itself is trading at $3 then there is also $1 of *extrinsic value*.

Suppose now, as an investor, you believe that the price of a stock is going to fall. It would be advisable to sell some of these stocks and repurchase them for lower prices, but you don't have these stocks and don't have enough funds to buys these stocks right now. It is still possible to sell these stocks without having them in an operation known as *short selling*. In this operation the investor borrows shares from a lender and immediately sells them, say for $100 each. The shares eventually become cheaper and the investor buys the same shares for $75 each. The investor returns the shares to the lender and profits $25 for each share. Had the price of the shares increased, the investor would rather incur in financial loss. On the other hand, the same operation could have been executed with options if the investor had purchased the right to sell the asset. Thus, options can be used to reduce the risk of financial operations.

The opposite operation, where the investor expects an increase in the price of the stock is known as *long buying*. Consider an *initial public offering* (IPO) where a company starts negotiating stocks. As an investor you expect that these stocks will start being negotiated at low values but you don't have enough funds to buy them. An investor can buy these stocks and sell them before having to pay for them. If the price really increases, the investor can profit with the operation.

1.3.1.1 Strategies with Options

There are some strategies that investors undertake to try to profit with options. Let's consider a stock trading at $40. An associated call option with a strike of $40 has a

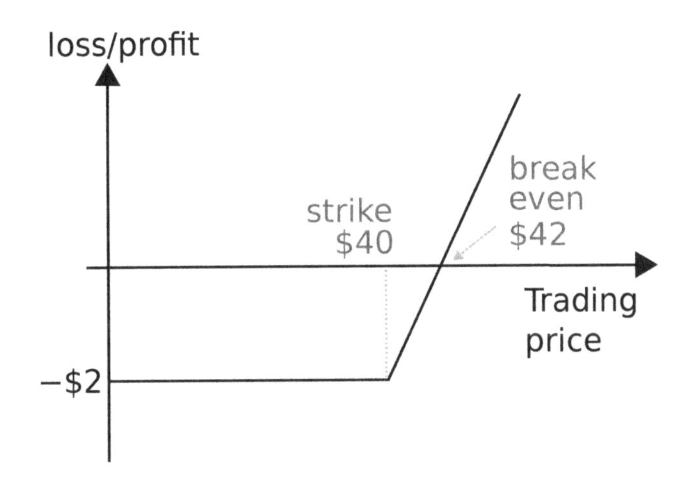

Figure 1.3: Payoff diagram for a long call operation

premium of $2. An investor can buy $200 in options, which gives the right to buy 100 shares in the exercise day. When this day comes, the stock is trading at $50. The investor executes the option and purchases a round lot of 100 stocks paying a total of $4,000. The investor immediately sells this lot for the current price, making $5,000. Therefore, the investor make $800 in profit.

This strategy, shown in Fig. 1.3 is known as *long call*. The investor starts off with a loss of $2 paid for the premium of the option. When the negotiable price of the stock reaches $40, the loss starts to decrease given the difference between this price and the strike. When the negotiable price reaches $42, the *break even point* is reached and the investor makes profit for any higher price.

On the other hand, if the investor believes that the trading price of this stock will fall, then $200 can be purchased in put options (same 2$ premium). If the stock is trading at $40 at the exercise day, the investor does not execute the right to sell the stocks and loses the premium. If, however, the trading price is lower than $38, say $28, the investor buys the stock at the spot market and executes the option to sell the stocks at $40. Therefore, the investor profits $40 - $28 - $2 = $10 per stock. This is known as *long put* and is shown in Fig. 1.4.

It is also possible to combine both strategies purchasing a put and a call options. If the trading price at the exercise day is higher than $42 or lower than $38 the investor executes one right ignoring the other. This strategy is known as *long straddle* and is shown in Fig. 1.5.

As this book was being written, something unusual happened. The popularization of e-commerce has turned many businesses based on physical stores obsolete. Following this tendency, many institutional investors have betted against GameStop (GME)[a] short selling it. The intention was of buying the shares back (*cover*) when

[a]A traditional video games store.

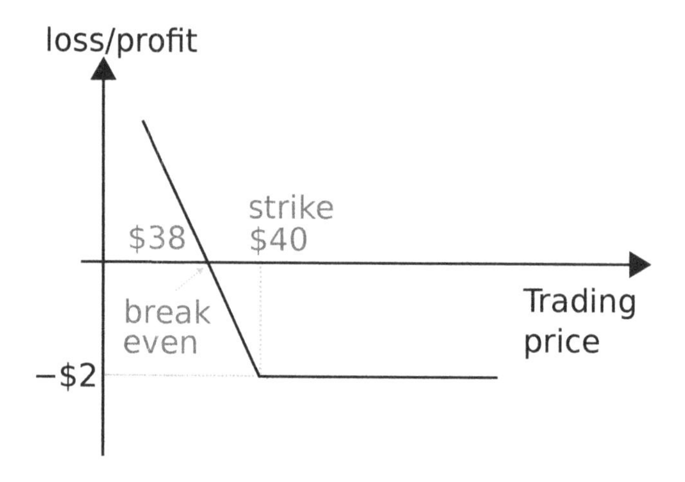

Figure 1.4: Payoff diagram for a long put operation

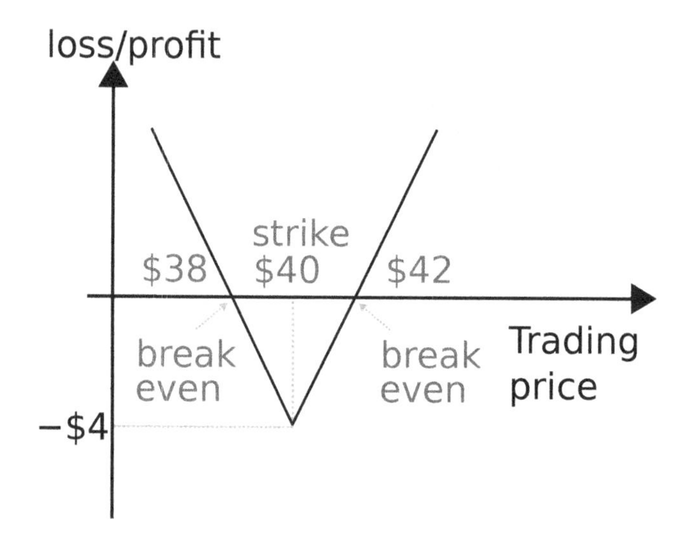

Figure 1.5: Payoff diagram for a long straddle operation

they would be negotiating at a lower price, then return the stocks to their original owners and pocket the difference. Many contrarian investors, however, felt that was an unfair game and coordinated themselves over the internet to start a rapid purchase of these stocks—a *short squeeze*. The price of GME increased over 5,000 % in a few days (see Fig. 1.6) forcing institutional investors to abandon their positions and accept considerable losses. Some hedge funds incurred in losses over 3 billion dollars as a result of this operation. This not only is a risky operation as it clearly invalidates the EMH.

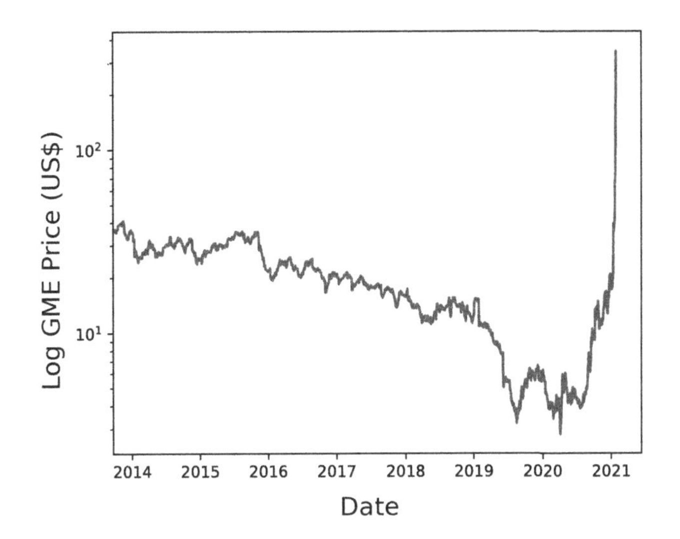

Figure 1.6: Negotiated price of GME over time obtained from Yahoo Finance

Notes

[1]Nicolaus Copernicus (1473–1543) Prussian astronomer.

[2]This is known today as Gresham's law in reference to Sir Thomas Gresham the Elder (1519–1579) English merchant.

[3]Lambert Adolphe Jacques Quetelet (1796–1874) Belgian astronomer, mathematician, and sociologist.

[4]Isidore Marie Auguste François Xavier Comte (1798–1857) French philosopher.

[5]Adam Smith (1723–1790) Scottish economist and philosopher.

[6]Isaac Newton (1643–1727) English natural philosopher.

[7]Harry Eugene Stanley (1941–) American physicist.

[8]Rosario Nunzio Mantegna (1960–) Italian Physicist.

[9]Maximilian Karl Emil Weber (1864–1920) German sociologist, philosopher, and political economist.

[10]Joseph Alois Schumpeter (1883–1950) Austrian economist, advisee of Eugen Böhm von Bawerk, adviser of James Tobin and Paul Samuelson among others.

[11]Carl Menger (1840–1921) Austrian economist.

[12]Term popularised by Robert King Merton (1910–2003) American sociologist.

[13]Friedrich August von Hayek (1899–1992) Austrian economist, Nobel laureate in 1974. Advisee of Friedrich Freiherr von Wieser.

[14]John Locke (1632–1704) English philosopher.

[15]Antoine Augustin Cournot (1801–1877) French mathematician and economist.

[16]Alfred Marshall (1842–1924) English economist.

[17]Pierre François Verhulst (1804–1849) Belgian mathematician, creator of the logistic function.

[18]Nicholas Kaldor (1908–1986) Hungarian economist, advisee of John Maynard Keynes and Gunnar Myrdal.

[19]John Fraser Muth (1930–2005) American economist.

[20]Irving Fisher (1867–1947) American economist and statistician, advisee of Josiah Willard Gibbs.

[21]Robert Emerson Lucas Jr. (1937–) American economist, Nobel laureate in 1995.

[22]Alban William Housego Philips (1914–1975) New Zealand economist.

[23]Amos Nathan Tversky (1937–1996) Israeli psychologist.

[24]Daniel Kahneman (1934–) Israeli psychologist and economist, Nobel laureate in 2002.

[25]Robert James Shiller (1946–2013) American economist, Nobel laureate in 2013.

[26]Richard H. Thaler (1945–) American economist, Nobel laureate in 2017.

[27]Proposed by Herbert Alexander Simon (1916–2001) American economist, political scientist and psychologist, Nobel laureate in 1978.

[28]Jules Augustin Frédéric Regnault (1834–1894) French stock broker.

[29]Louis Jean-Baptiste Alphonse Bachelier (1870–1946) French mathematician, advisee of Henri Poincaré.

[30]Benoit B. Mandelbrot (1924–2010) Polish mathematician. Adviser of Eugene Fama among others.

[31]Eugene Francis Fama (1939–) American economist, Nobel laureate in 2013.

[32]Paul Samuelson (1915–2009) American economist, Nobel laureate in 1970. Advisee of Joseph Schumpeter and adviser of Robert Cox Merton.

[33]Sanford Jay Grossman (1953–) American economist.

[34]Joseph Stiglitz (1943–) American economist, advisee of Robert Solow. Stiglitz was awarded the Nobel Memorial Prize in Economic Sciences in 2001.

2 Asset Return Statistics

The reported relationship between a security and another (typically fiat money) as a function of time is known as *ticker*. A "tick" then is understood as the price change the security undergoes and the ticker can include information about the volume negotiated, opening and closing prices among other information. A typical movement of the price of a security is shown in Fig. 2.1. The top left graph shows a situation where the price has an upward tendency and the closing price at the period is higher than the opening price. We call this tendency a *bull market*, whereas the opposite is known as a *bear market*. This happens in analogy with how these animals attack their preys either upwards or downwards. We should also add a situation where the opening and closing prices are very similar. Following the other analogies we shall call this a *crab market* since crabs walk sideways.

In order to better represent the information contained in a particular period we can use a candlesticks representation. Typically, the body part of a candlestick may indicate a bullish market period if it is hollow or green, or it can indicate a bearish market if it is solid black or red. Also the maximum and minimum prices reached during the period are indicated by the points of maximum and minimum of the vertical sticks on top and bottom of the body. A candlestick chart for longer periods can be created in a way similar to a renormalization group procedure.

We will start this chapter discussing return rates and will move towards a mathematical description of the fluctuations found in these returns along time. We will close this chapter studying time series and a set of *stylized facts* found in these series.

2.1 RETURN RATES

As an investor, would you be more willing to buy a security for 100 thousand dollars expecting to sell it in a few days for 101 thousand dollars or buy another security for 10 dollars expecting to sell it for 100 dollars? Certainly absolute prices are important, but in order for us to make better investment decisions, we must know what the expected *return rates* for the investments are. A linear return rate is given by:

$$
\begin{aligned}
r(t) &= \frac{p(t+\Delta) - p(t)}{p(t)} \\
&= \frac{p(t+\Delta)}{p(t)} - 1,
\end{aligned}
\tag{2.1}
$$

where $p(t)$ is the price of the security at time t and Δ is some period. For the example given, the first investment has a linear return rate of 1 % whereas the second investment has a linear return rate of 900 %. Although the absolute prices for the second investment are lower, the associated return is much higher than that of the first investment.

DOI: 10.1201/9781003127956-2

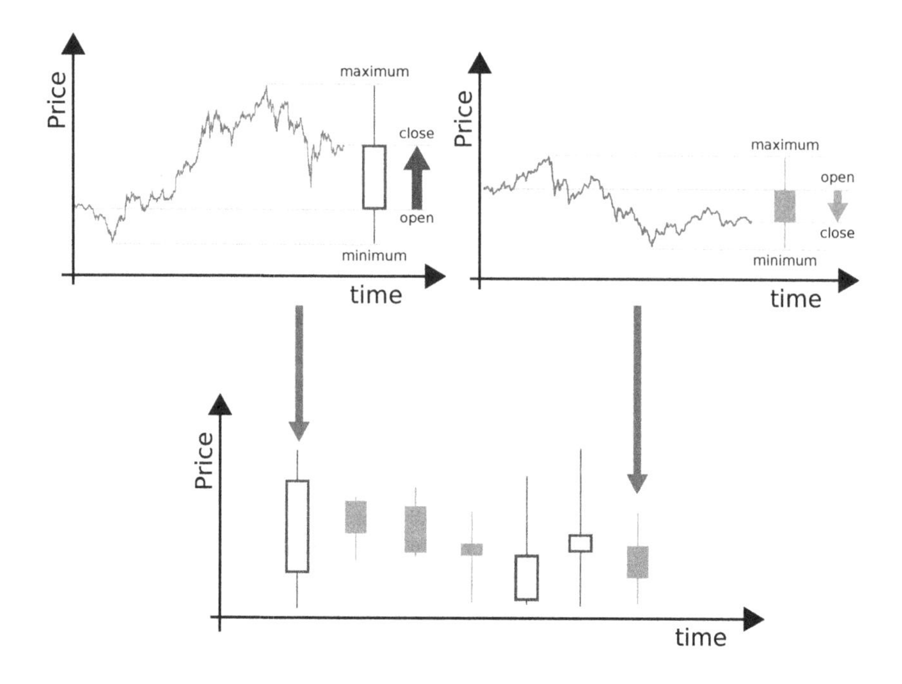

Figure 2.1: Examples of a bullish market (top left) and a bearish market (top right). A candlestick chart is shown in the bottom

This example also shows one problem. We are trying to compare two investments with returns with orders of magnitude of difference. One strategy that is used to deal with this situation is to use *logarithmic returns* instead. In order to obtain the log-return, let us first add one to Eq. 2.1 and then take its log:

$$\log(r(t)+1) = \log\left(\frac{p(t+\Delta)}{p(t)}\right). \tag{2.2}$$

The logarithmic is an analytic function[a] and so it can be represented by a Taylor[1] expansion around some $r(t)$:

$$\begin{aligned} \log(r(t)+1) &= \sum_{n=1}^{\infty}(-1)^{n+1}\frac{r^n(t)}{n} \\ &= r(t) - \frac{r^2(t)}{2} + \frac{r^3(t)}{3} - \dots \\ &\approx r(t) \text{ in a first order approximation.} \end{aligned} \tag{2.3}$$

[a]Basically, an infinitely differentiable function at any point in its domain.

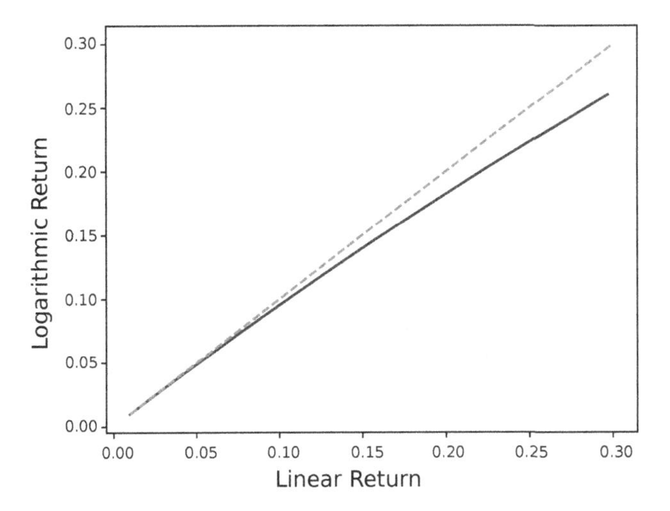

Figure 2.2: The logarithmic return (solid curve) as a function of the linear return. The dashed line shows the linear return as a function of itself

Therefore, in first order, we can approximate:

$$
\begin{aligned}
r(t) &\approx \log\left(\frac{p(t+\Delta)}{p(t)}\right) \\
&\approx \log\left(p(t+\Delta)\right) - \log\left(p(t)\right).
\end{aligned}
\tag{2.4}
$$

If we set Δ so that $t+\Delta$ is the time at the next immediate period of negotiation, we can write the log-return as a finite difference equation:

$$r_n \approx \log(p_{n+1}) - \log(p_n). \tag{2.5}$$

This is a first order approximation and so it applies better for small return values. Plotting the log-return as a function of the linear return as shown in Fig. 2.2 makes it easy to see that there is a growing divergence for large values of the linear return. Another way to look at log-returns is as a periodic compound rate. In order to observe this, let's rewrite the log-return as:

$$p_{n+1} = e^{r_n} p_n = \lim_{m\to\infty}\left(1 + \frac{r_n}{m}\right)^m p_n. \tag{2.6}$$

This is the total accumulated price for periodic compounding where the return was chopped into infinite sub-periods (a continuous).

Let us now consider one investment where we buy a security for 100 dollars expecting to sell it for 101 dollars and a second investment where we buy another security for 100,000 dollars expecting to sell it for 101,000 dollars. Both have exactly the same return rate but different prices. Therefore, if we are interested in studying the details of the return oscillations, it is a good idea to normalize the prices of different assets to the same basis. We then define a *normalized return* as:

$$\tilde{r}_n = \frac{r_n - \langle r \rangle}{\sqrt{\langle r^2 \rangle - \langle r \rangle^2}}. \tag{2.7}$$

The expected value in the last expression is for the whole period under analysis. Although many authors use $E[x]$ for the expected value, we will use the notation $\langle x \rangle$ here and throughout this book. This is because E can be easily confused for some variable and the angle brackets notation is commonly used in physics.

The following snippet computes the log-returns in Python:

```
import numpy as np

def ret(x):
        N = len(x)

        y = []
        for i in range(N-1):
                r = np.log(x[i+1])-np.log(x[i])
                y.append(r)

return y
```

The normalized returns can be computed with:

```
def nret(series):

        m = np.average(series)
        s = np.std(series)

        return [(r-m)/s for r in series]
```

2.2 PROBABILITY THEORY

As it was shown in Fig. 2.1, prices and returns display irregular movements along time. Here we will discuss the basics of random variables and study how we can use them to better understand the dynamics of returns. We will adopt the axiomatic approach introduced by Kolmogorov[2] in 1933 [27].

The first concept that has to be introduced is the *probability space* [28, 29]. The probability space is defined by the 3-uple (Ω, Σ, P). Ω is the *sample space*, a set of all possible results of an experiment. Σ is a set of events called σ-algebra. Finally, P is a *probability measure* that consists of probabilities assigned to the possible results.

Example 2.2.1. Finding the set of events for flipping a coin.

If the coin is not biased, we have $\Omega = \{H, T\}$, where H is a mnemonic for heads and T is a mnemonic for tails. The set of events for this example is: $\Sigma = \{\{\}, \{H\}, \{T\}, \{H, T\}\}$, $P(\{\}) = 0$, $P(\{H\}) = P(\{T\}) = 0.5$ e $P(\{H, T\}) = 1$. ■

2.2.1 PROBABILITY MEASURE

The *probability measure*, as the name implies, is a *measure*[3]. A measure, in turn, is a function $\mu : \Sigma \to \mathbb{R}$ that satisfies:

1. $\mu(E) \geq 0$, $\forall E \in \Sigma$, non-negativity
2. $\mu(\emptyset) = 0$, null empty set, and
3. $\mu\left(\bigcup_{k=1}^{\infty} E_k\right) = \sum_{k=1}^{\infty} \mu(E_k)$, $\forall E_k \in \Sigma$ countable additivity.

Example 2.2.2. Lebesgue measure.

A Lebesgue measure of a set is defined as:

$$\mathscr{L}^1(A) = \inf\left\{\sum_{n=1}^{\infty}(b_i - a_i) : A \subseteq \bigcup_{n=1}^{\infty}[a_n, b_n]\right\}, \tag{2.8}$$

where *inf* is the *infimum* defined as the biggest element of an ordered set T that is smaller or equal to all elements of the subset S for which it is being calculated. Following the same rationale, the *maximum* can be defined as the smallest element of T that is bigger or equal to all elements of S. For example, there is no minimum in the positive real numbers ($\mathbb{R}^+$) since a number can always be divided in smaller values. On the other hand, the infimum is 0. ■

The probability measure P is a measure that also satisfies:

1. $P : \Sigma \to [0, 1]$ and
2. $P(\Omega) = 1$.

The tuple (Ω, Σ) is known as *measurable space* (or *Borel*[4] *space*) whereas the tuple (Ω, Σ, μ) is known as *measure space*.

2.2.1.1 Conditional Probability

The probability for an event A to happen given that another event B has happened is written as $P(A|B)$. If $P(A|B) = P(A)$, the events are *independent*. In this case, having information about one event tells us nothing about the other. In the opposite case, we can write the probability as the ratio between the probability of both events happening together and the probability of event B happening [29, 30]:

$$P_r(A|B) = \frac{P_r(A \cap B)}{P_r(B)}. \tag{2.9}$$

One way of visualizing conditional probabilities is with an Euler[5] diagram as shown in Fig. 2.3.

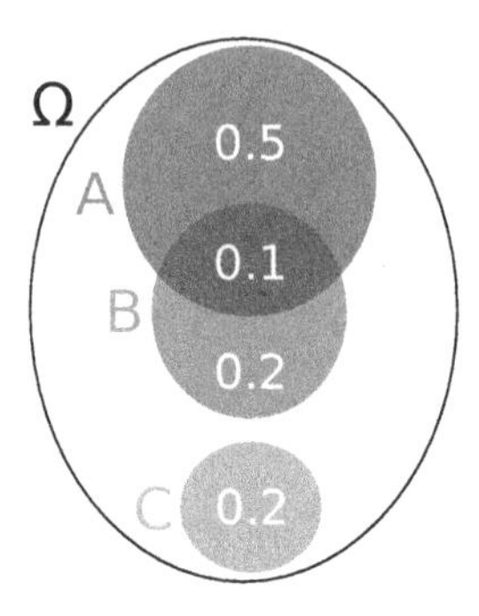

Figure 2.3: An example of an Euler diagram for probabilities

Example 2.2.3. Extract the probabilities from an Euler diagram.

Given the Euler diagram shown in Fig. 2.3 we can obtain the following probabilities:

$$\begin{aligned} P(A) &= 0.5 + 0.1 = 0.6 \\ P(B) &= 0.1 + 0.2 = 0.3 \\ P(A \cap B) &= 0.1 \\ P(A|B) &= 0.1/0.3 = 0.3333\ldots \end{aligned} \tag{2.10}$$

■

2.2.2 σ-ALGEBRA

If we know the probability of something happening we also have to know the probability of nothing happening. Therefore, a σ-algebra must contain the empty set and the sample space. Therefore, $\{\emptyset, \Omega\}$ is the smallest σ-algebra.

Let us consider the sample space $\Omega = \{H, T\}$ given in the example of Sec. 2.2 for flipping a coin. Considering a set of events $\Sigma = \{\emptyset, \Omega, \{H\}\}$ means that we know the probability of nothing happening, anything happening and the probability of obtaining heads. This, however, implies that we would not know the probability of obtaining tails, which is an absurd. How can we know the probability of obtaining head and not tails? Evidently, if we include H in the set of events, we must also include its complement T. Therefore, a σ-algebra must be closed under the complement.

Let us now consider a sample space $\Omega = \{1,2,3,4,5,6\}$ corresponding to the throw of a dice. Contemplating a set of events $\Sigma = \{\emptyset, \Omega, \{1\}, \{2\}\}$ would imply that we know the probability of obtaining 1 or 2, which is again an absurd. If 1 and 2 are in the set of events, then $\{1,2\}$ must also be present. Therefore, a σ-algebra must also be closed under countable union. Therefore, the biggest possible σ-algebra is the power set of the sample space. For the game of flipping a coin, that would be $\Sigma = \{\emptyset, \Omega, \{H\}, \{T\}\}$.

With these considerations, we can define a σ-algebra under a sample space Ω as the set Σ of the subsets of Ω that include the empty set and the sample space itself. Moreover, this set has to be closed under the complement and countable operations of union and intersection.

Example 2.2.4. Finding a σ-algebra.

For the sample space $\Omega = \{a,b,c,d\}$, one possible σ-algebra is $\Sigma = \{\emptyset, \{a,b\}, \{c,d\}, \{a,b,c,d\}\}$. ■

2.2.3 RANDOM VARIABLES

A random variable [28,31,32] can be considered as an observable output of an experiment that accounts for all of its outcomes according to certain probabilities if these outcomes are on a finite domain or to a density function if they are on an infinite domain.

More formally, a random variable in a probability space (Ω, Σ, P) is a function $X : \Omega \to \Omega'$ that is measurable in a Borel space (Ω', Σ'). This means that the inverse of a Borel set in Σ' is in Σ: $\{\omega : X(\omega) \in B\} \in \Sigma$, $\forall B \in \Sigma'$.

Example 2.2.5. The probability of finding exactly two consecutive heads when flipping three times the same coin.

The sample space in this case is $\Omega = \{(H,H,H), (H,H,T), (H,T,H), (H,T,T), (T,H,H), (T,H,T), (T,T,H), (T,T,T)\}$. Let X be a random variable associated with the number of H's in this game. This random variable X maps Ω to $\Omega' = \{0,1,2,3\}$, such that, for instance $X(\{T,H,H\}) = 2$ as shown in Fig. 2.4. Therefore, $P(X=2)$ is given by:

$$\begin{aligned} P(X=2) &= P\left(X^{-1}(2)\right) \\ &= P(\{(H,H,T),(T,H,H)\}) \\ &= 2/8 = 1/4. \end{aligned} \tag{2.11}$$

■

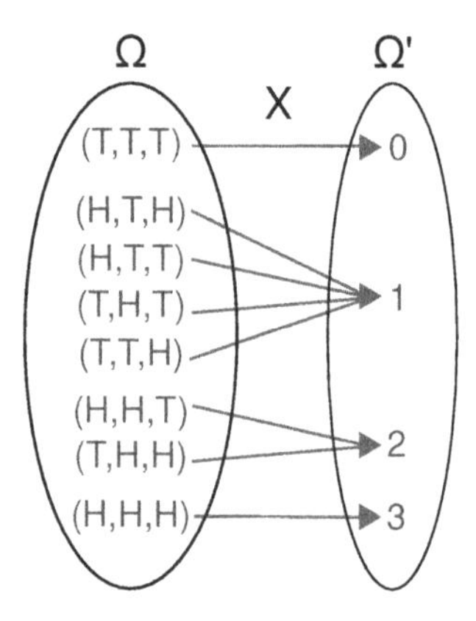

Figure 2.4: Sample space and measurable space for tossing a coin three times

2.2.3.1 Conditional Expectation

Let's consider six independent throws of two different and distinct coins P and Q: 1. (H,H), 2. (T,T), 3. (H,T), 4. (T,H), 5. (H,T), and 6.(T,H). This forms a sample space $\Omega = \{(H,H),(T,T),(H,T),(T,H)\}$ and a probability measure $P(H,H) = P(T,T) = 1/6$ and $P(H,T) = P(T,H) = 1/3$. Let's also consider a random variable $X : \Omega \to [0,1]$ such that $X(H) = 0$ and $X(T) = 1$. The expected value of X for coin P is clearly 1/2, but what is the expected value of X for coin P given that coin Q is tails?

Mathematically, consider a probability space (Ω, Σ, P) and let A, a set of disjoint events $\{A_1, A_2, \ldots\}$, be a measurable partition of Ω ($A_k = \{\omega \in \Omega : X(\omega) = x_k\}$). Let Σ_A be a σ-algebra generated by all unions and intersections of A. Then, the conditional expectation of $f(X)$ with respect to Σ_A is given by:

$$\langle f(X)|\Sigma_A\rangle(\omega) = \sum_i \langle f(X)|A_i\rangle \mathbf{1}_{A_i}(\omega), \tag{2.12}$$

where

$$\langle f(X)|A_i\rangle(\omega) = \frac{\langle f(X)\mathbf{1}_{A_i}\rangle}{P(A_i)}. \tag{2.13}$$

The latter can be interpreted as the center of mass of $f(X)$ on A_i.

In our example, the events where coin Q is tails are $\{(T,T),(H,T),(H,T)\}$. Therefore, a corresponding σ-algebra would be $\Sigma_A = \{\emptyset, \Omega, Y, \Omega \cap Y\}$, where $Y = \{(H,T),(T,T)\}$. The conditional expectation of X given that Q is tails is then:

$$\langle X|Y\rangle = \frac{\langle X\mathbf{1}_{(H,T),(T,T)}\rangle}{P((H,T),(T,T))} = \frac{0 \cdot 2/6 + 1 \cdot 1/6}{3 \cdot 1/6} = \frac{1}{3}. \tag{2.14}$$

It is also interesting to calculate the expected value of the conditional expectation:

$$
\begin{aligned}
\langle\langle X|Y\rangle\rangle &= \left\langle \sum_x xP(X=x|Y) \right\rangle \\
&= \sum_y \left(\sum_x xP(X=x|Y=y) \right) P(Y=y) \\
&= \sum_{y,x} xP(X=x, Y=y) \\
&= \sum_x x \sum_y P(X=x|Y=y) \leftarrow \text{marginalization} \\
&= \sum_x xP(X=x) = \langle X \rangle.
\end{aligned}
\tag{2.15}
$$

This is known as the *law of total expectation.*

2.2.4 DISTRIBUTION FUNCTIONS

Instead of a single probability measure, it is more common to work with distributions of probability [29]. The cumulative distribution function (CDF), for instance, is defined as:

$$F_X(x) := P[X \leq x] = P(\{\omega \in \Omega | X(\omega) \leq x\}). \tag{2.16}$$

When we calculate integrals with respect to a Lebesgue measure λ:

$$\int_B g d\lambda, \tag{2.17}$$

in a Borel set[b] we are calculating the area under the curve given by the function g in this set. This same integral can be written in all $\mathbb{R}$ if we include an *indicator function*:

$$\int_{\mathbb{R}} \mathbf{1}_B g d\lambda. \tag{2.18}$$

It is also possible to move this restriction to the measure itself:

$$\int_{\mathbb{R}} g d\mu. \tag{2.19}$$

In this case, μ is a measure that behaves as λ for the Borel set B but it returns 0 for Borel sets without an intersection with B. Thus, we can write:

$$\int_{\mathbb{R}} g d\mu = \int_{\mathbb{R}} \mathbf{1}_B g d\lambda. \tag{2.20}$$

Therefore, we have:

---

[b] Any set that can be created through the operations of complement and countable union or intersection. Therefore, the set of all Borel sets in Ω form a σ-algebra.

$$d\mu = f d\lambda, \tag{2.21}$$

where $f = \mathbf{1}_B$ is the density of μ known as the *Radon-Nikodym derivative*[6] with respect to the Lebesgue measure.

Following this procedure, if the CDF is a distribution function, we must have:

$$\int_\Omega dF = 1. \tag{2.22}$$

On the other hand, this equation can also be written as:

$$\int f(x)dx = 1. \tag{2.23}$$

If this density $f(x)$ exists, it is called *probability density function* (PDF) and is given by:

$$\begin{aligned} dF &= f(x)dx \\ f(x) &= \frac{dF(x)}{dx} \\ F(x) &= \int_{-\infty}^{x} f(t)dt. \end{aligned} \tag{2.24}$$

Therefore, loosely speaking, the PDF gives the probability that a random variable assumes a set of specific values. If the random variable is discrete, the PDF is called *probability mass function* (PMF).

Example 2.2.6. Finding the CDF of the Gaussian distribution.

The Gaussian distribution is given by:

$$f(x) = \frac{1}{\sqrt{\pi 2\sigma^2}} e^{-\frac{(x-\mu)^2}{2\sigma^2}}. \tag{2.25}$$

Therefore, its CDF is given by:

$$F(x) = \frac{1}{\sqrt{\pi 2\sigma^2}} \int_{-\infty}^{x} e^{-\frac{(y-\mu)^2}{2\sigma^2}} dy. \tag{2.26}$$

Substituting $z = (y-\mu)/\sqrt{2}\sigma \to dy = \sqrt{2}\sigma dz$, we get:

$$\begin{aligned} F(x) &= \frac{\sqrt{2}\sigma}{\sqrt{\pi 2\sigma^2}} \int_{-\infty}^{(x-\mu)/\sqrt{2}\sigma} e^{-z^2} dz \\ &= \frac{1}{\sqrt{\pi}} \left[\int_{-\infty}^{0} e^{-z^2} dz + \int_{0}^{(x-\mu)/\sqrt{2}\sigma} e^{-z^2} dz \right] \\ &= \frac{1}{2}\left[1 + \text{erf}\left(\frac{x-\mu}{\sqrt{2}\sigma}\right)\right]. \end{aligned} \tag{2.27}$$

■

2.2.4.1 Moments of a Distribution

The expected value of a continuous random variable is found by integrating its possible values weighted by the probability that they happen. Mathematically:

$$\langle X \rangle = \int_{-\infty}^{\infty} x f(x) dx. \tag{2.28}$$

We can define a *moment generating function* [31] as the expected value of e^{tX} for a given statistical distribution:

$$\begin{aligned} M_X(t) &= \langle e^{tX} \rangle \\ &= \left\langle 1 + tX + \frac{1}{2!} t^2 X^2 + \frac{1}{3!} t^3 X^3 + \ldots \right\rangle \\ &= \sum_{n=0}^{\infty} \langle X^n \rangle \frac{t^2}{n!}. \end{aligned} \tag{2.29}$$

With this definition we can find a specific moment:

$$m_n = \langle X^n \rangle = \left. \frac{d^n M_X}{dt^n} \right|_{t=0}. \tag{2.30}$$

This is particularly useful for calculating the distribution function of the sum of independent random variables. If we have $Z = X + Y$, for example, we can write the moment generating function for Z as:

$$\begin{aligned} M_Z(z) &= \left\langle e^{z(X+Y)} \right\rangle = \langle e^{zX} \rangle \langle e^{zY} \rangle \\ &= M_X(z) M_Y(z). \end{aligned} \tag{2.31}$$

Since the moment generating function can be interpreted as the bilateral Laplace[7] transform of the random variable, the inverse transform of the product is the convolution of the two distributions:

$$f_Z(z) = \int_{-\infty}^{\infty} f_Y(z - x) f_X(x) dx. \tag{2.32}$$

The nth *central moment* of a random variable X, another important definition, is given by:

$$\mu_n = \langle (X - \langle X \rangle)^n \rangle = \int_{-\infty}^{\infty} (x - \mu)^n f(x) dx. \tag{2.33}$$

Therefore, the 0th central moment is 1, the first central moment is zero and the third central moment is the variance.

The central moments $\forall c \in \mathbb{R}$ are translation invariant:

$$\mu_n(X + c) = \mu_n(X). \tag{2.34}$$

They are also homogeneous:

$$\mu_n(cX) = c^n \mu_n(X). \tag{2.35}$$

Furthermore, the three first central moments are additive for independent random variables:

$$\mu_n(X+Y) = \mu_n(X) + \mu_n(Y),\ n \in \{1,2,3\}. \tag{2.36}$$

The same procedure that we adopted to normalize returns can also be applied to moments so that we can define *standardized moments*. The nth standardized moment is created dividing the nth central moment by the nth power of the standard deviation:

$$\begin{aligned} \tilde{\mu}_n &= \frac{\mu_n}{\sigma^n}, \\ \sigma^n &= \left\langle (X - \langle X \rangle)^2 \right\rangle^{n/2}. \end{aligned} \tag{2.37}$$

The standardized moment of order 1 is always 0 and the second is always 1. The third standardized moment is a measure of asymmetry and the fourth standardized moment is a measure of kurtosis of the distribution. These standardized moments can be used to study the shape of the distribution of returns. For instance, the kurtosis of the normal distribution is 3 (*mesokurtic*). Therefore, distributions with a kurtosis smaller than 3 are known as *platykurtic*, whereas distributions with kurtosis greater than 3 are known as *leptokurtic*. This is a good indication about the chance of finding extreme events in the distribution as we shall see ahead.

2.2.5 LÉVY PROCESSES

One of the most used stochastic processes to model financial systems is the Lévy process [29,33–35][8]. A Lévy process is an indexed and ordered limited set of stochastic variables $\{X\}$ that obeys a set of properties[c]:

1. $P(X_0 = 0) = 1$;
2. The increments $X_t - X_s$ are independent;
3. The increments are also stationary, which implies that $X_t - X_s$ has the same distribution of X_{t-s} for $t > s$;
4. The probability measure related to this process is continuous. Therefore, $\lim_{\Delta\to 0} P(|X_{t+\Delta} - X_t| > \varepsilon) = 0,\ \forall\, \varepsilon > 0$ and $t > 0$;

If the distribution of increments is normal with zero mean and variance given by $\sigma^2 = t - s$, then this process is known as a *Wiener process* [36][9].

[c] A generalization of Lévy processes where the increments can have different distributions is known as an *additive process*.

2.2.5.1 Infinitely Divisible Processes

One interesting property of Lévy distributions is that they can be infinitely divided[10] [37,38]. Let's take one stochastic variable X_t that follows this process. For any integer N, according to property #2, this variable can be written as:

$$X_t = \sum_{k=1}^{n} \left(X_{\frac{kt}{n}} - X_{\frac{(k-1)t}{n}} \right), \ \forall n \in \mathbb{N}. \tag{2.38}$$

The Fourier transform of the distribution of the stochastic variables is called *characteristic function* and can simply be written as:

$$F_X(s) = \langle e^{isX} \rangle. \tag{2.39}$$

Thus, the characteristic function of X_t is:

$$F_X(s) = \left\langle e^{is\sum_{k=1}^{n}\left(X_{\frac{kt}{n}} - X_{\frac{(k-1)t}{n}} \right)} \right\rangle = \left\langle \prod_{k=1}^{n} e^{is\left(X_{\frac{kt}{n}} - X_{\frac{(k-1)t}{n}} \right)} \right\rangle. \tag{2.40}$$

Since the increments of a Lévy process are independent (property #2), we have:

$$F_X(s) = \prod_{k=1}^{n} \left\langle e^{is\left(X_{\frac{kt}{n}} - X_{\frac{(k-1)t}{n}} \right)} \right\rangle. \tag{2.41}$$

Yet, according to property #3, we have:

$$F_X(s) = \prod_{k=1}^{n} F_{X_{1/n}}(s) = \left(F_{X_{1/n}}(s) \right)^n. \tag{2.42}$$

Processes whose variables have the same distribution of the sum $\sum_{k=1}^{n} X_k$, $\forall n \in \mathbb{N}^+$ are called *infinitely divisible*.

2.2.5.2 Stable Distributions

Given two random variables with the same distribution, if the sum of these variables has the same distribution, then this distribution is called *stable*. As shown in Fig. 2.5, stable distributions are a subset of infinitely divisible processes. Let's see an example with Gaussian variables X_1 and X_2. The distribution of their sum can be computed using the characteristic function:

$$\begin{aligned} F_X(\omega) &= F_{X_1}(\omega) F_{X_2}(\omega) \\ &= e^{-\left(\sigma^2 \frac{\omega^2}{2} + i\mu\omega\right)} e^{-\left(\sigma^2 \frac{\omega^2}{2} + i\mu\omega\right)} \\ &= e^{-2\left(\sigma^2 \frac{\omega^2}{2} + i\mu\omega\right)}, \end{aligned} \tag{2.43}$$

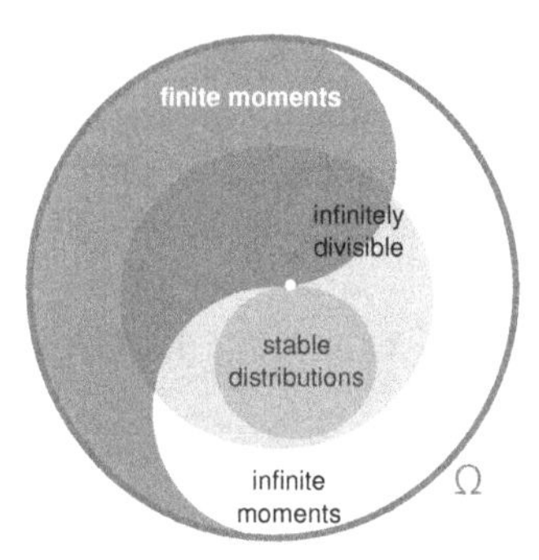

Figure 2.5: Venn diagram for distributions with finite/infinite moments, infinitely divisible processes and stable distribution. The point in the center shows the Gaussian distribution

Paul Lévy [39, 40] found in 1925 the whole set of distributions that are stable. These are given by:

$$\ln(F(\omega)) = \begin{cases} i\mu\omega - \gamma|\omega|^{\alpha}\left[1 - i\beta\frac{\omega}{|\omega|}\tan\left(\frac{\pi}{2}\alpha\right)\right] & \text{para } \alpha \neq 1, \\ i\mu\omega - \gamma|\omega|\left[1 + i\beta\frac{\omega}{|\omega|}\frac{2}{\pi}\ln|\omega|\right] & \text{para } \alpha = 1. \end{cases} \tag{2.44}$$

Some distributions are well known depending on specific values of α and β. For instance, when $\alpha, \beta = 1, 0$ we get the Cauchy-Lorentz distribution and when $\alpha = 2$ we get the Gaussian distribution.

When $\alpha \neq 1$ and $\beta = 0$, the zero centered stable distribution can be written as:

$$F(x) = \int_{-\infty}^{\infty} e^{ix\omega - \gamma|\omega|^{\alpha}} d\omega. \tag{2.45}$$

For the tail part of the distribution, the integral is dominated by small values of the conjugate variable ω. Therefore, it is possible to write a first non-trivial order approximation as:

$$\begin{aligned} F(x) &\approx \int_{-\infty}^{\infty} e^{ix\omega}\left(-\gamma|\omega|^{\alpha}\right) d\omega \\ &\propto \frac{\sin\left(\frac{\pi\alpha}{2}\right)\Gamma(\alpha+1)}{|x|^{\alpha+1}} \\ &\propto |x|^{-\alpha-1}. \end{aligned} \tag{2.46}$$

Thus, asymptotic Lévy distributions can follow a power law. One consequence is that Lévy distributions can have infinite moments. Indeed, the only stable distribution that has all its moments finite is the Gaussian distribution. Another consequence is the presence of self-similarity:

$$\begin{aligned} F(kx) &\approx |kx|^{-\alpha-1} \\ &\approx k^{-\alpha-1}|x|^{-\alpha-1} \text{ p/ } k>0 \\ &\approx k^H|x|^{-\alpha-1} \\ &\approx k^H F(x), \end{aligned} \tag{2.47}$$

where H is a self-similarity level also known as *Hurst coefficient*. Let's study the tails of some distributions in more details next.

2.2.6 DISTRIBUTION TAILS

Let's imagine a situation where we pick random people from a population. After we sample a large number of individuals, the height of each new individual contributes only slightly to the average height. Even if there is a giant in the population, the average height is not much affected because the height of most individuals are located around the population mean. In this case we say that the distribution has *light tails*.

Mathematically[d11], for a set of individuals $X_1, X_2, \ldots$, we must have[e]:

$$P\{\max(X_1, X_2, \ldots, X_N) > t)\} = o\left(P\left\{\sum_n^N X_n > t\right\}\right). \tag{2.48}$$

The Gaussian distribution is an example of a distribution that fulfills this requirement.

On the other hand, if we measure the average wealth of these individuals, we notice that this average is dominated by the wealth of a few individuals. We can think of a group of a thousand randomly picked individuals and some of them happens to be very rich. The average wealth of this group might be around some value until we include this rich person in the counting process. From this point forward, the average wealth is dominated by his or her wealth. In this case, extreme events are more frequent and we say that tail behavior of the distribution is *heavy*.

$$\begin{aligned} \lim_{t\to\infty} Pr\{\max(X_1, X_2, \ldots, X_N) > t)\} &\sim \lim_{t\to\infty} Pr\left\{\sum_n^N X_n > t\right\} \\ 1 - \lim_{t\to\infty} Pr\{\max(X_1, X_2, \ldots, X_N) \leq t\} &\sim \lim_{t\to\infty} Pr\left\{\sum_n^N X_n > t\right\} \\ 1 - \lim_{t\to\infty} Pr\{X_1 \leq t\} Pr\{X_2 \leq x\} \ldots Pr\{X_N \leq t\} &\sim \lim_{t\to\infty} Pr\left\{\sum_n^N X_n > t\right\}. \end{aligned}$$

---

[d]The small "o" notation means that if $f(x) = o(g(x))$, then $\lim_{x\to\infty} \frac{f(x)}{g(x)} = 0$.

[e]This is known by some authors as a *conspiracy principle* [41] since many things have to go wrong in a system so that we detect a problem.

For independent and identically distributed random variables (*i.i.d.*):

$$\begin{aligned}
1-\lim_{t\to\infty} Pr\{X_1\leq t\}^N &\sim \lim_{t\to\infty} Pr\left\{\sum_n^N X_n > t\right\}\\
1-\lim_{t\to\infty}(1-Pr\{X_1>t\})^N &\sim \lim_{t\to\infty} Pr\left\{\sum_n^N X_n > t\right\}\\
\lim_{t\to\infty} NF(t) &\sim \lim_{t\to\infty} F_N(t),
\end{aligned} \tag{2.49}$$

where we have used the binomial approximation, and F_N indicates the distribution of the sum. This is known as a *tail preservation criterium* [42]. This tells us that the distribution of the sum is preserved (keeps the same format) under finite addition and this defines a family of leptokurtic distributions. If the distribution of returns obeys this criterion, an investor can expect, for example, a higher probability for extreme gains and losses. Therefore, it implies more risk.

Heavy-tailed distributions can also be *fat*, *long* and *subexponential* as we shall see below. But before discussing these subclasses, we must discuss two mathematical tools that will help us analyze the tails: the Markov[12] inequality and the Chernoff[13] bound.

The expected value of a non-negative random variable can be calculated for two portions of a distribution, one closer to the origin and another corresponding to the tail:

$$\begin{aligned}
\langle X\rangle &= \int_{-\infty}^{\infty} x f_X(x)dx\\
&= \int_0^{\infty} x f_X(x)dx\\
&= \int_0^{a} x f_X(x)dx + \int_a^{\infty} x f_X(x)dx.
\end{aligned} \tag{2.50}$$

However, it is also true that:

$$\begin{aligned}
\langle X\rangle &= \int_0^{a} x f(x)dx + \int_a^{\infty} x f(x)dx \geq \int_a^{\infty} x f(x)dx\\
\langle X\rangle &\geq \int_a^{\infty} x f(x)dx \geq \int_a^{\infty} a f(x)dx = a.P(x\geq a).
\end{aligned} \tag{2.51}$$

Therefore,

$$P(x\geq a)\leq \langle X\rangle/a. \tag{2.52}$$

This is known as *Markov inequality* and gives us an important property of the distribution tail. To understand its importance, let's see how the exponential function of the tail behaves:

$$\begin{aligned}
Pr(X\geq a) &= Pr\left(e^{tX}\geq e^{ta}\right)\\
&\leq \frac{\langle e^{tX}\rangle}{e^{at}}\\
&\leq e^{-at}M_X(t),
\end{aligned} \tag{2.53}$$

This is known as the *Chernoff bound* for the distribution and imposes an upper bound on the tail in terms of the moment generating function.

2.2.6.1 Subexponential Distributions

In order to understand subexponential distributions, let's see what a *sub-Gaussian* distribution is first. The moment generating function for a zero centered Gaussian distribution is given by:

$$\begin{aligned} M_X(t) &= \frac{1}{\sqrt{2\pi\sigma^2}} \int_{-\infty}^{\infty} e^{tx} e^{-x^2/2\sigma^2} dx \\ &= \frac{1}{\sqrt{2\pi\sigma^2}} \int_{-\infty}^{\infty} e^{-(x-t\sigma^2)^2/2\sigma^2 + t^2\sigma^2/2} dx \\ &= e^{t^2\sigma^2/2}. \end{aligned} \tag{2.54}$$

Therefore, we can define sub-Gaussian distributions as those that obey:

$$\log(M_X(t)) \leq \frac{t^2\sigma^2}{2}. \tag{2.55}$$

Applying the Chernoff bound (Eq. 2.53) to this case, we find:

$$Pr(t \geq a) \leq e^{t^2\sigma^2/2 - ta}. \tag{2.56}$$

On the other hand, the Laplace distribution (double exponential distribution) has a moment generating function given by:

$$\begin{aligned} M_X(t) &= \frac{1}{2\lambda} \int_{-\infty}^{0} e^{x/\lambda} e^{tx} dx + \frac{1}{2\lambda} \int_{0}^{\infty} e^{-x/\lambda} e^{tx} dx \\ &= \frac{1}{2\lambda} \left[\int_{-\infty}^{0} e^{x(t+1/\lambda)} dx + \int_{0}^{\infty} e^{x(t-1/\lambda)} dx \right] \\ &= \frac{1}{\lambda} \left[\frac{e^{(t+1/\lambda)x}}{t+1/\lambda} \Bigg|_{-\infty}^{0} + \frac{e^{(t-1/\lambda)x}}{t-1/b} \Bigg|_{0}^{\infty} \right] \\ &= \frac{1}{2\lambda} \left[\frac{1}{t+1/\lambda} - \frac{1}{t-1/\lambda} \right], \text{ if } |t| < 1/\lambda \\ &= \frac{1}{2\lambda} \frac{-2/\lambda}{t^2 - 1/\lambda^2} \\ &= \frac{1}{1-\lambda^2 t^2} \\ &\leq \frac{1}{e^{-\lambda^2 t^2/2}}, \leftarrow 1^{\text{st}} \text{ order Taylor's expansion in reverse.} \\ &\leq e^{\lambda^2 t^2/2}. \end{aligned} \tag{2.57}$$

Therefore, proceeding as we did for the sub-Gaussian distributions, we define sub-exponential distributions as those that obey:

$$\log\left(M_X(t)\right) \leq \frac{t^2\lambda^2}{2}, \ \forall |t| < 1/\lambda. \tag{2.58}$$

Or applying the Chernoff bound:

$$Pr(x \geq a) \leq \exp\left\{\frac{\lambda^2 t^2}{2} - at\right\} = e^{g(t)}. \tag{2.59}$$

This, however, only applies to small values of a since the moment generating function is not defined for $t \geq 1/\lambda$.

It is possible to find a stricter limit by calculating the minimum of $g(t)$:

$$\begin{aligned} g(t) &= 1/2\lambda^2 t^2 - at \\ g'(t) &= \lambda^2 t - a = 0 \rightarrow t_{\min} = \frac{a}{\lambda^2} \\ g(t_{\min}) &= \frac{1}{2}\lambda^2\frac{a^2}{\lambda^4} - a\frac{a}{\lambda^2} = -\frac{1}{2}\frac{a^2}{\lambda^2}. \end{aligned} \tag{2.60}$$

This also imposes a bound for a:

$$\begin{aligned} &|t| < 1/\lambda \\ &\frac{1}{\lambda} > \frac{a}{\lambda^2} \rightarrow a < \lambda. \end{aligned} \tag{2.61}$$

At the boundary where $t = 1/\lambda$, we have:

$$\begin{aligned} g\left(t = \frac{1}{\lambda}\right) &= \frac{\lambda^2}{2\lambda^2} - \frac{a}{\lambda} \\ &\leq -\frac{a}{2\lambda}. \end{aligned} \tag{2.62}$$

Thus, according to Eq. 2.60, subexponential distributions resemble sub-Gaussian distributions near the origin. On the other hand, whereas sub-Gaussian distributions rapidly lose their tails, Eq. 2.62 shows that subexponential distributions have tails that fall exponentially without a defined moment generating function.

2.2.6.2 Long and Fat Tails

A distribution is said to have a *long tail* if its tail probability have a tendency to become constant for large values. Mathematically, we can say:

$$
\begin{aligned}
\lim_{x\to\infty} P[X > x+\Delta | X > x] = 1,\ \Delta > 0 \\
\lim_{x\to\infty} \frac{P((X > x+\Delta) \cap (X > x))}{P(X > x)} = 1 \\
\lim_{x\to\infty} \frac{P(X > x+\Delta)}{P(X > x)} = 1 \\
\lim_{x\to\infty} P(X > x+\Delta) \sim \lim_{x\to\infty} P(X > x), \\
\therefore \lim_{x\to\infty} f(x+\Delta) \sim \lim_{x\to\infty} f(x).
\end{aligned}
\tag{2.63}
$$

Let's see, for instance, a distribution function such as:

$$
f(x) = \frac{1}{\log(x)},\ x \geq 1. \tag{2.64}
$$

Applying the definition of a long tail distribution we get:

$$
\begin{aligned}
\lim_{x\to\infty} \frac{\log(x)}{\log(x+\Delta)} &= \lim_{x\to\infty} \frac{\log(x)}{\log\left(x\left(1+\frac{\Delta}{x}\right)\right)} \\
&= \lim_{x\to\infty} \frac{\log(x)}{\log(x)+\log\left(1+\frac{\Delta}{x}\right)} \\
&= \frac{\log(x)}{\log(x)} = 1.
\end{aligned}
\tag{2.65}
$$

This means that for large values, the tail tends to flatten out and the expected fluctuations resulting from this tail can be big.

Fat tails, on the other hand, are those distributions whose tails can be modeled with a power law. One example of such distributions is the Pareto distribution that is discussed next. We close this discussion by showing the tails of the Gaussian, the exponential and the Pareto distributions in Fig. 2.6. It is instructive to note how the tail of the Gaussian distribution falls rapidly and the tail of the Pareto distribution is not limited by the exponential distribution.

2.2.7 PARETO DISTRIBUTION

We close this section studying the Pareto[14] distribution, which is of particular importance in econophysics. For instance, it is a distribution commonly used to describe the wealth distribution in a society.

The domain of this distribution is $[x_0, \infty)$, where x_0 is a constant. Its PDF is given by:

$$
f(x) = \alpha \frac{x_0^{\alpha}}{x^{\alpha+1}}. \tag{2.66}
$$

Its CDF is then given by:

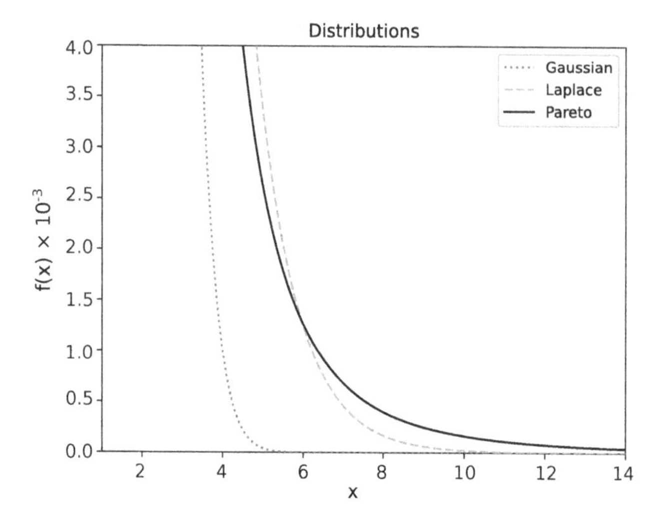

Figure 2.6: The Gaussian distribution (light tail), the Lorenz distribution (exponential), and the Pareto distribution (heavy tail). All distributions are calculated to have the same variance

$$\begin{aligned} P(Z \leq z) &= \int_{x_0}^{z} \alpha x_0^{\alpha} x^{-(\alpha+1)} dx \\ &= -\frac{\alpha}{\alpha} x_0^{\alpha} x^{-(\alpha+1)+1} \Big|_{x_0}^{z} \\ &= -x_0^{\alpha} z^{-\alpha} + 1 \\ &= 1 - \left(\frac{x_0}{z}\right)^{\alpha} \end{aligned} \tag{2.67}$$

and its complementary CDF (CCDF) is given by:

$$\bar{P}(z) = P(Z > z) = 1 - P(Z \leq z) = \left(\frac{x_0}{z}\right)^{\alpha}. \tag{2.68}$$

The CCDF of the Pareto distribution leads to a special property. If we take the logarithm of a Pareto distributed random variable:

$$Y = \log\left(\frac{X}{x_0}\right), \tag{2.69}$$

we find for its CDF:

$$\begin{aligned} Pr(Y > y) &= 1 - Pr\left(\ln\left(\frac{X}{x_0}\right) \leq y\right) \\ &= 1 - Pr(X \leq x_0 e^y) \\ &= \left(\frac{x_0}{x_0 e^y}\right)^{\alpha} \\ &= e^{-\alpha y}. \end{aligned} \tag{2.70}$$

Therefore, the logarithm of a Pareto distributed random variable is exponentially distributed.

The *survival function* (or *reliability function*) of a Pareto distributed random variable holds another interesting property related to its expectation beyond a specific value z. This is given by:

$$S(z) = \frac{\langle X\mathbf{1}_{X>z}\rangle}{\bar{P}(z)}. \tag{2.71}$$

Its numerator is given by:

$$\begin{aligned} \langle X\mathbf{1}_{X>z}\rangle &= \int_z^{\infty} x\alpha \frac{x_0^{\alpha}}{x^{\alpha+1}} dx = \alpha x_0^{\alpha} \int_z^{\infty} x^{-\alpha} dx \quad = \alpha x_0^{\alpha} \left.\frac{x^{1-\alpha}}{1-\alpha}\right|_z^{\infty} \\ &= \frac{\alpha}{\alpha - 1} x_0^{\alpha} z^{1-\alpha} \text{ if } \alpha > 1. \end{aligned} \tag{2.72}$$

Therefore, the survival function is given by:

$$\begin{aligned} S(z) &= \frac{\alpha}{\alpha - 1} x_0^{\alpha} z^{1-\alpha} \left(\frac{z}{x_0}\right)^{\alpha} \\ &= \frac{\alpha}{\alpha - 1} z. \end{aligned} \tag{2.73}$$

The additional expected survival is then:

$$\begin{aligned} S(z) - z &= \frac{\alpha}{\alpha - 1} z - z \\ &= \frac{z}{\alpha - 1}. \end{aligned} \tag{2.74}$$

For $\alpha = 2$, for example, when the process survives z, it is expected that it will survive another value z. This kind of behavior is known as *Lindy effect*[f].

The kth standardized moment for a Pareto distributed random variable is not always defined:

[f]Not related to any scientist but to a restaurant in New York where comedians used to meet and make predictions about how much longer their shows would last.

$$\begin{aligned}
\left\langle x^k \right\rangle &= \alpha x_0^\alpha \int_{x_0}^{\infty} dx \frac{x^k}{x^{\alpha+1}} \\
&= \alpha x_0^k \int_{x_0}^{\infty} dx x^{k-\alpha-1} \\
&= \alpha x_0^k \frac{x^{k-\alpha}}{k-\alpha}\bigg|_{x_0}^{\infty} \\
&= \alpha x_0^\alpha \frac{x_0^{k-\alpha}}{\alpha-k}, \text{ for } k<\alpha \\
&= \frac{\alpha}{\alpha-k} x_0^k.
\end{aligned} \tag{2.75}$$

It is possible to estimate the tail index α of the distribution using maximum likelihood estimation. Its likelihood function is given by:

$$\mathscr{L} = \alpha^N x_0^{N\alpha} \prod_n x_n^{-(\alpha+1)} \tag{2.76}$$

and its log is:

$$\log \mathscr{L} = N\log(\alpha) + N\alpha\log(x_0) - (\alpha+1)\sum_n \log(x_n). \tag{2.77}$$

Minimizing it with respect to α we get:

$$\begin{aligned}
\frac{\partial \log \mathscr{L}}{\partial \alpha} &= \frac{N}{\alpha} + N\log(x_0) - \sum_n \log(x_n) \\
0 &= \frac{N}{\alpha} + \sum_n \log\frac{x_n}{x_0} \\
\hat{\alpha} &= \frac{N}{\sum_n \log\left(\frac{x_n}{\hat{x}_0}\right)},
\end{aligned} \tag{2.78}$$

where $\hat{x}_0 = \min_i x_i$. This is known as the *Hill estimator*[15] [43].

2.2.7.1 Lorenz Curve

As we stated before, the Pareto distribution is widely used to model the wealth distribution in a society. A graphical representation of this distribution is given by the Lorenz[16] curve [44]. For a PDF given by $f(x)$ and its equivalent CDF $F(x)$, the Lorenz curve is given by:

$$L(F(x)) = \frac{\int_{-\infty}^{x(F)} t f(t) dt}{\int_{-\infty}^{\infty} t f(t) dt} = \frac{\int_0^F x(G) dG}{\int_0^1 x(G) dG}, \tag{2.79}$$

where $x(F)$ is the inverse of the CDF. Thus, the Lorenz curve shows the normalized expected value of a random variable as a function of its cumulative value. Usually, economists use this curve to show the cumulative share of wealth as a function of the cumulative share of people. A straight diagonal line implies perfect equality, whereas the bending of the curve implies some level of inequality. For instance, if the tail index is ≈ 1.16, approximately 80 % of the wealth would be held by 20 % of the society, which is known as *Pareto principle*. Countries with the highest inequalities in the world have tail indexes not smaller than 1.3, though.

For the Pareto distribution we get:

$$F(x) = 1 - \left(\frac{x_0}{x}\right)^{\alpha} \rightarrow x(F) = \frac{x_0}{(1-F)^{1/\alpha}}. \tag{2.80}$$

Thus, the Lorenz curve becomes:

$$L(F) = \frac{\int_0^F \frac{x_0}{(1-G)^{1/\alpha}} dG}{\int_0^1 \frac{x_0}{(1-G)^{1/\alpha}} dG}. \tag{2.81}$$

The integral in the numerator is:

$$\begin{aligned} \int_0^F \frac{x_0}{(1-G)^{1/\alpha}} dG &= x_0 \frac{(1-G)^{1-1/\alpha}}{1-1/\alpha}\bigg|_0^F \\ &= x_0 \frac{(1-F)^{1-1/\alpha}}{1-1/\alpha} - \frac{x_0}{1-1/\alpha} \\ &= \frac{x_0}{1-1/\alpha}\left[(1-F)^{1-1/\alpha} - 1\right]. \end{aligned} \tag{2.82}$$

Thus, the Lorenz curve becomes:

$$\begin{aligned} L(F) &= \frac{(1-F)^{1-1/\alpha} - 1}{-1} \\ &= 1 - (1-F)^{1-1/\alpha}. \end{aligned} \tag{2.83}$$

This curve is shown in Fig. 2.7 and can be plotted in Python using:

```
F = np.linspace(0,1,100)
L = [1-(1-Fn)**(1-1.0/alfa) for Fn in F]
pl.plot(F,L)
```

2.2.7.2 Gini Index

As discussed previously, a homogeneous wealth distribution would imply that x % individuals would hold x % of the total wealth. Therefore, the Lorenz curve would be a diagonal straight line with an inclination of $\pi/4$ rads. In order to quantify how unequal the wealth distribution is, Gini[17] developed the following measure [45]:

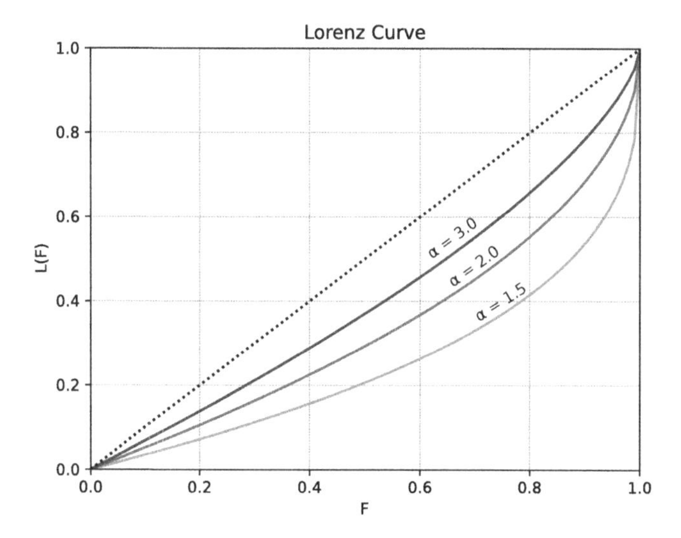

Figure 2.7: Lorenz curve for different tail indexes. The dashed line corresponds to $\alpha \to \infty$

$$G = 1 - 2\left(\int_0^1 L(F)dF\right). \tag{2.84}$$

For the Pareto distribution, the Gini index is given by:

$$\begin{aligned} G &= 1 - 2\left(\int_0^1 1 - (1-F)^{1-1/\alpha} dF\right) \\ &= 1 - 2\left(F|_0^1 - \left.\frac{(1-F)^{2-1/\alpha}}{2-1/\alpha}\right|_0^1\right) \\ &= 1 - 2\left(1 + \frac{1}{2-1/\alpha}\right) \\ &= \frac{1}{2\alpha - 1}. \end{aligned} \tag{2.85}$$

Typically, countries with high inequality have Gini indexes around 60 %, whereas countries with low inequalities have Gini indexes around 25 %. In these extreme cases, the tail indexes are:

$$\begin{aligned} \alpha &= \frac{1}{2}\left(1 + \frac{1}{G}\right) \\ \alpha_{\text{min}} &\approx 1.33, \\ \alpha_{\text{max}} &\approx 2.50. \end{aligned} \tag{2.86}$$

2.3 MODELS FOR FINANCIAL TIME SERIES

A time series is an indexed dataset ordered in time. Mathematically, we describe a time series with a set $X = \{x_m | m \in T\}$, where T is an indexed set given by $T = \{1,\ldots,M\}$. The same way we can use a time series to describe the rainfall amount along a set of days, we can use a time series to model the price evolution of a financial asset. It is possible to model a time series in different ways. Here we will study the autoregressive model (AR), the moving average (MA) and the combination of both, the autoregressive moving average (ARMA) [31].

2.3.1 AUTOREGRESSIVE MODEL (AR)

Let's describe the price of an asset x_t as:

$$x_t = c + \sum_{i=1}^{p} \phi_i x_{t-i} + \varepsilon_t, \tag{2.87}$$

where c is a bias, ϕ_i are parameters that we can adjust to obtain a nice fit, and ε_t is a component related to random shocks (or *innovations*). This random variable typically has a zero expected value and unitary variance. A graphical representation of this process is shown in Fig. 2.8

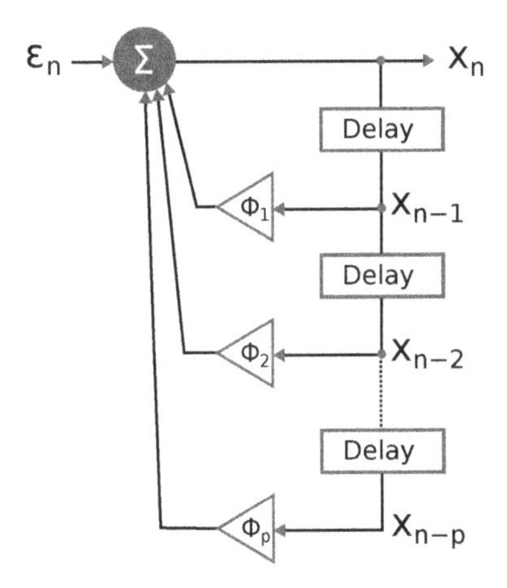

Figure 2.8: Graphical representation of an autoregressive process

This is the equivalent of an infinite impulse response (IIR) filter, where the current price depends explicitly on previous values.

If we ignore the price history such that $\phi_i = 0$, $\forall i$, then the current price depends only on the random shock:

$$x_t = c + \varepsilon_t. \tag{2.88}$$

Since we used no information about the past, this model is known as AR(0). Its expected value is given by:

$$\langle x_t \rangle = c. \tag{2.89}$$

The variance of an AR(0) process is given by:

$$\begin{aligned} \mathrm{VAR}(x_t) &= \mathrm{VAR}(c) + \mathrm{VAR}(\varepsilon_t) \\ \sigma_x^2 &= \sigma_e^2. \end{aligned} \tag{2.90}$$

If we take only the previous price information, we obtain the AR(1) process given by:

$$x_t = c + \phi x_{t-1} + \varepsilon_t. \tag{2.91}$$

If $\phi = 1$, then the AR(1) model is a random walk model around a value c. Let's explore a few properties of this model. For instance, the expected value of the AR(1) process is:

$$\begin{aligned} \langle x_t \rangle &= \langle c \rangle + \langle \phi x_{t-1} \rangle + \langle \varepsilon_t \rangle \\ \mu &= c + \phi \langle x_{t-1} \rangle + \langle \varepsilon_t \rangle. \end{aligned} \tag{2.92}$$

Considering that the price of the asset is a *wide-sense stationary process* ($\langle x_{t+n} \rangle = \langle x_t \rangle$, see more in Sec. 2.4.1), then:

$$\mu = c + \phi \mu \rightarrow \mu = \frac{c}{1-\phi}. \tag{2.93}$$

The variance of an AR(1) process is given by:

$$\begin{aligned} \sigma_x^2 &= \mathrm{VAR}(c) + \mathrm{VAR}(\phi x_{t-1}) + \mathrm{VAR}(\varepsilon_t) \\ &= 0 + \phi^2 \sigma_x^2 + \sigma_e^2 \\ \sigma_x^2 &= \frac{\sigma_e^2}{1-\phi^2}. \end{aligned} \tag{2.94}$$

Since the variance is always positive, we have that $|\phi| < 1$.

The autocovariance is given by:

$$\begin{aligned} \gamma_s = \mathrm{COV}(x_t, x_{t-s}) &= \langle (x_{t-s} - \langle x_{t-s} \rangle)(x_t - \langle x_t \rangle) \rangle \\ &= \langle (x_{t-s} - \langle x_{t-s} \rangle)\left(c + \phi x_{t-1} + \varepsilon_t - c - \phi \langle x_{t-1} \rangle\right) \rangle \\ &= \langle (x_{t-s} - \mu)\left(\phi (x_{t-1} - \mu) + \varepsilon_t\right) \rangle \\ &= \phi \langle (x_{t-s} - \mu)(x_{t-1} - \mu) \rangle \\ &= \phi \gamma_{s-1}. \end{aligned} \tag{2.95}$$

Therefore:

$$\gamma_s = \phi \gamma_{s-1} = \phi^2 \gamma_{s-2} = \phi^3 \gamma_{s-3} = \ldots = \phi^s \gamma_0 = \phi^s \sigma_x^2. \tag{2.96}$$

The autocorrelation is given then by:

$$\rho = \frac{\mathrm{COV}(x_t, x_{t-s})}{\mathrm{STD}(x_t)STD(x_{t-s})} = \frac{\gamma_s}{\sigma_x^2} = \phi^s. \tag{2.97}$$

Interestingly, this can also be written as:

$$\sigma_{xx}^2 = \phi^s = e^{s \log \phi} = e^{-s/\tau}, \tag{2.98}$$

where $\tau = -1/\log \phi$ is a *correlation time*.

Finally, we can use the Wiener-Khinchin theorem (Sec. 2.4.1.1) to compute the spectral density from the autocovariance:

$$\begin{aligned} S(\omega) &= \sum_{s=-\infty}^{\infty} \frac{\sigma_e^2}{1-\phi^2} \phi^{|s|} e^{-j\omega s} \\ &= \frac{\sigma_e^2}{1-\phi^2} \left[1 + \sum_{s=1}^{\infty} \left(e^{-j\omega s} \phi^s + e^{j\omega s} \phi^s \right) \right] \\ &= \frac{\sigma_e^2}{1-\phi^2} \left[1 + \sum_{s=1}^{\infty} \left(e^{-j\omega} \phi \right)^s + \sum_{s=1}^{\infty} \left(e^{j\omega} \phi \right)^s \right]. \end{aligned} \tag{2.99}$$

To proceed let's use the result:

$$\sum_{n=1}^{\infty} x^n = \sum_{n=0}^{\infty} x^n - 1 = \frac{1}{1-x} - 1 = \frac{x}{1-x}. \tag{2.100}$$

Therefore:

$$\begin{aligned} S(\omega) &= \frac{\sigma_e^2}{1-\phi^2} \left[1 + \frac{e^{-j\omega}\phi}{1-e^{-j\omega}\phi} + \frac{e^{j\omega}\phi}{1-e^{j\omega}\phi} \right] \\ &= \frac{\sigma_e^2}{1-\phi^2} \frac{(1-e^{-j\omega}\phi)(1-e^{j\omega}\phi) + \left(1-e^{j\omega}\phi\right) e^{-j\omega}\phi + \left(1-e^{-j\omega}\phi\right) e^{j\omega}\phi}{\left(1-e^{-j\omega}\phi\right)\left(1-e^{j\omega}\phi\right)} \\ &= \frac{\sigma_e^2}{1-\phi^2} \frac{1-e^{j\omega}\phi - e^{-j\omega}\phi + \phi^2 + e^{-j\omega}\phi - \phi^2 + e^{j\omega}\phi - \phi^2}{\left(1-e^{-j\omega}\phi\right)\left(1-e^{j\omega}\phi\right)} \\ &= \frac{\sigma_e^2}{1-\phi^2} \frac{1-\phi^2}{\left(1-e^{-j\omega}\phi\right)\left(1-e^{j\omega}\phi\right)} \\ &= \frac{\sigma_e^2}{1-e^{j\omega}\phi - e^{-j\omega}\phi + \phi^2} = \frac{\sigma_e^2}{1-2\phi\cos(\omega)+\phi^2}. \end{aligned} \tag{2.101}$$

The autocorrelation and the spectral density for an AR(1) process are shown in Fig. 2.9. It is interesting to note that positive ϕ produces a nicely decaying autocorrelation function and a spectral density that is dominated by low frequency components. On the other hand, negative coefficients produce alternating autocorrelations and spectral densities that are dominated by high frequency components. Thus, negative ϕ produces signals in the time domain that are rougher than those signals produced by positive ϕ.

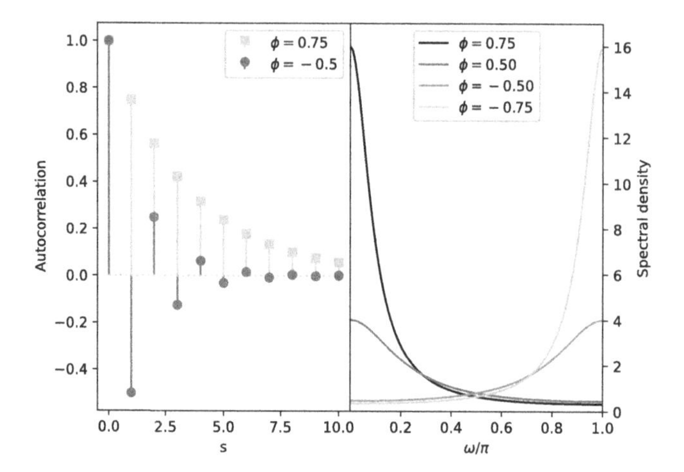

Figure 2.9: Left: Autocorrelation function and Right: Spectral density for an AR1 process

2.3.2 MOVING AVERAGE (MA)

The AR(1) process can be recursively modified such that it becomes an infinite history. In order to distinguish it from the standard AR process, let's write the adjustment parameter as θ:

$$\begin{aligned} x_t &= c+\theta x_{t-1}+\varepsilon_t \\ &= c+\theta\left(c+\theta x_{t-2}+\varepsilon_{t-1}\right)+\varepsilon_t \\ &= c+\theta\left(c+\theta\left(c+\theta x_{t-3}+\varepsilon_{t-2}\right)+\varepsilon_{t-1}\right)+\varepsilon_t \\ &= c+\theta c+\theta^2 c+\theta^3 x_{t-3}+\theta^2\varepsilon_{t-2}+\phi\varepsilon_{t-1}+\varepsilon_t \end{aligned} \tag{2.102}$$

If we continue this process indefinitely towards the past and considering that $|\theta|<1$ and the first element of this series is not much relevant, we get:

$$\begin{aligned} x_t &= \sum_{n=0}^{\infty}\theta^n c+\sum_{n=0}^{\infty}\theta^n\varepsilon_{t-n} \\ &= \frac{c}{1-\theta}+\sum_{n=0}^{\infty}\theta^n\varepsilon_{t-n} \\ x_t &= \mu+\sum_{n=0}^{\infty}\theta^n\varepsilon_{t-n} \end{aligned} \tag{2.103}$$

This way, θ appears as a kernel and the price signal is a response to random shocks at different time period. This is known as a *moving average process*. The graphical representation of this process is shown in Fig. 2.10. Although it can be a more complicated process to work with because of the shocks, it is always stable.

It is often possible to make the opposite travel from an MA to an AR process. For instance, for the MA(1) process we have:

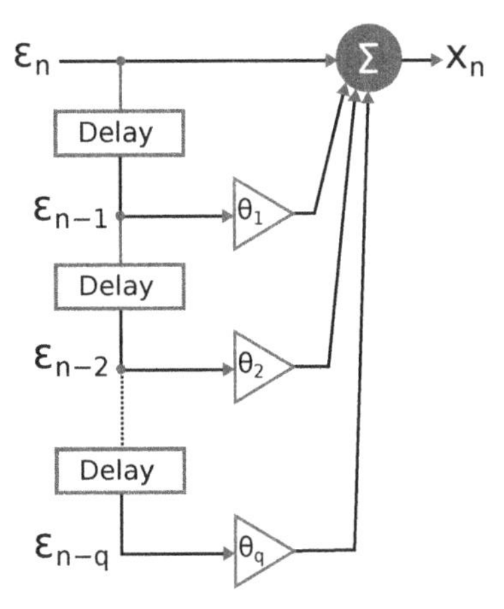

Figure 2.10: Graphical representation of a moving average process

$$\begin{aligned} x_t &= \varepsilon_t + \theta\varepsilon_{t-1} \\ x_{t-1} &= \varepsilon_{t-1} + \theta\varepsilon_{t-2} \\ x_{t-2} &= \varepsilon_{t-2} + \theta\varepsilon_{t-3} \\ x_{t-3} &= \varepsilon_{t-3} + \theta\varepsilon_{t-4} \\ &\vdots \end{aligned} \tag{2.104}$$

Therefore,

$$\begin{aligned} x_t &= \varepsilon_t + \theta x_{t-1} - \theta^2\varepsilon_{t-2} \\ &= \varepsilon_t + \theta x_{t-1} - \theta^2 x_{t-2} + \theta^3\varepsilon_{t-3} \\ &= \varepsilon_t + \theta x_{t-1} - \theta^2 x_{t-2} + \theta^3 x_{t-3} - \theta^4\varepsilon_{t-4} \\ x_t &= \sum_{n=1}^{\infty} (-1)^{n-1}\theta^n x_{t-n} + \varepsilon_t. \end{aligned} \tag{2.105}$$

The equivalence of adjustment parameters in this case is:

$$\phi_n = (-1)^{n-1}\theta^n. \tag{2.106}$$

Like the AR(0) process, the MA(0) is also trivial. Its expected value is just c and its variance is σ_e^2. The MA(1) process is given by:

$$x_t = \mu + \varepsilon_t + \theta\varepsilon_{t-1}. \tag{2.107}$$

Its expected value is given by:

$$\langle x_t \rangle = \langle \mu \rangle + \langle \varepsilon_t \rangle + \langle \theta\varepsilon_{t-1} \rangle = \mu. \tag{2.108}$$

The variance of this process is:

$$\begin{aligned}\mathrm{VAR}(x_t) &= \mathrm{VAR}(\mu) + \mathrm{VAR}(\varepsilon_t) + \mathrm{VAR}(\theta\varepsilon_{t-1}) \\ &= \sigma_e^2 + \theta^2\sigma_e^2 \\ &= (1+\theta^2)\sigma_e^2.\end{aligned} \tag{2.109}$$

Its autocovariance is given by:

$$\begin{aligned}\gamma_s &= \langle (x_t - \langle x_t\rangle)(x_{t-s} - \langle x_{t-s}\rangle)\rangle = \langle (\varepsilon_t + \theta\varepsilon_{t-1})(\varepsilon_{t-s} + \theta\varepsilon_{t-s-1})\rangle \\ &= \left\langle \varepsilon_t\varepsilon_{t-s} + \theta\varepsilon_t\varepsilon_{t-s-1} + \theta\varepsilon_{t-1}\varepsilon_{t-s} + \theta^2\varepsilon_{t-1}\varepsilon_{t-s-1}\right\rangle \\ &= \delta(s)\sigma_e^2 + \theta\sigma_e^2\delta(s-1) + \theta^2\sigma_e^2\delta(s) \\ &= \sigma_e^2\left[(1+\theta^2)\delta(s) + \theta\delta(s-1)\right].\end{aligned} \tag{2.110}$$

Therefore, its autocorrelation is:

$$\begin{aligned}\rho_x &= \frac{\sigma_e^2\left[(1+\theta^2)\delta(s) + \theta\delta(s-1)\right]}{(1+\theta^2)\sigma_e^2} \\ &= \delta(s) + \frac{\theta}{1+\theta^2}\delta(s-1).\end{aligned} \tag{2.111}$$

The spectral density can be obtained once again using the Wiener-Khinchin theorem with the covariance:

$$\begin{aligned}S(\omega) &= \sum_{s=-\infty}^{\infty} e^{-j\omega s}\sigma_e^2\left((1+\theta^2)\delta(s) + \theta\delta(|s|-1)\right) \\ &= \sigma_e^2\left(1+\theta^2+\theta(e^{-j\omega}+e^{j\omega})\right) \\ &= \sigma_e^2\left(1+\theta^2+2\theta\cos(\omega)\right).\end{aligned} \tag{2.112}$$

Therefore, when θ is positive and the autocorrelation is positive, the spectrum is dominated by low frequency components. In the opposite case, when θ is negative, the spectrum is dominated by high frequency components. This behavior is shown in Fig. 2.11.

2.3.3 AUTOREGRESSIVE MOVING AVERAGE (ARMA)

It is possible to combine an autoregressive process with a moving average to compose an *autoregressive moving average* process:

$$y_t = \sum_{n=1}^{N} \phi_n y_{t-n} + \varepsilon_n + \sum_{m=1}^{M} \theta_m \varepsilon_{t-m}. \tag{2.113}$$

A graphical representation of an ARMA process is shown in Fig. 2.12.

The ARMA model is particularly appealing since a wide-sense stationary time series with zero mean can be modeled as a sum of a stochastic (linear combination of lags of a white noise process) component and an uncorrelated deterministic (linear

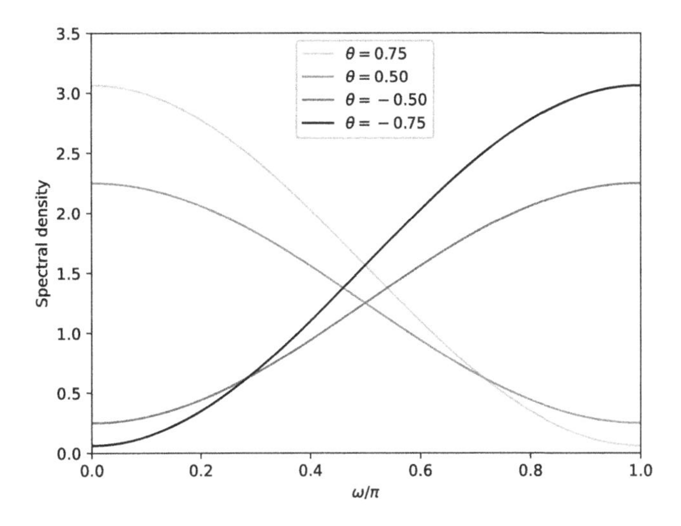

Figure 2.11: The spectral density for an MA1 process

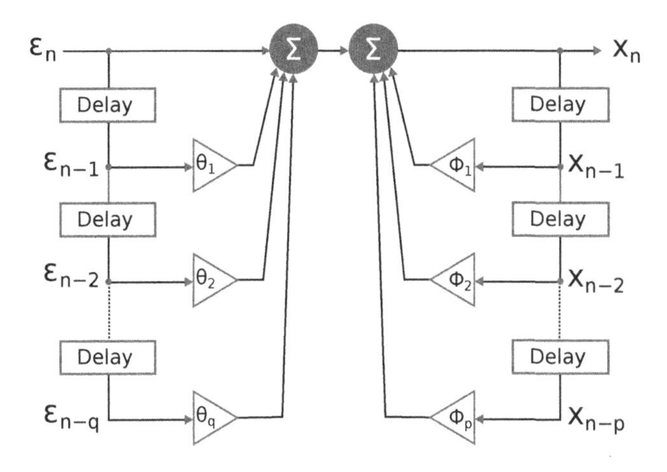

Figure 2.12: Graphical representation of an ARMA process

combination of lags of a signal) component. This is known as *Wold's decomposition* [46][18]. In other words, Wold's decomposition states that any wide-sense stationary time series has an ARMA representation.

ARMA(p,0) processes are just AR(p) processes and ARMA(0,q) processes are just MA(q) processes. For an ARMA(1,1) process we have:

$$y_t = \mu + \phi y_{t-1} + \varepsilon_t + \theta \varepsilon_{t-1}. \tag{2.114}$$

The expected value of y_t is:

$$\langle y_t \rangle = \mu + \phi \langle y_{t-1} \rangle \rightarrow \langle y_t \rangle = \frac{\mu}{1-\phi}. \tag{2.115}$$

The variance can be found by first computing $y_t y_t$ and $y_t y_{t-1}$:

$$\begin{aligned} \langle y_t^2 \rangle &= \phi \langle y_{t-1} y_t \rangle + \langle \varepsilon_t y_t \rangle + \theta \langle \varepsilon_{t-1} y_t \rangle \\ \langle y_t y_{t-1} \rangle &= \phi \langle y_{t-1} y_{t-1} \rangle + \langle \varepsilon_t y_{t-1} \rangle + \theta \langle \varepsilon_{t-1} y_{t-1} \rangle. \end{aligned} \tag{2.116}$$

In order to solve these equations, first we need:

$$\begin{aligned} \langle y_t \varepsilon_t \rangle &= \phi \langle y_{t-1} \varepsilon_t \rangle + \langle \varepsilon_t \varepsilon_t \rangle + \theta \langle \varepsilon_{t-1} \varepsilon_t \rangle \\ &= \sigma_e^2, \end{aligned} \tag{2.117}$$

given that $\langle y_{t-1} \varepsilon_t \rangle = 0$, since the signal is not correlated with the noise. We also need:

$$\begin{aligned} \langle \varepsilon_{t-1} y_t \rangle &= \phi \langle \varepsilon_{t-1} y_{t-1} \rangle + \langle \varepsilon_{t-1} \varepsilon_t \rangle + \theta \langle \varepsilon_{t-1} \varepsilon_{t-1} \rangle \\ &= \phi \langle \varepsilon_t y_t \rangle + \theta \sigma_e^2 \\ &= \sigma_e^2 (\phi + \theta), \end{aligned} \tag{2.118}$$

With these results we can rewrite Eqs. 2.116:

$$\begin{aligned} \gamma_0 &= \phi \gamma_1 + \sigma_e^2 + \theta \sigma_e^2 (\phi + \theta) \\ \gamma_1 &= \phi \gamma_0 + \theta \sigma_e^2. \end{aligned} \tag{2.119}$$

Therefore,

$$\begin{aligned} \gamma_0 &= \phi(\phi \gamma_0 + \theta \sigma_e^2) + \sigma_e^2 + \theta \sigma_e^2 (\phi + \theta) \\ &= \phi^2 \gamma_0 + \phi \theta \sigma_e^2 + \sigma_e^2 + \phi \theta \sigma_e^2 + \theta^2 \sigma_e^2 \\ &= \frac{1 + 2\phi\theta + \theta^2}{1 - \phi^2} \sigma_e^2. \end{aligned} \tag{2.120}$$

2.3.4 BOX-JENKINS APPROACH

Box[19] and Jenkins[20] created a popular approach to identify the parameters of ARMA processes [47]. This approach is divided in three steps: i) **identification**: In this step, the sizes of the AR and MA series are estimated, ii) **estimation**: here the ϕ and θ parameters are estimated, and iii) **diagnostics**: in this last step, the residues of the model are analyzed. All steps are undertaken in a cyclic fashion until convergence is obtained.

In the identification phase, the autocorrelation function (ACF) and the partial autocorrelation function (PACF) of the time series are plotted. The presence of a long persistence in the ACF may indicate that a low frequency trend has to be removed or the series could be differentiated. If the ACF decays, but the PACF drops quickly after some value p, then an AR(p) model can be used. If the opposite situation occurs with the PACF decaying and the ACF drops quickly after some value q, then an

MA(q) model can be used. On the other hand, if both functions decay without any abrupt cut, then a mixed model has to be considered.

In the estimation phase, the ϕ and θ parameters are estimated using techniques such as least squares, the Yule-Walker equations and maximum likelihood. We will study some of these techniques next.

After the parameters are estimated, a diagnostic is performed doing a portmanteau test with the ACF and the PACF of the residues. If the test fails, then we go back to the first step. One test that is commonly used is the Box-Pierce test [48]. The null hypothesis for this test is that the residues are i.i.d. The following statistics is computed:

$$Q(m) = n \sum_{l=1}^{m} r_l^2, \tag{2.121}$$

where m is the number of lags in the ACF, n is the number of terms in the series, and

$$r_l^2 = \frac{\sum_{t=l+1}^{n} a_t a_{t-l}}{\sum_{t=1}^{n} a_t^2}, \tag{2.122}$$

where a_n are the residues.

In the Box-Pierce test $Q(m)$ is chi-square distributed with k degrees of liberty if the residues are random. Therefore, the null hypothesis is rejected if $Q > \chi_{1-\alpha,k}$ with level of significance α and $k = m - p - q$ degrees of freedom.

Another test is the Box-Ljung [49], where the statistics Q is modified to:

$$Q(m) = n(n+2) \sum_{l=1}^{m} \frac{r_l^2}{n-l}. \tag{2.123}$$

The Box-Ljung test is known to produce a Q statistics with a distribution closer to the χ^2 distribution.

2.3.4.1 Partial Autocorrelation

The correlation function for a time series x is simply given by:

$$ACF(\tau) = \frac{\langle x_t x_{t-\tau} \rangle}{\sigma^2}. \tag{2.124}$$

The *partial autocorrelation function* is a conditional correlation that takes into account what we already known about previous values of the time series. For instance, for the PACF(2) we get:

$$\rho_x(2) = \frac{\mathrm{cov}(x_t, x_{t-2} | x_{t-1})}{\sigma(x_t | x_{t-1}) \sigma(x_{t-2} | x_{t-1})} \tag{2.125}$$

and the PACF(3) is given by:

$$\rho_x(3) = \frac{\mathrm{cov}(x_t, x_{t-3} | x_{t-1}, x_{t-2})}{\sigma(x_t | x_{t-1}, x_{t-2}) \sigma(x_{t-3} | x_{t-1}, x_{t-2})}. \tag{2.126}$$

In order to compute this conditional correlation, let's write the elements of the time series as a linear combination of lagged elements:

$$\begin{aligned}\hat{x}_t &= \alpha_{11}x_{t-1}+e_t\\ &= \alpha_{21}x_{t-1}+\alpha_{22}x_{t-2}+e_t\\ &= \alpha_{31}x_{t-1}+\alpha_{32}x_{t-2}+\alpha_{33}x_{t-3}+e_t\\ &\vdots\\ &= \sum_{i=1}^{k}\alpha_{ki}x_{t-i}\end{aligned} \tag{2.127}$$

Equation 2.124 shows that the ACF is the angular coefficient of a linear regression of x_t for an independent $x_{t-\tau}$. Therefore, the α_{kk} coefficients are the PACF(k) since they correspond to the correlation between x_t and x_{t-k}. It is possible to find the α coefficients minimizing the square error:

$$\begin{aligned}0 &= \frac{\partial\left(x_t-\hat{x}_t\right)^2}{\partial\alpha_p}\\ &= \frac{\partial\left(x_t-\sum_{i=1}^{k}\alpha_{ki}x_{t-k}\right)^2}{\partial\alpha_{kp}}\\ &= \frac{\partial\left(x_t^2-2\sum_{i=1}^{k}\alpha_{ki}x_tx_{t-i}+\sum_{ij}\alpha_{ki}\alpha_{kj}x_{t-i}x_{t-j}\right)}{\partial\alpha_{kp}}\\ &= -2x_tx_{t-p}+2\sum_{i=1}^{k}\alpha_{ki}x_{t-i}x_{t-p}.\end{aligned} \tag{2.128}$$

Taking the expected value:

$$\begin{aligned}\langle x_tx_{t-p}\rangle &= \sum_{i=1}^{k}\alpha_{ki}\langle x_{t-i}x_{t-p}\rangle\\ \mathrm{cov}(p) &= \sum_{i=1}^{k}\alpha_{ki}\mathrm{cov}(i-p),\ \text{ divide by }\mathrm{cov}(0)\\ \rho_x(p) &= \sum_{i=1}^{k}\alpha_{ki}\rho_x(i-p).\end{aligned} \tag{2.129}$$

In matrix format:

$$\begin{bmatrix}\rho_x(1)\\ \rho_x(2)\\ \vdots\\ \rho_x(k)\end{bmatrix} = \begin{bmatrix}\rho_x(0) & \rho_x(1) & \dots & \rho_x(k-1)\\ \rho_x(-1) & \rho_x(0) & \dots & \rho_x(k-2)\\ \vdots & \vdots & \ddots & \vdots\\ \rho_x(k-1) & \rho_x(k-2) & \dots & \rho_x(0)\end{bmatrix}\begin{bmatrix}\alpha_{k1}\\ \alpha_{k2}\\ \vdots\\ \alpha_{kk}\end{bmatrix}. \tag{2.130}$$

Or in matrix notation:

$$\mathbf{r}_k = \mathbf{R}_k \mathbf{a}_{k,k}. \tag{2.131}$$

Once the PACF is calculated, a *correlogram* is computed as shown in Fig. 2.13. The value after which the PACF shows a quick drop is an indication about the order of the model. Under the supposition that the coefficients follow a normal distribution, the limits of significance are given by $\pm 1.96/\sqrt{N}$ for a level of 5%, where N is the size of the time series.

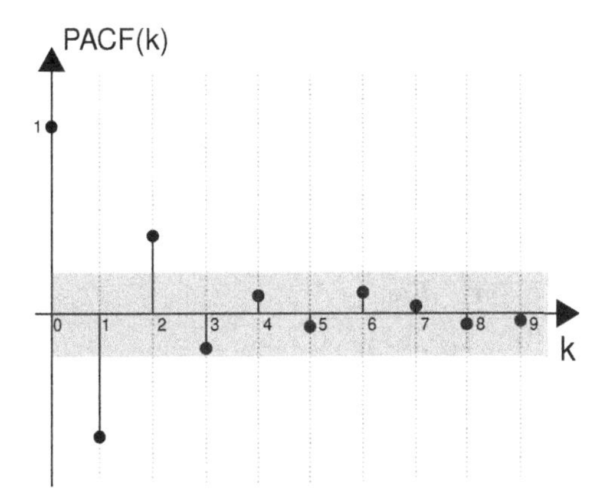

Figure 2.13: Example of PACF where the limit of significance is given by the shaded area. In this example, a second order model could be tested

2.3.4.2 Durbin-Levinson Recursive Method

According to Eq. 2.130 we must do a matrix inversion for each value of k and then find the last element of $\mathbf{a}_{k,k}$. There is, however, a faster method to do this computation devised by Durbin[21] and Levinson[22] [50, 51].

For a matrix:

$$\mathbf{E} = \begin{bmatrix} 0 & 0 & \dots & 0 & 1 \\ 0 & 0 & \dots & 1 & 0 \\ \vdots & \vdots & \ddots & \vdots & \vdots \\ 0 & 1 & \dots & 0 & 0 \\ 1 & 0 & \dots & 0 & 0 \end{bmatrix}, \tag{2.132}$$

it is possible to write Eq. 2.130 as:

$$\begin{bmatrix} \mathbf{r}_{k-1} \\ \rho_x(k) \end{bmatrix} = \begin{bmatrix} \mathbf{R}_{k-1} & \mathbf{E}\mathbf{r}_{k-1} \\ \mathbf{r}_{k-1}^T \mathbf{E} & \rho_x(0) \end{bmatrix} \begin{bmatrix} \mathbf{a}_{k-1,k} \\ \alpha_{kk} \end{bmatrix}. \tag{2.133}$$

The first line of this matrix equation (using $\mathbf{R}_k^{-1}\mathbf{r}_k = \mathbf{a}_{k,k}$) is given by:

$$\begin{aligned}
\mathbf{r}_{k-1} &= \mathbf{R}_{k-1}\mathbf{a}_{k-1,k} + \mathbf{E}\mathbf{r}_{k-1}\alpha_{kk} \\
\mathbf{R}_{k-1}^{-1}\mathbf{r}_{k-1} &= \mathbf{a}_{k-1,k} + \mathbf{R}_{k-1,k}^{-1}\mathbf{E}\mathbf{r}_{k-1}\alpha_{kk} \\
\mathbf{a}_{k-1,k-1} &= \mathbf{a}_{k-1,k} + \mathbf{E}\mathbf{R}_{k-1}^{-1}\mathbf{r}_{k-1}\alpha_{kk} \\
&= \mathbf{a}_{k-1,k} + \mathbf{E}\mathbf{a}_{k-1,k-1}\alpha_{kk} \\
\rightarrow \mathbf{a}_{k-1,k} &= \mathbf{a}_{k-1,k-1} - \mathbf{E}\mathbf{a}_{k-1,k-1}\alpha_{kk}.
\end{aligned} \tag{2.134}$$

The second equation gives:

$$\begin{aligned}
\rho_x(k) &= \mathbf{r}_{k-1}^T\mathbf{E}\mathbf{a}_{k-1,k} + \rho_x(0)\alpha_{kk} \\
&= \mathbf{r}_{k-1}^T\mathbf{E}\left(\mathbf{a}_{k-1,k-1} - \mathbf{E}\mathbf{a}_{k-1,k-1}\alpha_{kk}\right) + \rho_x(0)\alpha_{kk} \\
&= \mathbf{r}_{k-1}^T\mathbf{E}\mathbf{a}_{k-1,k-1} - \mathbf{r}_{k-1}^T\mathbf{E}\mathbf{E}\mathbf{a}_{k-1,k-1}\alpha_{kk} + \rho_x(0)\alpha_{kk} \\
&= \mathbf{r}_{k-1}^T\mathbf{E}\mathbf{a}_{k-1,k-1} + \left(\rho_x(0) - \mathbf{r}_{k-1}^T\mathbf{a}_{k-1,k-1}\right)\alpha_{kk} \\
\rightarrow \alpha_{kk} &= \frac{\rho_x(k) - \mathbf{r}_{k-1}^T\mathbf{E}\mathbf{a}_{k-1,k-1}}{\rho_x(0) - \mathbf{r}_{k-1}^T\mathbf{a}_{k-1,k-1}}.
\end{aligned} \tag{2.135}$$

Thus, it is possible to find the coefficients $\alpha_{k,k}$ recursively in a more computationally effective way.

2.3.4.3 Yule-Walker Equations

Yule[23] and Walker[24] have independently developed a general method [52,53] to find the parameters of time series. This method is based on computing the ACF. Let's describe how this technique works applying it to an AR process:

$$x_i = \sum_{j=1}^{p} \phi_j x_{i-j} + \varepsilon_i. \tag{2.136}$$

Let's multiply it for an element with a delay k:

$$x_{i-k}x_i = \sum_{j=1}^{p} \phi_j x_{i-k}x_{i-j} + x_{i-k}\varepsilon_i, \tag{2.137}$$

and let's take its expected value:

$$\langle x_{i-k}x_i\rangle = \sum_{j=1}^{p} \phi_j \langle x_{i-k}x_{i-j}\rangle + \langle x_{i-k}\varepsilon_i\rangle. \tag{2.138}$$

Since the innovations are now correlated with the signal, we have that:

$$\sigma_{xx}^2(k) = \sum_{j=1}^{p} \phi_j \sigma_{xx}^2(j-k). \tag{2.139}$$

Dividing the whole expression by $\sigma_{xx}^2(0)$:

$$r_{xx}(k) = \sum_{j=1}^{p} \phi_j r_{xx}(j-k). \tag{2.140}$$

If we remember that $r_{xx}(-a) = r_{xx}(a)$ e $r_{xx}(0) = 1$ we can write the previous equation in matrix format:

$$\begin{bmatrix} r_{xx}(1) \\ r_{xx}(2) \\ \vdots \\ r_{xx}(p-1) \\ r_{xx}(p) \end{bmatrix} = \begin{bmatrix} 1 & r_{xx}(1) & \dots & r_{xx}(p-2) & r_{xx}(p-1) \\ r_{xx}(1) & 1 & \dots & r_{xx}(p-3) & r_{xx}(p-2) \\ \vdots & \vdots & & \vdots & \vdots \\ r_{xx}(p-2) & r_{xx}(p-1) & \dots & 1 & r_{xx}(1) \\ r_{xx}(p-1) & r_{xx}(p-2) & \dots & r_{xx}(1) & 1 \end{bmatrix} \begin{bmatrix} \phi_1 \\ \phi_2 \\ \vdots \\ \phi_{p-1} \\ \phi_p \end{bmatrix} \tag{2.141}$$

Or in matrix notation:

$$\mathbf{r} = \mathbf{R}\Phi \tag{2.142}$$

Matrix $\mathbf{R}$ is known as *dispersion matrix*. It is a symmetric matrix with complete rank. Therefore, it is invertible and we can write:

$$\Phi = \mathbf{R}^{-1}\mathbf{r}. \tag{2.143}$$

Thus we can directly obtain the values of the parameters ϕ_n.

2.3.5 HETEROSCEDASTICITY

So far we have only considered the case where time series have the same finite variance along their whole periods. This case is known as *homoscedasticity*. In financial series, however, it is easy to find situations where this does not happen. As we will see ahead in Sec. 2.4.2, variance tends to cluster in time and show some inertia. This change in variance at different time periods is known as *heteroscedasticity*. Therefore, not only it is important to develop models for the variance in time series as it is also important to consider their histories. One way of detecting heteroscedasticity in time series is with the Breusch-Pagan[25] test where the variance of the residuals of a linear regression depends on the independent variable.

In order to model heteroscedasticity, Engle[26] proposed an *autoregressive conditional heteroskedasticity* (ARCH) [26, 54, 55] process, for which he was awarded the Nobel prize in 2003. Consider a stochastic variable for the log-returns given by $r_t = \sigma_t \varepsilon_t$, where ε_t is normally distributed $\sim \mathscr{N}(0,1)$. An ARCH process is then given by:

$$\sigma_t^2 = \alpha_0 + \sum_{n=1}^{q} \alpha_n r_{t-n}^2. \tag{2.144}$$

For the ARCH(1) process, given a history $\mathscr{H}_t = r_t, r_{t-1}, r_{t-2}, \ldots$, we have:

$$\begin{aligned}\mathrm{VAR}(r_t|\mathscr{H}_{t-1}) &= \langle r_t^2|\mathscr{H}_{t-1}\rangle \\ &= \langle \sigma_t^2 \varepsilon_t^2|\mathscr{H}_{t-1}\rangle \\ &= \sigma_t^2 \langle \varepsilon_t^2|\mathscr{H}_{t-1}\rangle \\ &= \sigma_t^2 = \alpha_0 + \alpha_1 r_{t-1}^2.\end{aligned} \tag{2.145}$$

The unconditional variance, though, can be found using the law of total expectation:

$$\sigma_u^2 = \langle \mathrm{VAR}(r_t|\mathscr{H}_{t-1})\rangle = \alpha_0 + \alpha_1 \sigma_u^2 \rightarrow \alpha_0 = (1-\alpha_1)\sigma_u^2. \tag{2.146}$$

Therefore, the conditional variance can also be written as:

$$\sigma_t^2 = (1-\alpha_1)\sigma_u^2 + \alpha_1 r_{t-1}^2 = \sigma_u^2 + \alpha_1 (r_{t-1}^2 - \sigma_u^2). \tag{2.147}$$

From this equation we see that the conditional variance of the ARCH(1) model is a linear combination of the unconditional variance and a deviation of the square error.

The ARCH(1) process has zero conditional expectation, though:

$$\langle r_t|\mathscr{H}_{t-1}\rangle = \langle \sigma_t \varepsilon_t|\mathscr{H}_{t-1}\rangle = \sigma_t \langle \varepsilon_t|r_{t-1}, r_{t-2}, \ldots\rangle = 0. \tag{2.148}$$

Using the law of total expectation again we see that the unconditional expectation is also zero:

$$\langle\langle r_t|\mathscr{H}_{t-1}\rangle\rangle = \langle 0\rangle = 0. \tag{2.149}$$

We can use these results to obtain the covariance:

$$\begin{aligned}\langle r_t r_{t-1}\rangle &= \langle\langle r_t r_{t-1}|\mathscr{H}_{t-1}\rangle\rangle \\ &= \langle r_{t-1}\langle r_t|\mathscr{H}_{t-1}\rangle\rangle \\ &= \langle r_{t-1} 0\rangle = 0.\end{aligned} \tag{2.150}$$

Since one cannot predict r_t based on its history $\mathscr{H}_{t-1}$, the ARCH(1) model obeys the EMH (see Sec. 1.3).

2.3.5.1 Generalized ARCH Model

A generalization of ARCH models (GARCH) in the form of:

$$\sigma_t^2 = \alpha_0 + \sum_{n=1}^{p} \alpha_n r_{t-n}^2 + \sum_{m=1}^{q} \beta_n \sigma_{t-m}^2 \tag{2.151}$$

was proposed in 1986 by Bollerslev[27] and Taylor[28] [56, 57].

The GARCH(1,1) process is given by:

$$\mathrm{VAR}(r_t|\mathscr{H}_{t-1}) = \sigma_t^2 = \alpha_0 + \alpha_1 r_{t-1}^2 + \beta \sigma_{t-1}^2. \tag{2.152}$$

This can be written as an infinite ARCH process if we rewrite it back substituting it in itself:

$$\begin{aligned}\sigma_t^2 &= \alpha_0 + \alpha_1 r_{t-1}^2 + \beta(\alpha_0 + \alpha_1 r_{t-2}^2 + \beta\sigma_{t-2}^2)\\ &= \alpha_0 + \beta\alpha_0 + \alpha_1 r_{t-1}^2 + \alpha_1\beta r_{t-2}^2 + \beta^2\sigma_{t-2}^2\\ &= \alpha_0 + \beta\alpha_0 + \alpha_1 r_{t-1}^2 + \alpha_1\beta r_{t-2}^2 + \beta^2(\alpha_0 + \alpha_1 r_{t-3}^2 + \beta\sigma_{t-3}^2)\\ &\to \frac{\alpha_0}{1-\beta} + \alpha_1\sum_{n=1}^{\infty}\beta^{n-1}r_{t-n}^2, \text{ if } |\beta|<1.\end{aligned} \tag{2.153}$$

Its unconditional variance is given by:

$$\begin{aligned}\sigma_u^2 &= \langle\sigma_t^2\rangle = \langle\alpha_0 + \alpha_1\sigma_{t-1}^2\varepsilon_{t-1}^2 + \beta\sigma_{t-1}^2\rangle\\ &= \alpha_0 + \alpha_1\sigma_u^2 + \beta\sigma_u^2\\ \sigma_u^2 &= \frac{\alpha_0}{1-(\alpha_1+\beta)}.\end{aligned} \tag{2.154}$$

We can use the unconditional variance to compute the unconditional kurtosis. In order to do this, first we must calculate the expected value of the fourth power of $\langle\varepsilon\rangle$:

$$\begin{aligned}\langle\varepsilon^4\rangle &= \frac{1}{\sqrt{2\pi}}\int_{-\infty}^{\infty}\varepsilon^4 e^{-\varepsilon^2/2}d\varepsilon\\ &= -\frac{1}{\sqrt{2\pi}}\int_{-\infty}^{\infty}\varepsilon^3\frac{d}{d\varepsilon}e^{-\varepsilon^2/2}d\varepsilon\\ &= -\frac{1}{\sqrt{2\pi}}\left(\varepsilon^3 e^{-\varepsilon^2/2}\Big|_{-\infty}^{\infty} - 3\int_{-\infty}^{\infty}x^2\frac{d}{d\varepsilon}e^{-\varepsilon^2/2}d\varepsilon\right)\\ &= 3\langle\varepsilon^2\rangle = 3.\end{aligned} \tag{2.155}$$

The expected value of the fourth power of σ_u is:

$$\begin{aligned}\sigma_u^4 = \langle\sigma_t^4\rangle &= \langle(\alpha_0 + \alpha_1 r_{t-1}^2 + \beta\sigma_{t-1}^2)^2\rangle\\ &= \langle(\alpha_0 + \alpha_1\varepsilon_{t-1}^2\sigma_{t-1}^2 + \beta\sigma_{t-1}^2)^2\rangle\\ &= \langle\alpha_0^2 + (\alpha_1\varepsilon_{t-1}^2 + \beta)^2\sigma_{t-1}^4 + 2\alpha_0(\alpha_1\varepsilon_{t-1}^2 + \beta)\sigma_{t-1}^2\rangle\\ &= \langle\alpha_0^2 + (\alpha_1^2\varepsilon_{t-1}^4 + \beta^2 + 2\beta\alpha_1\varepsilon_{t-1}^2)\sigma_{t-1}^4 + 2\alpha_0(\alpha_1\varepsilon_{t-1}^2 + \beta)\sigma_{t-1}^2\rangle\\ &= \alpha_0^2 + (3\alpha_1^2 + \beta^2 + 2\alpha_1\beta)\sigma_u^4 + 2\alpha_0(\alpha_1+\beta)\sigma_u^2\\ &= \alpha_0^2 + (3\alpha_1^2 + \beta^2 + 2\alpha_1\beta)\sigma_u^4 + \frac{2\alpha_0^2(\alpha_1+\beta)}{1-(\alpha_1+\beta)}\\ &= \frac{\alpha_0^2 - (\alpha_1+\beta)\alpha_0^2 + 2\alpha_0^2(\alpha_1+\beta)}{1-(\alpha_1+\beta)} + (3\alpha_1^2 + \beta^2 + 2\alpha_1\beta)\sigma_u^4\\ &= \frac{(1+\alpha_1+\beta)\alpha_0^2}{(1-\alpha_1-\beta)(1-3\alpha_1^2-\beta^2-2\alpha_1\beta)},\end{aligned} \tag{2.156}$$

given that $1-3\alpha_1^2-\beta^2-2\alpha_1\beta>0$.

The kurtosis of the GARCH(1,1) model is thus:

$$\begin{aligned}
K &= \frac{\langle r_t^4\rangle}{\langle r_t^2\rangle^2} = \frac{\langle \sigma_t^4\varepsilon_t^4\rangle}{\langle \sigma_t^2\varepsilon_t^2\rangle} = 3\frac{\sigma_u^4}{\sigma_u^2} \\
&= 3\frac{(1+\alpha_1+\beta)\alpha_0^2}{(1-\alpha_1-\beta)(1-3\alpha_1^2-\beta^2-2\alpha_1\beta)}\left(\frac{1-(\alpha_1+\beta)}{\alpha_0}\right)^2 \\
&= 3\frac{(1+\alpha_1+\beta)(1-\alpha_1-\beta)}{1-3\alpha_1^2-\beta^2-2\alpha_1\beta} = 3\frac{1-(\alpha_1+\beta)^2}{1-(\alpha_1+\beta)^2-2\alpha_1^2} \\
&= 3+\frac{6\alpha_1^2}{1-3\alpha_1^2-\beta^2-2\alpha_1\beta}.
\end{aligned} \tag{2.157}$$

Note that since $1-3\alpha_1^2-\beta^2-2\alpha_1\beta > 0$, the excess kurtosis is always positive. Therefore, the GARCH(1,1) model can model heavy tails.

ARCH and GARCH models can be easily handled in python. The following snippet, for instance, fits previously obtained log-returns with a GARCH(1,1) model and displays a summary of information related to this fitting.

```
from arch import arch_model

model = arch_model(log_returns,vol='GARCH',p=1,q=1,dist='Normal')
model_fit = model.fit()
estimation = model_fit.forecast()
print(model_fit.summary())
```

2.4 STYLIZED EMPIRICAL FACTS OF FINANCIAL TIME SERIES

Econophysics as a science has the huge ambition of discovering specific dynamics of economics systems that do not depend directly on the human will. This, however, creates a huge problem. Given that economics deals with human beings and they act subjectively, the conclusion that the human action is unpredictable is striking. Indeed many sociologists have pointed out that even if one embarks on a mission to forecast the human action, the very act of trying to forecast it would impact the measurement itself [58]. Despite this burden, many regularities have been empirically observed in the economic and social human activities when one deals with summaries of relevant facts. These regularities have been called by Kaldor[29] as *stylized facts* [59, 60], but even before him, general tendencies of the economic activity had already been observed [61].

It is important to point out that these stylized facts are just general tendencies and may hide several details. For instance, it is observed a tendency for salaries to correlate with the level of education. Nonetheless, it is not uncommon to find graduate students having low incomes during their period in graduate school.

Thus, stylized facts behave for the social scientist the same way physical measurements behave for the natural scientist. The econophysicist observes these regularities as natural phenomena and tries to explain them using the mathematical framework developed to explain similar physical phenomena. Furthermore, the existence of these stylized facts implies that these regularities can be extracted and separated from more holistic frameworks. This agrees with the approach of Brentano[30] where a phenomenon could be studied and understood apart from the whole. Moreover, according to his view, specific historical and psychological details were not determinant for the study of economic systems. This view influenced a legion of economists such as Menger that used it to elaborate his theory of marginal utility.

Since stylized facts deal with tendencies, they are best analyzed with statistical tools. However, in order to use many statistical tools, the time series has to be *stationary* and *ergodic*. If the dataset is not stationary, the statistical properties change over time and no generalization can be obtained. Also, we expect that statistical averages of the time series estimate real averages. Furthermore, finite size effects have to be taken care of when necessary. Many statistical analyses are performed on small time intervals and this may not represent a general tendency for longer periods.

2.4.1 STATIONARITY

A *stochastic process* is defined as an indexed collection of random variables $\{X_1, X_2, X_3, \ldots, X_N\}$. If the unconditional joint probability of this process is time invariant, then this process is *stationary* (or *strictly* or *strongly* stationary). Mathematically:

$$F_X(x_{t_1+\tau}, \ldots, x_{t_N+\tau}) = F_X(x_{t_1}, \ldots, x_{t_N}). \tag{2.158}$$

Therefore, statistical parameters such as the average and variance are also time independent.

This is, nonetheless, a very strict definition that can often complicate computations. We can relax this restriction by imposing that only the average and the autocovariance are time invariant and the second moment is finite. Under these restrictions, the stochastic process is said to be *wide-sense stationary* (or *covariance* stationary). Mathematically:

$$\mu = \langle x(t) \rangle = \langle x(t+\Delta) \rangle, \ \forall \Delta \in \mathbb{R}. \tag{2.159}$$

The autocovariance is a function that tells us how much similar a function is to itself when compared at two different time points. Mathematically it is given by:

$$\begin{aligned} C_{XX}(t_1, t_2) &= \langle (x(t_1) - \langle x(t_1) \rangle)(x(t_2) - \langle x(t_2) \rangle) \rangle \\ &= \langle (x(t_1) - \mu)(x(t_2) - \mu) \rangle \\ &= C_{XX}(t_1 - t_2, 0) = C_{XX}(\Delta, 0) = C_{XX}(\Delta). \end{aligned} \tag{2.160}$$

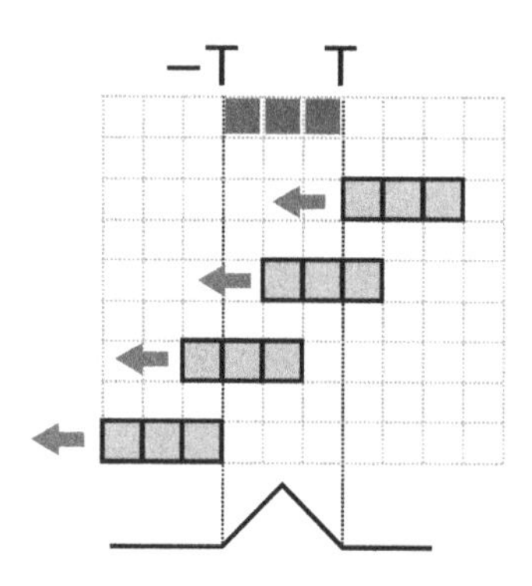

Figure 2.14: Graphical representation of an autocovariance process. A copy of a signal is translated in time while the product between them is computed at every time shift

Indeed, we see that if we make:

$$\begin{aligned} C_{XX}(t_1-t_2,t_2-t_2) &= \langle (x(t_1-t_2)-\langle x(t_1-t_2)\rangle)(x(t_2-t_2)-\langle x(t_2-t_2)\rangle)\rangle \\ &= \langle (x(t_1-t_2)-\mu)(x(0)-\mu)\rangle \\ &\propto C_{XX}(t_1-t_2,0) = C_{XX}(\Delta,0), \end{aligned} \tag{2.161}$$

since $\mu = \langle x(0)\rangle$ is an average and $\neq x(0)$ is the value at that point. Thus, we can define an autocorrelation function indexed only by a time difference as graphically shown in Fig. 2.14.

In time series, the stationarity of a stochastic process can be verified by the presence of a unit root. To see how the importance of the unit root, let's take the MA representation of an AR1 process:

$$\begin{aligned} y_t &= a\cdot y_{t-1}+\varepsilon_t \\ &= a(a\cdot y_{t-2}+\varepsilon_{t-1})+\varepsilon_t \\ &= a\left[a(a\cdot y_{t-3}+\varepsilon_{t-2})+\varepsilon_{t-1}\right]+\varepsilon_t \\ &\to \sum_{n=0}^{\infty} a^n\varepsilon_{t-n}. \end{aligned} \tag{2.162}$$

The variance of this process is given by:

$$\begin{aligned} var\{y_t\} &= \sum_{n=0}^{\infty} var\{a^n\varepsilon_{t-n}\} \\ \sigma_y^2 &= \sum_{n=0}^{\infty} a^{2n} var\{\varepsilon_{t-n}\} \\ &= \frac{\sigma_e^2}{(1-a^2)}, \text{ if } |a|<1. \end{aligned} \tag{2.163}$$

On the other hand, if $a = 1$, then

$$\sigma_y^2 = t^2 \sigma_e^2 \tag{2.164}$$

and the variance not only depends but increases with time.

A popular method to check for the unit root in time series is using the Dickey-Fuller[31] test. This test considers an AR1 process:

$$\begin{aligned} y_t - y_{t-1} &= (a-1)y_{t-1} + \varepsilon_t \\ \Delta y_t &= \beta y_{t-1} + \varepsilon_t. \end{aligned} \tag{2.165}$$

Coefficient β is obtained with a linear regression between Δy_t and y_{t-1}. The process is stationary when $\beta < 0$.

2.4.1.1 Fourier Transform

The autocorrelation function is actually a very important tool to access many stylized facts in time series. Although the definition given above is accurate, it is much simpler to calculate it using a Fourier[32] transform. The Fourier transform pair used in this book is given by:

$$\begin{aligned} F(\omega) &= \mathscr{F}\{f(t)\} = \int_{-\infty}^{\infty} f(t) e^{-j\omega t} dt \\ f(t) &= \mathscr{F}^{-1}\{F(\omega)\} = \frac{1}{2\pi} \int_{-\infty}^{\infty} F(\omega) e^{j\omega t} d\omega. \end{aligned} \tag{2.166}$$

Numerically, it is estimated by the discrete Fourier transform (DFT) given by:

$$X_m = \sum_{n=0}^{N-1} x_n e^{-j\omega_m n}, \tag{2.167}$$

where

$$\omega_m = \frac{2\pi m}{M}, \ m \in \{0, 1, \ldots, M-1\}. \tag{2.168}$$

The variable m going from $M/2$ to M produces an angular frequency that goes from π to 2π, which is equivalent of it going from $-\pi$ back to 0. Therefore the first $M/2$ points correspond to positive frequencies whereas the last $M/2$ points correspond to reflected negative frequencies. This mechanism is shown in Fig. 2.15.

The DFT is computed, though, using efficient algorithms such as *Cooley-Tuckey*[33]. When using such algorithms, the DFT is known as *fast Fourier transform* (FFT).

The DFT assumes a periodic signal. If we simply ignore it, the Fourier transform can show spectral leakage and have a reduced dynamical reserve. In order to circumvent this problem, the signal is typically multiplied by a window function such as the Hann window:

$$w_n = \sin^2\left(\frac{n\pi}{N-1}\right). \tag{2.169}$$

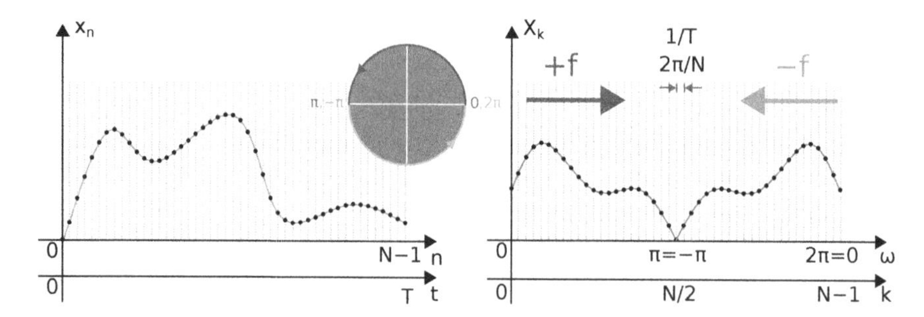

Figure 2.15: Left: A discretized signal containing N points within a period T. Right: Its discrete Fourier transform. The inset depicts the Fourier circle showing the equivalence between going from 0π to 2π and from $-\pi$ back to 0

The autocorrelation function can nicely be computed from the Fourier transform and take advantage from these fast algorithms. For this, let's consider a signal given by its Fourier transform:

$$x(t) = \frac{1}{2\pi}\int_{-\infty}^{\infty} X(\omega)e^{-j\omega t}d\omega. \tag{2.170}$$

The autocovariance function can be written as:

$$\begin{aligned}C_{XX}(\tau) &= \lim_{T\to\infty}\frac{1}{T}\int_{-T/2}^{T/2}\left(\frac{1}{2\pi}\int_{-\infty}^{\infty}X(\omega')e^{-j\omega'(t+\tau)}d\omega'\right)\left(\frac{1}{2\pi}\int_{-\infty}^{\infty}X^*(\omega)e^{j\omega t}d\omega\right)dt\\ &= \frac{1}{(2\pi)^2}\int_{-\infty}^{\infty}\int_{-\infty}^{\infty}X(\omega')X^*(\omega)\left(\lim_{T\to\infty}\frac{1}{T}\int_{-T/2}^{T/2}e^{-j(\omega'-\omega)t}dt\right)e^{-j\omega'\tau}\,d\omega\,d\omega'.\end{aligned} \tag{2.171}$$

It is easy to show that the integral inside the parentheses gives a delta function $\delta(\omega-\omega')$. We are then left with:

$$\begin{aligned}C_{XX}(\tau) &= \frac{1}{(2\pi)^2}\int_{-\infty}^{\infty}\int_{-\infty}^{\infty}R(\omega)R^*(\omega')\delta(\omega-\omega')e^{-j\omega'\tau}d\omega\,d\omega'\\ &= \frac{1}{(2\pi)^2}\int_{-\infty}^{\infty}|R(\omega)|^2e^{-j\omega\tau}d\omega\\ &= \frac{1}{2\pi}\mathscr{F}^{-1}\{R(\omega)R^*(\omega)\}.\end{aligned} \tag{2.172}$$

The argument of the inverse Fourier transform is known as *power spectral density* (PSD) and the main result is known as the *Wiener-Khinchin theorem*[34].

This can be implemented in Python according to the following snippet:

```
N = len(data)

# Hanning window
s = [w*d for w,d in zip(np.hanning(N), data)]

# FFT
F = np.fft.fft(s)
S = [f*np.conj(f)/N for f in F]

# Autocorrelation
z = np.fft.ifft(S).real
C = np.append(z[N/2:N], z[0:N/2])
```

2.4.2 COMMON EMPIRICAL FACTS

Figure 2.16 shows the price of Bitcoin[g] and its normalized returns for approximately 7000 minutes from a random initial date.

Money

Money is the most liquid asset in an economy that: i) can be generally accepted as a medium of exchange, ii) can be used as a unit of accounting, and iii) can store value.

For instance, precious metals satisfy all these three requirements and for a long time gold was used as currency. After the adoption of the Bretton Woods system of monetary management in 1944, most economies adopted the *gold standard* monetary system where the value of a paper money is based on a fixed amount of gold. Since 1973 most economies (specially the US with the 'Nixon shock') dropped the gold standard in favor of *fiat money*—a government issued currency not backed by any precious metal.

Since Bitcoin does not meet the standard definition of money, we prefer to call it just *digital good.*

The distribution of returns shown in Fig. 2.17 is clearly not Gaussian, but fits much better with a heavy tail distribution such as the t-student distribution. This is

[g] A popular digital good based on blockchain technology.

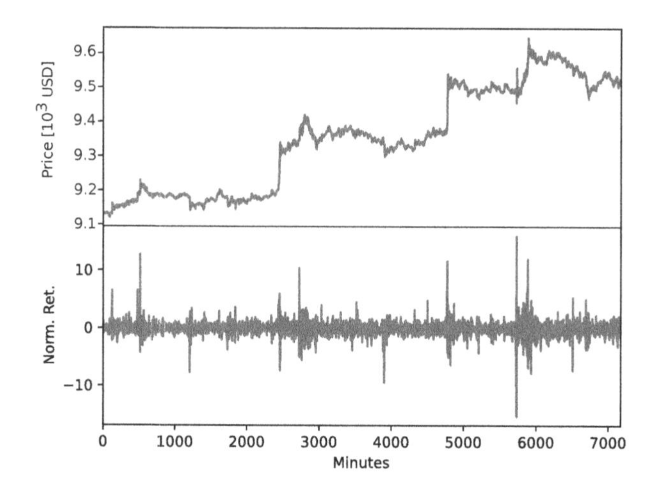

Figure 2.16: Top: Price of Bitcoin in US$ for approximately 7000 consecutive minutes obtained at a random initial data, and Bottom: Respective normalized returns for the same period

a known stylized fact: the distribution of log returns for many financial assets has a tendency to form fat tails for high frequency data. It is important to emphasize that this tendency occurs for high frequency data. If the returns are calculated for longer time steps, this tendency disappears and the distribution approaches a Gaussian distribution. This is another stylized fact known as *aggregational Gaussianity*.

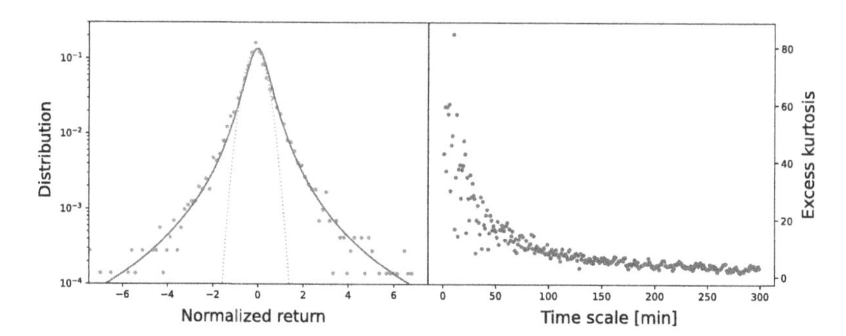

Figure 2.17: Left: The solid disks correspond to the estimated distribution of normalized returns for the time series shown in Fig. 2.16. The dotted line is a Gaussian fitting of the data, whereas the solid line is a fitting with a t-student distribution. Right: Excess kurtosis for the distribution as a function of the time scale used to calculate the log-returns

It is widely reported that price returns in liquid markets[h] have either no correlation or correlations that decay quickly to zero [62]. Since no periodicity can be found

[h]A market with a significant number of buyers and sellers and relatively low transaction costs.

and no strategy can be developed using this information, the absence of correlation is usually considered an indication of an efficient market. Although returns do not show any significant correlation, the volatility[i] usually does and is known as *volatility clustering*. Therefore, volatility exhibits a memory effect and price variations tend to show persistence over finite periods. This behavior is visualized in Fig. 2.18. Another way to access whether the time series shows volatility clustering is fitting the time series with a GARCH process.

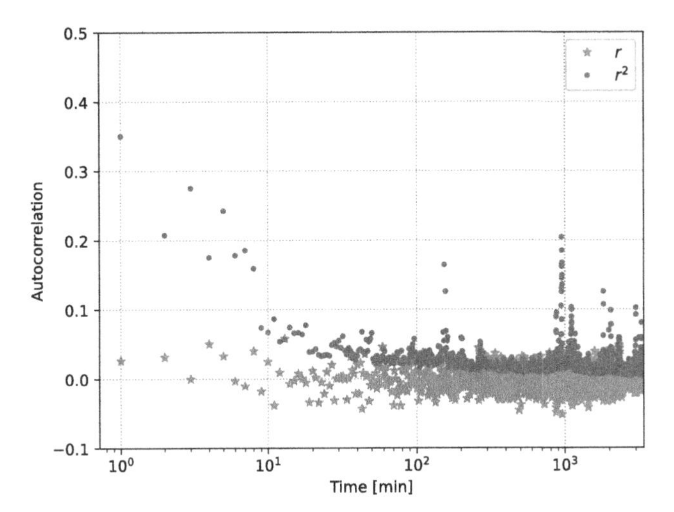

Figure 2.18: The autocorrelation function of returns (stars) and returns squared (circles). Whereas the former does not show any significative correlation, the latter explicitly shows some memory

Financial assets, including Bitcoin, have many other stylized facts [62] such as positive correlation between volume and volatility and *fluctuation scaling* [63]. The latter is characterized by a power law between the average return and the corresponding volatility. This is known in biology as *Taylor's law*[35].

Notes

[1]Brook Taylor (1685–1731) English mathematician.

[2]Andrey Nikolaevich Kolmogorov (1903–1987) Russian mathematician.

[3]Originally defined by Henri Léon Lebesgue (1875–1941) French mathematician; advisee of Émile Borel.

[4]Félix Édouard Justin Émile Borel (1871–1956) French mathematician advisor of Henri Lebesgue among others.

[5]Leonhard Euler (1707–1783) Swiss mathematician; advisee of Johann Bernoulli and advisor of Joseph-Louis Lagrange, among others.

[6]Johann Karl August Radon (1887–1956) Austrian mathematician and Otto Marcin Nikodym (1887–1974) polish mathematician.

[i]The variance within a given period.

[7]Pierre-Simon, marquis de Laplace (1749–1827) French polymath. Advisee of Jean d'Alembert and advisor of Siméon Denis Poisson.

[8]Paul Pierre Lévy (1886–1971) French mathematician, advisee of Jacques Hadamard and Vito Volterra. Adviser of Benoît Mandelbrot.

[9]Norbert Wiener (1894–1964) North American mathematician and philosopher. Advisee of Josiah Royce and adviser of Amar Bose.

[10]Identified in 1929 by Bruno de Finetti (1906–1985) Italian statistician.

[11]This notation was introduced in 1909 by Edmund Georg Hermann Landau (1877–1938) German mathematician; advisee of Georg Frobenius and adviser of Niels Bohr's brother Harald Bohr among others.

[12]Andrey Andreyevich Markov (1856–1922) Russian mathematician, advisee of Pafnuty Lvovich Chebyshev and adviser of Aleksandr Lyapunov and Abraham Besicovitch among others.

[13]Herman Chernoff (1923–) American mathematician, advisee of Abraham Wald.

[14]Vilfredo Federico Damaso Pareto (1848–1923) italian polymath.

[15]Bruce M Hill (1944–2019) American statistician.

[16]Max Otto Lorenz (1876–1959) American economist.

[17]Corrado Gini (1884–1965) Italian statistician.

[18]Shown in 1938 para Herman Ole Andreas (1908–1992) Swedish statistician.

[19]George Edward Pelham Box (1919–2013) British statistician, advisee of Egon Sharpe Pearson, son of Karl Pearson. Box created the famous aphorism: "All models are wrong, but some are useful."

[20]Gwilym Meirion Jenkins (1932–1982) British statistician.

[21]James Durbin (1923–2012) British statistician.

[22]Norman Levinson (1912–1975) American mathematician, advisee of Norbert Wiener.

[23]George Udny Yule (1871–1951) British statistician.

[24]Sir Gilbert Thomas Walker (1868–1958) English physicist and statistician.

[25]Trevor Stanley Breusch (1953–) Australian econometrician, and Adrian Rodney Pagan (1947–) Australian economist.

[26]Robert Fry Engle III (1942–) North American statistician.

[27]Tim Peter Bollerslev (1958–) Danish economist, advisee of Robert F. Engle.

[28]Stephen John Taylor (1954–) British economist.

[29]Nicholas Kaldor (1908–1986) Hungarian economist, advisee of Lionel Robins.

[30]Franz Clemens Honoratus Hermann Brentano (1838–1917) Prussian philosopher and psychologist, adviser of Edmund Husserl and Sigmund Freud among others.

[31]David Alan Dickey (1945–) advisee of Wayne Arthur Fuller (1931–) American statisticians.

[32]Jean-Baptiste Joseph Fourier (1768–1830) French mathematician, advisee of Joseph-Louis Lagrange and adviser of Peter Gustav Lejeune Dirichlet and Claude-Louis Navier among others.

[33]James William Cooley (1926–2016) and John Wilder Tukey (1915–2000) American mathematicians.

[34]Norbert Wiener (1894–1964) American mathematician, adviser of Amar Bose among others. Aleksandr Yakovlevich Khinchin (1894–1959) Russian mathematician.

[35]Lionel Roy Taylor (1924–2007) British ecologist.

3 Stochastic Calculus

The book value of a company is defined as its assets subtracted from its liabilities. Accountants use this information to build a consistent economic history of the company allowing investors to make forecasts about its future value.

In 2013, though, Apple had one of its highest sales ever, its balances were in good shape and everything pointed to a bright future. Investors, on the other hand, had a pessimistic view about the company and were trading its stocks at low prices. This kind of mismatch between the economic health of a company and the negotiated price of its stocks led regulators to push companies to use a *fair-value* type accounting based on the value a company would worth if it were to be sold. One of this accounting strategies, mark-to-market (MTM), is even part of the accounting standards adopted by the U.S. Securities and Exchange Commission (SEC).

On the other hand, this kind of accounting, for example, pushed Enron, a company spotted six consecutive years as the most innovative in America, to adopt fraudulent accounting practices. For instance, it would build a power plant and adopt the projected profit as real and actual profit in order to elevate its public image. Had the profit not materialized, it would transfer the asset to an off-the-books corporation and hide the loss. After the fraudulent behavior was discovered, the company's price per share dropped from \$90 to less than \$1 in about one year, the company went bankrupt, its employees lost \$2 billions from their pensions, and shareholders lost \$74 billions.

What is a fair game? What is the probability for an event like this or a change of expectations to happen? This chapter deals with random processes and their calculations. We will begin studying fair games, move to stochastic chains, mean reverting processes, and then close with point processes.

3.1 MARTINGALES AND FAIR GAMES

Martingales [31, 64] are stochastic processes whose name is borrowed from the vocabulary of gamblers from France. It was first used by Ville[1] in 1939 [65], but the theory dates back to 1934 with the works of Paul Lévy [66]. In a dÁlembert's[2] martingale, for instance, a gambler starts with one coin. He loses a bet and doubles the next bet for two coins. He loses this bet again and accumulates a loss of three coins. Once again, the gambler doubles the next bet for four coins but this time the gambler wins. Thus, the gambler wins four coins and accumulates a net result of one coin. As long as the gambler keeps this strategy, it is possible to end up with the net result of one coin. For the nth bet, the gambler bets $b_n = 2^{n-1}$ coins. If he stops at this bet, he accumulates a result of:

$$S_n = b_n - \sum_{k=1}^{n-1} b_k = 2^{n-1} - \sum_{k=1}^{n-1} 2^{k-1} = 2^{n-1} - \frac{1-2^{n-1}}{1-2} = 1.$$

DOI: 10.1201/9781003127956-3

Martingales are associated with the concept of a *fair game*. This means that for such game the chance of winning a certain amount is the same of losing this same amount. If we apply this concept to the market we say that the current price cannot depend on its history, the market price has no memory. Therefore, there is no opportunity for arbitrage and, according to Fama (Sec. 1.3), this models an efficient market. We can say that, independent of its history $X_1,\ldots,X_n$, the expected value for the next value of this stochastic process is exactly the present value. Mathematically, this is described by:

$$\langle X_{n+1}|X_1,\ldots,X_n\rangle = X_n, \tag{3.1}$$

given that the expected present value is finite:

$$\langle |X_n|\rangle < \infty. \tag{3.2}$$

The best way to grasp the mathematical usefulness of martingales is by studying a set of examples. Let's consider first a *martingale bet* (not to be confused with a martingale process) where a random variable X can assume the value $+1$ if the gambler wins or -1 if the gambler loses. Let's also define a random variable B for the current bet. For the nth bet, the wealth of the gambler is described by the random variable Y as:

$$Y_n = Y_0 + \sum_{i=1}^{n} B_i X_i, \tag{3.3}$$

where Y_0 is the initial wealth. The conditional expected value for the wealth at the next bet is given by:

$$\begin{aligned}
\langle Y_{n+1}|Y_0,Y_1,\ldots,Y_n\rangle &= \left\langle Y_0 + \sum_{i=1}^{n+1} B_i X_i \middle| Y_0,\ldots,Y_n \right\rangle \\
&= \left\langle Y_0 + \sum_{i=1}^{n} B_i X_i + B_{n+1}X_{n+1} \middle| Y_0,\ldots,Y_n \right\rangle \\
&= \left\langle Y_0 + \sum_{i=1}^{n} B_i X_i \middle| Y_0,\ldots,Y_n \right\rangle + \langle B_{n+1}X_{n+1}|Y_0,\ldots,Y_n\rangle \\
&= \langle Y_n|Y_0,\ldots,Y_n\rangle + \frac{1}{2}(B_{n+1} - B_{n+1}) \\
&= Y_n.
\end{aligned} \tag{3.4}$$

The fact that the conditional expected future value does not depend on the history of the random variable is known as *martingale property*. It is also possible to define processes in which current value of the random variable works as a limiter for the expected conditional future value. If it works as a limiter for the lower bound:

$$\langle X_{n+1}|X_1,\ldots,X_n\rangle \geq X_n, \tag{3.5}$$

then this is known as a *submartingale*. On the other hand, if it is a limiter for the upper bound:

$$\langle X_{n+1}|X_1,\ldots,X_n\rangle \leq X_n, \tag{3.6}$$

then this process is known as a *supermartingale*.

Let's now consider a fair game where the gambler tosses an unbiased coin and wins a constant amount Δ depending on the side that is showing up when it lands. If the wealth is described by the random variable X_n for the nth bet, the expected conditional value is $X_n+\Delta$ with a 50 % chance and $X_n-\Delta$ with the same chance. Therefore, $\langle X_{n+1}|X_1\ldots X_n\rangle = 1/2(X_n+\Delta)+1/2(X_n-\Delta)=X_n$. However, if the coin is biased towards winning, we have a supermartingale. Otherwise, it is a submartingale.

In a broader sense, martingales can be defined in terms of *filtration*[3]. Consider $\mathscr{F}_t$ as the set of all measurement elements up to time t (a σ-algebra). The finite sequence $\mathscr{F}_{0\leq t\leq T}$ such that $\mathscr{F}_0\subseteq\mathscr{F}_1\subseteq\ldots\subseteq\mathscr{F}_t\subseteq\ldots\subseteq\mathscr{F}_T$ is considered a filtration of a sample space Ω. If the filtration is composed of σ-algebras generated by $\{X_1,\ldots,X_n\}$, then $(\mathscr{F}_n)$ is the *natural filtration* of the sequence (X_n). Thus, a random variable is said to be *adapted to the filtration* $\mathscr{F}_n : n\in\mathbb{N}$ if $X_n\in\mathscr{F}_n$.

For instance, let's assume a binary market where we only consider if the price of an asset went up (+) or down (−). For a two period binary market, the sample space is $\Omega=\{++,+-,-+,--\}$. The smallest σ-algebra is given by $\{\emptyset,\Omega\}$ and will be denoted by $\mathscr{F}_0$. For the first period, in addition to the trivial σ-algebra, there are two more possibilities: $F_+=\{++,+-\}$ e $F_-\{-+,--\}$. Therefore, $\mathscr{F}_1=\{\mathscr{F}_t,F_+,F_-\}$. For the second period, we have all subsets of Ω.

With the definition of filtration, we can define a martingale $X_n : n\in\mathbb{N}$ with respect to the filtration $\mathscr{F}_n : n\in\mathbb{N}$:

1. $\langle|X_n|\rangle<\infty$;
2. X_n is adapted to the filtration $\mathscr{F}_n$, i.e. X_n is $\mathscr{F}_n$-measurable;
3. $\langle X_{n+1}|\mathscr{F}_n\rangle = X_n, \forall n\in\mathbb{N}$.

Another widely used type of martingale is the *Doob martingale* [36,64][4]. For any bounded function $f:\Omega^n\to\mathbb{R}, f(X_1,\ldots,X_n)$ the Doob martingale is given by:

$$\begin{aligned} Z_n &= \langle f(X)|\mathscr{F}_n\rangle \\ Z_0 &= \langle f(X)\rangle. \end{aligned} \tag{3.7}$$

Such Doob process is a martingale because:

$$\begin{aligned} \langle Z_n|\mathscr{F}_{n-1}\rangle &= \langle\langle f(X)|\mathscr{F}_n\rangle|\mathscr{F}_{n-1}\rangle \\ &= \langle f(X)|\mathscr{F}_{n-1}\rangle, \text{ law of total expectation} \\ &= Z_{n-1}. \end{aligned} \tag{3.8}$$

3.1.1 RANDOM WALKS

Let's analyze here some random walks. For instance, let X be a random variable such that $\langle X \rangle = \mu$. The position of a particle after n steps is given by $Z_n = \sum_{i=1}^{n}(X_i - \mu)$. The expected conditional value of its future position is given by:

$$
\begin{aligned}
\langle Z_{n+1}|Z_1 \dots Z_n \rangle &= \left\langle \sum_{i=1}^{n+1}(X_i - \mu) \middle| Z_1 \dots Z_n \right\rangle \\
&= \left\langle \sum_{i=1}^{n}(X_i - \mu) + X_{n+1} - \mu \middle| Z_1 \dots Z_n \right\rangle \\
&= \langle Z_n + X_{n+1} - \mu | Z_1 \dots Z_n \rangle \\
&= \langle Z_n | Z_1 \dots Z_n \rangle + \langle X_{n+1} - \mu | Z_1 \dots Z_n \rangle \\
&= Z_n + \langle X_{n+1} - \mu \rangle \\
&= Z_n.
\end{aligned}
\tag{3.9}
$$

Therefore, the simple random walk is an example of a martingale.

Bachelier (Sec. 1.3) in his seminal work [21] assumes an efficient market where successive prices are statistically independent. Also, he assumes a complete market where there are buyers and sellers for assets marked at any price. Given these hypotheses, he assumes that the prices undergo a movement composed of a deterministic and a random component. The random component is a random walk that can be described by a martingale, i.e. its expected future value does not depend on its history.

Let's now consider the geometric random walk given by $Z_n = \prod_{i=1}^{n} X_i$ where the random variables X_i are independent such that $\langle X_i \rangle = 1$ and $X_i > 0$. Its conditional expected value is:

$$
\begin{aligned}
\langle Z_{n+1}|Z_1 \dots Z_n \rangle &= \left\langle \prod_{i=1}^{n+1} X_i \middle| Z_1 \dots Z_n \right\rangle \\
&= \left\langle X_{n+1} \prod_{i=1}^{n} X_i \middle| Z_1 \dots Z_n \right\rangle \\
&= \langle X_{n+1} \rangle \langle Z_n | Z_1 \dots Z_n \rangle \\
&= Z_n.
\end{aligned}
\tag{3.10}
$$

Therefore, it is also a martingale.

The exponential random walk is another process with a multiplicative structure that finds applications in stochastic finance where price changes are described with respect to the current price. Let's take $Z_n = e^{\lambda \sum_{i=1}^{n} X_i}$ where the random variables X_i are independent and can only assume the values ± 1. The conditional expected future value is given by:

$$
\begin{aligned}
\langle Z_{n+1}|Z_1\ldots Z_n\rangle &= \left\langle e^{\lambda\sum_{i=1}^{n+1}X_i}\middle|Z_1\ldots Z_n\right\rangle \\
&= \left\langle e^{\lambda X_{n+1}}e^{\lambda\sum_{i=1}^{n}X_i}\middle|Z_1\ldots Z_n\right\rangle \\
&= \left\langle e^{\lambda X_{n+1}}\right\rangle Z_n \\
&= 1/2\left(e^{\lambda}+e^{-\lambda}\right)Z_n \\
&= \cosh(\lambda)Z_n \ \geq Z_n.
\end{aligned} \tag{3.11}
$$

Therefore, it is a submartigale. Multiplicative processes have been used, for example, to study the distribution of wealth in a society [67].

3.1.2 PÓLYA PROCESSES

A Pólya[5] process is one where balls of different colors (red and green) are randomly picked from an urn, returned, and additional c balls of the same color are added to the urn. If $c = 0$, then we have a Bernoulli[6] process with replacement, or if $c = -1$ then we get a hypergeometric process without replacement.

At the nth trial, there are a red balls and b green balls in the urn. For the next trial, there are two possibilities for the fraction Y of red balls. We can get $(a+c)/(a+b+c)$ with probability $a/(a+b)$ or a fraction $a/(a+b+c)$ with probability $b/(b+c)$:

$$
\begin{aligned}
\langle Y_{n+1}|Y_1,\ldots,Y_n\rangle &= \frac{a}{a+b}\frac{a+c}{a+b+c}+\frac{b}{a+b}\frac{a}{a+b+c} \\
&= \frac{a(a+c)+ab}{(a+b)(a+b+c)} \\
&= \frac{a(a+c+b)}{(a+b)(a+b+c)} \\
&= \frac{a}{a+b} = Y_n.
\end{aligned} \tag{3.12}
$$

Therefore, this fraction is a martingale.

This process has a very interesting property. In order to arrive at this property, let's calculate the probability of drawing green balls in the first and second trials, and a red ball in the third trial:

$$
\begin{aligned}
P(G_1G_2R_3) &= P(G_1)P(G_2|G_1)P(R_3|G_1G_2) \\
&= \frac{b}{(a+b)}\frac{(b+c)}{(a+b+c)}\frac{a}{(a+b+2c)}.
\end{aligned} \tag{3.13}
$$

Let's compare this probability with that of picking first a red ball then two green balls:

$$
\begin{aligned}
P(R_1G_2G_3) &= P(R_1)P(G_2|R_1)P(G_3|R_1G_2) \\
&= \frac{a}{(a+b)}\frac{b}{(a+b+c)}\frac{(b+c)}{(a+b+2c)}.
\end{aligned} \tag{3.14}
$$

Finally, let's also compare those with the probability of obtaining a green ball followed by a red and another green balls:

$$\begin{aligned} P(G_1R_2G_3) &= P(G_1)P(R_2|G_1)P(G_3|G_1R_2) \\ &= \frac{b}{(a+b)}\frac{a}{(a+b+c)}\frac{(b+c)}{(a+b+2c)}. \end{aligned} \tag{3.15}$$

Note that the denominator is always the same and we have a permutable sequence. This is valid for any choice of balls, since the number of balls increases linearly with the period. Therefore, the probability of obtaining a sequence with k red balls and $N-k$ green balls is given by:

$$\begin{aligned} p &= \frac{a(a+c)\dots(a+(k-1)c)b(b+c)\dots(b+(N-k-1)c)}{(a+b)(a+b+c)(a+b+2c)\dots(a+b+(N-1)c)} \\ &= \frac{a^{(c,k)}b^{(c,N-k)}}{(a+b)^{(c,N)}}, \end{aligned} \tag{3.16}$$

where

$$r^{(s,j)} = r(r+s)(r+2s)\dots(r+(j-1)s). \tag{3.17}$$

It is possible to rewrite the probability p using a shorter notation:

$$\begin{aligned} p &= \frac{\prod_{i=1}^{k}(a+(i-1)c)\prod_{i=1}^{N-k}(b+(i-1)c)}{\prod_{i=1}^{N}(a+b+(i-1)c)} \\ &= \frac{\prod_{i=1}^{k}\left(\frac{a}{c}+i-1\right)\prod_{i=1}^{N-k}\left(\frac{b}{c}+1-1\right)}{\prod_{i=1}^{N}\left(\frac{a+b}{c}+i-1\right)} \\ &= \frac{\Gamma\left(\frac{a}{c}+k\right)\Gamma\left(\frac{b}{c}+N-k\right)\Gamma^{-1}\left(\frac{a}{c}\right)\Gamma^{-1}\left(\frac{b}{c}\right)}{\Gamma\left(\frac{a+b}{c}+N\right)\Gamma^{-1}\left(\frac{a+b}{c}\right)} \\ &= \frac{\Gamma\left(\frac{a}{c}+k\right)\Gamma\left(\frac{b}{c}+N-k\right)}{\Gamma\left(\frac{a}{c}+\frac{b}{c}+N\right)}\frac{\Gamma\left(\frac{a}{c}+\frac{b}{c}\right)}{\Gamma\left(\frac{a}{c}\right)\Gamma\left(\frac{b}{c}\right)} \\ &= \frac{B\left(\frac{a}{c}+k,\frac{b}{c}+N-k\right)}{B\left(\frac{a}{c},\frac{b}{c}\right)} \end{aligned} \tag{3.18}$$

Since we have a permutable sequence, there are $\binom{N}{k}$ possibilities. Therefore, the PDF of the sum $Z_n = \sum_i^N X_i$ is:

$$P(Z_n = k) = \binom{N}{k}\frac{B\left(\frac{a}{c}+k,\frac{b}{c}+N-k\right)}{B\left(\frac{a}{c},\frac{b}{c}\right)}. \tag{3.19}$$

which can also be written as:

$$P(Z_n = k) = \frac{\Gamma(N+1)}{\Gamma(k+1)\Gamma(N-k+1)}\frac{1}{B\left(\frac{a}{c},\frac{b}{c}\right)}\frac{\Gamma\left(\frac{a}{c}+k\right)\Gamma\left(\frac{b}{c}+N-k\right)}{\Gamma\left(\frac{a}{c}+\frac{b}{c}+N\right)} \tag{3.20}$$

Using Stirling's[7] approximation[a] we get for $x, k \to \infty$:

$$\begin{aligned}\frac{\Gamma(x+a)}{\Gamma(x+b)} &= \frac{\Gamma((x+a-1)+1)}{\Gamma((x+b-1)+1)} \\ &\approx \frac{e^{-x-a+1}(x+a-1)^{1/2+x+a-1}}{e^{-x-b+1}(x+b-1)^{1/2+x+b-1}} \\ &\approx \frac{x^x x^{-1/2} x^a}{x^x x^{-1/2} x^b} \\ &\approx x^{a-b}.\end{aligned} \tag{3.21}$$

Using this result in Eq. 3.20 we get a beta distribution:

$$\begin{aligned}P(Z_n = k) &= \frac{1}{B\left(\frac{a}{c}, \frac{b}{c}\right)} k^{a/c-1}(N-k)^{b/c-1} N^{1-a/c-b/c} \\ P\left(\frac{Z_n}{N} = u\right) &= \frac{1}{B\left(\frac{a}{c}, \frac{b}{c}\right)} u^{a/c-1}(1-u)^{b/c-1}, \ u = k/N.\end{aligned} \tag{3.22}$$

Therefore, Pólya's processes can potentially lead to distributions with heavy tails. This finds parallels in many socioeconomic phenomena. For instance, Merton (Sec. 1) observed that fame and status can lead to cumulative advantage [68]. This is known in social sciences as *Matthew effect*. In economics, it is often observed that the growth rate of firms is independent of their sizes. This is basically because the concentration of capital can lead to progressive more investments. This is known as the *law of proportionate effect* or *Gibrat's law* [69][8].

3.1.3 STOPPING TIMES

From the definition of a martingale, we saw that for a stochastic process $X = \{X_n : n \geq 0\}$:

$$\begin{aligned}\langle X_{n+1} | X_0, X_1, \ldots, X_n \rangle &= X_n \\ \langle X_{n+1} - X_n | X_0, X_1, \ldots, X_n \rangle &= 0.\end{aligned} \tag{3.23}$$

This implies that for a fair game, the expected values of the additions $\Delta X = X_{n+1} - X_n$ are independent on the history of the game. Thus, we infer that after N bets, the expected value of X_N must be the same as that at the beginning of the process. Mathematically:

$$\begin{aligned}\langle X_n | \mathscr{F}_{n-1} \rangle &= X_{n-1} \\ \langle \langle X_n | \mathscr{F}_{n-1} \rangle | \mathscr{F}_{n-2} \rangle &= \langle X_{n-1} | \mathscr{F}_{n-2} \rangle \\ \langle X_n | \mathscr{F}_{n-1} \rangle &= \langle X_{n-1} | \mathscr{F}_{n-2} \rangle\end{aligned} \tag{3.24}$$

Continuing this procedure until we reach the first element of the series:

[a] $\Gamma(x+1) \approx \sqrt{2\pi} e^{-x} x^{1/2+x}$.

$$\langle X_n|\mathscr{F}_{n-1}\rangle = \langle X_{n-1}|\mathscr{F}_{n-2}\rangle = \langle X_{n-2}|\mathscr{F}_{n-3}\rangle = \ldots = \langle X_1|\mathscr{F}_0\rangle = X_0. \tag{3.25}$$

But what happens if the process is suddenly interrupted after some period T? In order to verify what happens, we need to use *Lebesgue's dominated convergence theorem*:

Lebesgue's Dominated Convergence Theorem

The Lebesgue dominated convergence theorem [64] states that for a set of measurable functions $\{f_n : \mathbb{R} \to \mathbb{R}\}$ that converge to a function f dominated by some other integrable function g, $|f_n(x)| \leq g(x), \forall x \in \mathbb{R}$, then f is integrable and:

$$\lim_{n\to\infty} \int_{\mathbb{R}} f_n d\mu = \int_{\mathbb{R}} f d\mu, \tag{3.26}$$

for a measure μ.

Thus, we must check whether the stochastic process that we are considering is dominated. Let's write it as a telescopic sum:

$$X_{t\wedge T} = X_0 + \sum_{n=1}^{t\wedge T} (X_{n+1} - X_n), \; t \in \mathbb{N}_0, \tag{3.27}$$

where $X_{t\wedge T}$ indicates that the process is stopped at an instant T.

If $X_{t\wedge T}$ is limited, then:

$$|X_{t\wedge T} - X_0| = \left|\sum_{n=0}^{t\wedge T} (X_{n+1} - X_n)\right| \leq T \sup_{n\in\mathbb{N}} |X_{n+1} - X_n| < \infty, \tag{3.28}$$

if the increments are limited.

Applying, Lebesgue's theorem:

$$\begin{gathered} \lim_{t\to\infty} \int X_{t\wedge T} dP = \int X_0 dP \\ \lim_{t\to\infty} \langle X_{t\wedge T}\rangle = \langle X_0\rangle. \end{gathered} \tag{3.29}$$

Since this result is valid for any t we arrive at the *optional stopping theorem* [31,36]:

$$\langle X_T\rangle = \langle X_0\rangle. \tag{3.30}$$

Example 3.1.1. Probability of winning after a stop

Let's consider X_i as the gain that a gambler obtains in a bet, and Z_i the accumulated gain until the ith bet. If the gambler begins with a total of 0 coins and ends with W gains or L losses, the gambler stops betting. What is the probability q of ending up with W? Applying Eq. 3.30:

$$\begin{aligned}\langle Z_T\rangle &= \langle Z_0\rangle = 0\\ \langle Z_T\rangle &= qW-(1-q)L = 0\\ q(W+L)-L &= 0\\ q &= \frac{L}{L+W}.\end{aligned} \tag{3.31}$$

■

3.1.4 WALD'S EQUATION

Let's consider the following martingale:

$$Z_n = \sum_{j=1}^{n}(X_j - \langle X\rangle) \tag{3.32}$$

with $\langle Z_1\rangle = 0$. Its expected value at T is:

$$\begin{aligned}\langle Z_T\rangle &= \left\langle \sum_{j=1}^{T}(X_j-\langle X\rangle)\right\rangle\\ &= \left\langle \sum_{j=1}^{T}X_j - T\langle X\rangle\right\rangle\\ &= \left\langle \sum_{j=1}^{T}X_j\right\rangle - \langle T\rangle\langle X\rangle.\end{aligned} \tag{3.33}$$

However, according to the stopping time theorem we have:

$$\begin{aligned}\langle Z_T\rangle &= \langle Z_0\rangle = 0\\ \left\langle \sum_{j=1}^{T}X_j\right\rangle - \langle T\rangle\langle X\rangle &= 0\\ \left\langle \sum_{j=1}^{T}X_j\right\rangle &= \langle T\rangle\langle X\rangle.\end{aligned} \tag{3.34}$$

This last expression is known as *Wald's equation* [31][9].

Example 3.1.2. Gambler's ruin

Let's consider a bet where the gambler starts with an amount x_0 and wins according to a random variable ξ_n such that $X_n = \sum_i^n \xi_i$. Thus, at the nth bet, the gambler has an accumulated wealth of $W_n = x_0 + X_n$. The gambler stops betting if he or she is ruined obtaining $W_n = 0 \rightarrow X_n = -x_0$ or if the gambler reaches a certain amount $W_n = a \to X_n = a - x_0$. The probability of the former is ρ_0 whereas the probability of the later is ρ_a. According to Wald's equation:

$$\begin{aligned} \langle X_T \rangle &= \langle T \rangle \langle \xi \rangle \\ -x_0\rho_0 + (a - x_0)\rho_a &= \langle T \rangle \cdot 0 = 0 \end{aligned} \tag{3.35}$$

We also have that $\rho_0 + \rho_a = 1$. Thus:

$$\begin{aligned} \begin{bmatrix} 1 & 1 \\ -x_0 & a - x_0 \end{bmatrix} \begin{bmatrix} \rho_0 \\ \rho_a \end{bmatrix} &= \begin{bmatrix} 1 \\ 0 \end{bmatrix} \\ \begin{bmatrix} \rho_0 \\ \rho_a \end{bmatrix} &= \frac{1}{a - x + x} \begin{bmatrix} a - x_0 & -1 \\ x_0 & 1 \end{bmatrix} \begin{bmatrix} 1 \\ 0 \end{bmatrix} \\ &= \frac{1}{a} \begin{bmatrix} a - x_0 \\ x_0 \end{bmatrix}. \end{aligned} \tag{3.36}$$

From this last equation we see that the greedier the gambler is, the more likely it is for him or her to be ruined. ■

Example 3.1.3. Unbiased dice

We throw a six faced dice until obtaining an odd number. What is the expected number of fives? To solve this problem we can use a random variable X that assumes 0 if the number drawn is even or 1 if it is odd. The game stops when the accumulated sum of outcomes of X is 1. According to Wald's equation:

$$\begin{aligned} \left\langle \sum_{n=1}^{T} X_n \right\rangle &= \langle T \rangle \langle X \rangle \\ 1 &= \langle T \rangle \frac{1}{2} \\ \langle T \rangle &= 2. \end{aligned} \tag{3.37}$$

It takes on average two bets to end the game. Let's now make X assume 1 when we reach number 5:

$$\begin{aligned} \left\langle \sum_{n=1}^{T} X_n \right\rangle &= \langle T \rangle \langle X \rangle \\ &= 2 \times \frac{1}{6} \\ &= \frac{1}{3}. \end{aligned} \tag{3.38}$$

■

Example 3.1.4. Sum of outcomes

A dice is thrown repeatedly until number 1 is reached. What is the expected value of the sum of the numbers obtained until the game is stopped? We can once again create a random variable X that assumes 1 when number 1 is drawn. Applying Wald's equation:

$$\begin{aligned}\left\langle \sum_{n=1}^{T} X_n \right\rangle &= \langle T \rangle \frac{1}{6} \\ \langle T \rangle &= 6.\end{aligned} \tag{3.39}$$

Let's make now X be a random variable that is exactly the result of the dice:

$$\begin{aligned}\left\langle \sum_{n=1}^{T} X_n \right\rangle &= 6 \times \frac{1}{6}(1+2+3+4+5+6) \\ &= 21.\end{aligned} \tag{3.40}$$

■

3.1.5 GALTON-WATSON PROCESS

A branching process[10] [31, 70] is a stochastic process that starts with a single node. A random variable X_n that admits only positive values is drawn and its result corresponds to the number of sub-nodes connected to this node. Thus, the number of nodes in a step n of this process is given by:

$$Z_n = \sum_{k=1}^{Z_{n-1}} X_n. \tag{3.41}$$

This process is illustrated in Fig. 3.1. This kind of process is used to study, for example, the evolution in the number of firms. Every innovation can be understood as an opportunity to the development of other products and processes that lead to the formation of new businesses (see, for instance [71]).

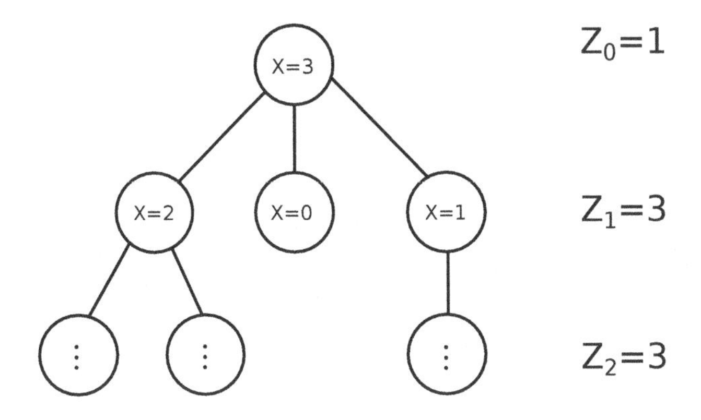

Figure 3.1: An example of a Galton-Watson ramification process

The expected number of nodes is calculated as:

$$Z_n = X_1 + X_2 + \ldots + X_{Z_{n-1}}. \tag{3.42}$$

Thus, it is possible to use Wald's equation for the stopping time Z_{n-1}:

$$\langle Z_n \rangle = \langle Z_{n-1} \rangle \langle X \rangle \tag{3.43}$$

However,

$$\langle Z_{n-1} \rangle = \langle Z_{n-2} \rangle \langle X \rangle \tag{3.44}$$

which leads to:

$$\langle Z_n \rangle = \langle Z_{n-2} \rangle \langle X \rangle^2. \tag{3.45}$$

This can be repeated n times leading to:

$$\begin{aligned} \langle Z_n \rangle &= \langle Z_{n-n} \rangle \langle X \rangle^n \\ &= \mu^n, \end{aligned} \tag{3.46}$$

given that we start with one node ($Z_0 = 1$).

This calculation can also be written as:

$$\begin{aligned} \langle Z_n | Z_1, Z_2 \ldots, Z_{n-1} \rangle &= \langle Z_{n-1} | Z_1, Z_2, \ldots, Z_{n-1} \rangle \langle X \rangle \\ &= Z_{n-1}\mu. \end{aligned} \tag{3.47}$$

Since $\mu \geq 1$, $Z_{n-1}\mu \geq Z_n$ this is a submartingale.

It is possible now to create a stochastic process:

$$W_n = \frac{X_n}{\mu^n}, \tag{3.48}$$

which is a martingale because:

$$\begin{aligned} \langle W_{n+1} | W_1, W_2, \ldots, X_n \rangle &= \left\langle \frac{X_{n+1}}{\mu^{n+1}} \middle| W_1, W_2, \ldots, X_n \right\rangle \\ &= \frac{1}{\mu} \frac{1}{\mu^n} \langle X_{n+1} | W_1, W_2, \ldots, W_n \rangle \\ &= \frac{1}{\mu} \frac{X_n}{\mu^n} \mu = \frac{X_n}{\mu^n} = Z_n. \end{aligned} \tag{3.49}$$

3.1.5.1 Extinction

Let's consider a stochastic process given by $Z = \sum_{i=1}^{N} X_i$, where X are i.i.d random variables. The stopping time N is also a random variable. The probability generating function for this process is given by:

$$G_Z(s) = \sum_{k=0}^{\infty} P(Z = k) s^k. \tag{3.50}$$

The probability that appears in this equation can be written as:

$$
\begin{aligned}
P(Z=k) &= \sum_{n=0}^{\infty} P(Z=k \cap N=n) \\
&= \sum_{n=0}^{\infty} P(Z=k|N=n)\,P(N=n).
\end{aligned}
\tag{3.51}
$$

Therefore, the probability generating function can be written as:

$$
\begin{aligned}
G_Z(s) &= \sum_{n=0}^{\infty} Pr(N=n) \sum_{k=0}^{\infty} Pr(Z=k|N=n)\,s^k \\
&= \sum_{n=0}^{\infty} Pr(N=n) G_X(s)^n \\
&= G_N\left(G_X(s)\right).
\end{aligned}
\tag{3.52}
$$

Using G_n for the probability generating function for the Z_n process:

$$
\begin{aligned}
G_n(s) &= G_{n-1}\left(G_X(s)\right) \\
G_{n-1}(s) &= G_{n-2}\left(G_X(s)\right) \\
G_{n-1}\left(G_X(s)\right) &= G_{t-2}\left(G_X\left(G_X(s)\right)\right) \\
G_n(s) &= G_{n-2}\left(G_X^{(2)}(s)\right) \\
G_n(s) &= G_{n-n}\left(G_X^{(n)}(s)\right) \\
&= G_X^{(n)}(s),
\end{aligned}
\tag{3.53}
$$

where $G_X^{(n)}$ is the nth interaction of G_X. Going backwards we get:

$$
\begin{aligned}
G_n(s) &= G_X^{(n)}(s) \\
&= G_X\left(G_X^{(n-1)}(s)\right) \\
G_{n-1}(s) &= G_X^{(n-1)}(s) \\
G_n(s) &= G_X\left(G_{n-1}(s)\right).
\end{aligned}
\tag{3.54}
$$

The probability of extinction after the nth generation is:

$$e_n = Pr(X_n = 0). \tag{3.55}$$

This probability has the property that $e_n \leq 1$. Moreover, since $X_n = 0 \rightarrow X_{n+1} = 0$, then $e_n \leq e_{n+1}$. Therefore, the sequence $\{e_n\}$ is limited. The ultimate probability of extinction is given by:

$$e = \lim_{n \to \infty} e_n. \tag{3.56}$$

However,

$$\begin{aligned} e_n &= Pr(X_n = 0) \\ &= G_n(0) \\ &= G_X\left(G_{n-1}(0)\right) \\ &= G_X\left(e_{n-1}\right). \end{aligned} \tag{3.57}$$

Taking the limit:

$$\begin{aligned} e &= \lim_{n\to\infty} e_n \\ &= \lim_{n\to\infty} G_X\left(e_{n-1}\right) \\ &= G_X\left(e\right). \end{aligned} \tag{3.58}$$

For any nonnegative root r of this equation:

$$\begin{aligned} e_1 &= G(e_0) = G(0) \leq G(r) = r \\ e_2 &= G(e_1) = e_1 \leq r \\ &\vdots \\ e_n &\leq r. \\ e &= \lim_{n\to\infty} \leq r. \end{aligned} \tag{3.59}$$

Therefore, e is the smallest nonnegative root.

Since G is continuous, not decremental and convex, there are one or two fixed points for $e = G_X(e)$. If $G'(1) > 1$, then there are two fixed points, whereas if $G'(1) \leq 1$ then there is only one solution. Nonetheless, $G'(1) = \langle X \rangle$. Therefore, there is only one fixed point if the mean is smaller or equal to unity.

Thus, this process has two regimes. If $\langle X \rangle < 1$, then $e = 1$ and this is known as a *subcritical process*. On the other hand, if $\langle X \rangle > 1$, then $e < 1$ and this is known as a *supercritical process*. These two regimes are shown in Fig. 3.2.

3.1.6 AZUMA-HOEFFDING THEOREM

Let's finish our discussion about martingales studying a concentration inequality known as *Azuma-Hoeffding inequality* [31][11].

Knowing that the exponential function is convex[b] and given a random variable that satisfies the Lipschitz[12] condition $|X_n - X_{n-1}| \leq c_n$, it is possible to write:

---

[b] A function $f : A \to \mathbb{R}$ is convex if $\forall x, y \in A$ and $\forall t \in [0,1]$, $f(tx + (1-t)y) \leq tf(x) + (1-t)f(y)$.

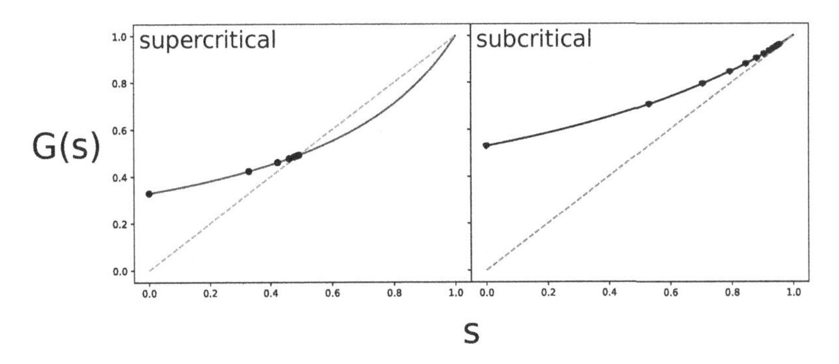

Figure 3.2: A supercritical process (left) and a subcritical (right) process. The dashed line indicates the trivial case $G(s) = s$ and the dots are interaction points that converge to the fixed points

$$\begin{aligned}
\exp(tX_n) &= \exp\left\{\frac{1}{2}\left[\left(\frac{X_n}{c_n}+1\right)c_nt + \left(1-\frac{X_n}{c_n}\right)(-c_nt)\right]\right\} \\
&\leq \frac{1}{2}\left(\frac{X_n}{c_n}+1\right)e^{c_nt} + \frac{1}{2}\left(1-\frac{X_n}{c_n}\right)e^{-c_nt}, \ \forall t \in [0,1] \\
&\leq \frac{e^{c_nt}+e^{-c_nt}}{2} + \frac{e^{c_nt}-e^{-c_nt}}{2}\frac{X_n}{c_n} \\
&\leq \cosh(c_nt) + \frac{X_n}{c_n}\sinh(c_nt) \\
&\leq \sum_{k=0}^{\infty}\frac{(c_nt)^{2k}}{(2k)!} + \frac{X_n}{c_n}\sinh(c_nt) \ \leftarrow \text{Taylor expansion} \\
&\leq \sum_{k=0}^{\infty}\frac{(1/2(c_nt)^2)^k}{k!} + \frac{X_n}{c_n}\sinh(c_nt), \ \text{since } 2^k k! \leq (2k)! \\
&\leq e^{\frac{(c_nt)^2}{2}} + \frac{X_n}{c_n}\sinh(c_nt) \ \leftarrow \text{Taylor expansion.}
\end{aligned} \tag{3.60}$$

Now, according to the Chernoff bound (Eq. 2.53):

$$\begin{aligned}
P(X_n - X_0 \geq a) &\leq e^{-at}\langle e^{t(X_n-X_0)}\rangle = e^{-at}\left\langle \exp\left\{t\sum_{k=1}^{n}(X_k - X_{k-1})\right\}\right\rangle \\
&= e^{-at}\left\langle e^{t(X_n-X_{n-1})}\exp\left\{t\sum_{k=1}^{n-1}(X_k - X_{k-1})\right\}\right\rangle.
\end{aligned} \tag{3.61}$$

Using the law of total expectation (Eq. 2.15) and considering that X_k for $k \leq n-1$ is measurable with respect to a filtration $\mathscr{F}_{n-1}$:

$$
\begin{aligned}
P(X_n - X_0 \geq a) &\leq e^{-at} \left\langle \left\langle e^{t(X_n - X_{n-1})} \exp\left\{ t \sum_{k=1}^{n-1} (X_k - X_{k-1}) \right\} \middle| \mathscr{F}_{n-1} \right\rangle \right\rangle \\
&\leq e^{-at} \left\langle \left\langle e^{t(X_n - X_{n-1})} \middle| \mathscr{F}_{n-1} \right\rangle \exp\left\{ t \sum_{k=1}^{n-1} (X_k - X_{k-1}) \right\} \right\rangle \\
&\leq e^{-at} \left\langle \exp\left\{ \sum_{k=1}^{n-1} (X_k - X_{k-1}) \right\} \right. \\
&\quad \left. \left(e^{(c_n t)^2/2} + \frac{1}{c_n} \sinh(c_n t) \left\langle X_n - X_{n-1} | \mathscr{F}_{n-1} \right\rangle \right) \right\rangle,
\end{aligned}
\tag{3.62}
$$

where we have used Eq. 3.60 in the last line. Also, the martingale on the right hand side of the equation must be zero. Therefore:

$$
P(X_n - X_0 \geq a) \leq e^{-at} e^{(c_n t)^2/2} \left\langle \exp\left\{ \sum_{k=1}^{n-1} (X_k - X_{k-1}) \right\} \right\rangle. \tag{3.63}
$$

If we repeat the same procedure n times, we end up with:

$$
P(X_n - X_0 \geq a) \leq e^{-at} \exp\left\{ \sum_{k=1}^{n} e^{(c_k t)^2/2} \right\}. \tag{3.64}
$$

We can now choose the value of t that produces the lowest bound:

$$
\begin{aligned}
&\frac{d}{dt} e^{-at} \exp\left\{ \sum_{k=1}^{n} e^{(c_k t)^2/2} \right\} = \left(-a + t \sum_{k=1}^{n} c_k^2 \right) e^{-at} \exp\left\{ \sum_{k=1}^{n} e^{(c_k t)^2/2} \right\} \\
&0 = \left(-a + t \sum_{k=1}^{n} c_k^2 \right) \rightarrow t = \frac{a}{\sum_{k=1}^{n} c_k^2}.
\end{aligned}
\tag{3.65}
$$

Therefore,

$$
P(X_n - X_0 \geq a) \leq \exp\left\{ -\frac{a^2}{2 \sum_{k=1}^{n} c_k^2.} \right\}. \tag{3.66}
$$

Doing the same calculations for a negative t:

$$
P(X_n - X_0 \leq -a) \leq \exp\left\{ -\frac{a^2}{2 \sum_{k=1}^{n} c_k^2.} \right\}. \tag{3.67}
$$

Combining both results:

$$
P(|X_n - X_0| \geq a) \leq 2 \exp\left\{ -\frac{a^2}{2 \sum_{k=1}^{n} c_k^2.} \right\}. \tag{3.68}
$$

Example 3.1.5. Binary Bets

Let's consider a binary option (also known as *fixed return options*) where the gambler only wins a fixed amount if a call option expires in the money. Let's represent the outcomes of this bet by $X_1, X_2, \ldots, X_n$. What is the chance that the gambler finds a sequence $B_1, B_2, \ldots, B_k$ within X?

The number of possibilities of including a string of size k within another of size n is given by:

$$N = n - k + 1. \tag{3.69}$$

The probability of a sequence of size k in X be exactly B is:

$$p = \left(\frac{1}{2}\right)^k. \tag{3.70}$$

Thus, the expected number of occurrences of B in X is:

$$\langle F \rangle = Np = (n - k + 1)\left(\frac{1}{2}\right)^k. \tag{3.71}$$

Let's now create a Doob martingale:

$$\begin{aligned} Z_n &= \langle F | \mathscr{F}_n \rangle, \\ Z_0 &= \langle F \rangle. \end{aligned} \tag{3.72}$$

The function F is Lipschitz limited, since each character of X cannot be in more than $c_n = k$ matches. The probability of $|Z_k - Z_0|$ being greater than some value a is then given by:

$$P(|Z_k - Z_0| \geq a) \leq 2\exp\left\{-\frac{a^2}{2nk^2}\right\}. \tag{3.73}$$

This equation states that this probability is concentrated on the expected value of F, since the probability of finding excess values greater than a certain amount decays exponentially fast.

■

3.2 MARKOV CHAINS

Markov[13] chain [29–32, 72, 73] is a process widely used to describe the likelihood of future events given a present state. In this process, the transition probability from the current state to another depends only on the present state, a memoryless principle known as *Markov property*. More formally, given a probability space $(\Omega, \mathscr{F}, P)$, where Ω is discrete, finite or countable, a stochastic process X_n adapted to $\mathscr{F}_n$ is a Markov chain if:

$$P(X_n \in F|\mathscr{F}_s) = P(X_n \in F|X_s), \; F \in \mathscr{F}, \; \forall n > s \tag{3.74}$$

almost surely. An example of a Markov chain is illustrated by the toy model in Fig. 3.3.

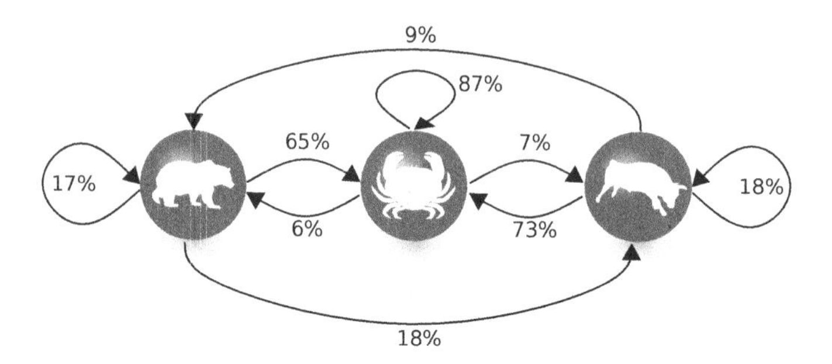

Figure 3.3: A Markov chain as a toy model for the stock market. A *crab market* is defined here as a market where the difference between the close and open prices is not greater than ± 5 % of the open price. A *bull/bear* market is one where the difference is greater/smaller then such value

Path Dependence

Many (if not most) economic and social phenomena, however, not only have memory as they also show *path dependence*. For instance, many inferior standards are still used today because of the legacy they build in a positive feedback fashion [74–76].

The path dependence of the final state of a physical system in parameter space is known as *nonholonomy*. Nonholonomic systems are also called *nonintegrable* (but may be solvable, nonetheless) since there are less constants of motion than functions that govern their dynamics. Hence, these systems may show interesting phenomena such as chaotic behavior [77–79].

3.2.1 TRANSITION FUNCTION

For a probability space $(\Omega, \mathscr{F}, P)$, the function $P : \Omega \times \mathscr{F} \rightarrow \mathbb{R}$ is known as *probability transition function* if $P(x, \cdot)$ is a probability measure in the space $(\Omega, \mathscr{F})$ for all $x \in \Omega$, and for all $F \in \mathscr{F}$, the map $s \rightarrow P(s, F)$ is $\mathscr{F}$-measurable.

For a discrete Markov chain that starts in a state i, the probability that it is in a state j after an interval $n + m$ is given by:

$$\begin{aligned}
P_{ij}^{n+m} &= P(X_{n+m} = j|X_0 = i) = P(X_{n+m} = j \cap \Omega|X_0 = i) \\
&= P\left(X_{n+m} = j \cap \bigcup_k X_m = k|X_0 = i\right) \\
&= \sum_k P(X_{n+m} = j \cap X_m = k|X_0 = i) \\
&= \sum_k P(X_{n+m} = j \cap X_m = k \cap X_0 = i)/P(X_0 = i) \\
&= \sum_k P(X_{n+m} = j|X_m = k \cap X_0 = i) P(X_m = k \cap X_0 = i)/P(X_0 = i) \\
&= \sum_k P(X_{n+m} = j|X_m = k \cap X_0 = i) P(X_m = k|X_0 = i) \\
&= \sum_k P(X_{n+m} = j|X_m = k) P(X_m = k|X_0 = i) \\
&= \sum_k P(X_n = j|X_0 = k) P(X_m|X_0 = i) \\
&= \sum_k P_{jk}^n P_{ki}^m.
\end{aligned} \tag{3.75}$$

For a continuous Markov chain, we have the equivalent formulation:

$$P^{n+1}(x,F) = \int_\Omega P^n(y,F)P(x,dy), \tag{3.76}$$

where the integration is performed over the measure $P(x,\cdot)$. This equation is known as *Chapman-Kolmogorov equation* [29–32,36,64,73,80,81][14].

Example 3.2.1. Markov chain for the stock market

The transition matrix for the Markov chain in Fig. 3.3 is given by:

$$\begin{aligned}
\mathbf{P} &= \begin{bmatrix} \mathscr{P}_r(\text{crab} \to \text{crab}) & \mathscr{P}_r(\text{crab} \to \text{bear}) & \mathscr{P}_r(\text{crab} \to \text{bull}) \\ \mathscr{P}_r(\text{bear} \to \text{crab}) & \mathscr{P}_r(\text{bear} \to \text{bear}) & \mathscr{P}_r(\text{bear} \to \text{bull}) \\ \mathscr{P}_r(\text{bull} \to \text{crab}) & \mathscr{P}_r(\text{bull} \to \text{bear}) & \mathscr{P}_r(\text{bull} \to \text{bull}) \end{bmatrix} \\
&= \begin{bmatrix} 0.87 & 0.06 & 0.07 \\ 0.65 & 0.17 & 0.18 \\ 0.73 & 0.09 & 0.18 \end{bmatrix}
\end{aligned} \tag{3.77}$$

Had any element $p_{ii} = 1$ and $p_{ij} = 0$, $\forall i \neq j$, then it would be impossible for the chain to leave this state and it would be called an *absorbent state*. ■

3.2.2 STATIONARY DISTRIBUTION

For any initial condition $\mathbf{x}^0$, after one transition we reach:

$$\mathbf{x}^1 = \mathbf{x}^0\mathbf{P}. \tag{3.78}$$

For the next interactions we get:

$$\begin{aligned} \mathbf{x}^2 &= \mathbf{x}^1\mathbf{P} \\ &= \mathbf{x}^0\mathbf{P}\mathbf{P} \\ &= \mathbf{x}^0\mathbf{P}^2. \end{aligned} \tag{3.79}$$

For the nth operation, the state of the chain is:

$$\mathbf{x}^n = \mathbf{x}^0\mathbf{P}^n. \tag{3.80}$$

At the limit, the transition probability is:

$$\mathbf{P}_\infty = \lim_{N\to\infty} \mathbf{P}^N. \tag{3.81}$$

For the transition matrix of the previous example, after only five interactions we get:

$$\mathbf{P}_\infty \to \begin{bmatrix} 0.842 & 0.070 & 0.087 \\ 0.842 & 0.070 & 0.087 \\ 0.842 & 0.070 & 0.087 \end{bmatrix}. \tag{3.82}$$

If the initial state was $\mathbf{x}^0 = \begin{bmatrix} 1 & 0 & 0 \end{bmatrix}$, for example, the limit distribution (*stationary distribution* [30, 31, 73]) would be $\mathbf{x}^0 = \begin{bmatrix} 0.842 & 0.070 & 0.087 \end{bmatrix}$, which tells us that 84 % of the days the market would be *crabish*, 7 % bearish, and 8.7 % bullish.

It is possible to obtain the stationary distribution π in a more elegant way using the Chapman-Kolmogorov equation:

$$\begin{aligned} p_{ij}^{n+1} &= \sum_k p_{ik}^n p_{kj} \\ p_{ij}^{\infty} &= \sum_k p_{ik}^{\infty} p_{kj} \\ \pi_j &= \sum_k \pi_k p_{kj} \\ \pi &= \pi\mathbf{P}, \end{aligned} \tag{3.83}$$

which tells us that π is an invariant measure for $\mathbf{P}$. This is also a matrix formulation of the *detailed balance principle*: $\lambda_i p_{ij} = \lambda_j p_{ji}$, $\forall i, j$, where λ is any measure. This can also be written as:

$$\begin{aligned} \pi &= \pi\mathbf{P} \\ \pi^T &= (\pi\mathbf{P})^T = \mathbf{P}^T\pi^T \\ \mathbf{P}^T\pi^T &= \pi^T. \end{aligned} \tag{3.84}$$

Therefore, the stationary distribution is the eigenvector of $\mathbf{P}^T$ that corresponds to the eigenvalue 1, if it exists.

Since π is a distribution, it also has to satisfy $\pi\mathbf{J} = \mathbf{e}$, where $\mathbf{J}$ is a unit matrix[c] where all its elements are 1, and $\mathbf{e}$ is a unit vector. Doing some algebra:

$$\begin{aligned} \pi\mathbf{P} - \pi\mathbf{J} &= \pi - \mathbf{e} \\ \pi &= \mathbf{e}\left(\mathbf{I} + \mathbf{J} - \mathbf{P}\right)^{-1}. \end{aligned} \tag{3.85}$$

Example 3.2.2. Stationary distribution for the Markov chain in Fig. 3.3

$$\begin{aligned} \pi &= \begin{bmatrix} 1 & 1 & 1 \end{bmatrix} \left(\begin{bmatrix} 1 & 0 & 0 \\ 0 & 1 & 0 \\ 0 & 0 & 1 \end{bmatrix} + \begin{bmatrix} 1 & 1 & 1 \\ 1 & 1 & 1 \\ 1 & 1 & 1 \end{bmatrix} - \begin{bmatrix} 0.87 & 0.06 & 0.07 \\ 0.65 & 0.17 & 0.18 \\ 0.73 & 0.09 & 0.18 \end{bmatrix} \right)^{-1} \\ &= \begin{bmatrix} 0.84229209 & 0.07036004 & 0.08734787 \end{bmatrix} \\ &\approx \begin{bmatrix} 84.23\% \text{ crab} & 7.04\% \text{ bear} & 8.73\% \text{ bull} \end{bmatrix}. \end{aligned} \tag{3.86}$$

■

3.2.2.1 Detailed Balance

Let's study the detailed balance in more detail. Let $(X_n, n \geq 0)$ be a Markov chain and consider a process that moves back in time: $X_n \to X_{n-1} \to \ldots \to X_0$. The probability of finding the chain in a state x_n given that it was in the states $x_{n+1}, x_{n+2}, \ldots$ is given by:

$$\begin{aligned} &P(X_n = x_n | X_{n+1} = x_{n+1} \cap X_{n+2} = x_{n+2} \cap \ldots) = \\ &= \frac{P(X_n = x_n \cap X_{n+1} = x_{n+1} \cap X_{n+2} = x_{n+2} \cap \ldots)}{P(X_{n+1} = x_{n+1} \cap X_{n+2} = x_{n+2} \cap \ldots)} \\ &= \frac{P(X_{n+2} = x_{n+2} \cap \ldots | X_{n+1} = x_{n+1} \cap X_n = x_n) P(X_{n+1} = x_{n+1} \cap X_n = x_n)}{P(X_{n+2} = x_{n+2} \cap \ldots | X_{n+1} = x_{n+1}) P(X_{n+1} = x_{n+1})}. \end{aligned} \tag{3.87}$$

According to the Markov property:

$$\begin{aligned} &P(X_n = x_n | X_{n+1} = x_{n+1} \cap X_{n+2} = x_{n+2} \cap \ldots) = \\ &= \frac{P(X_{n+2} = x_{n+2} \cap \ldots | X_{n+1} = x_{n+1})}{P(X_{n+2} = x_{n+2} \cap \ldots | X_{n+1} = x_{n+1})} \frac{P(X_{n+1} = x_{n+1} \cap X_n = x_n)}{P(X_{n+1} = x_{n+1})} \\ &= P(X_n = x_n | X_{n+1} = x_{n+1}). \end{aligned} \tag{3.88}$$

Therefore this process is also a Markov chain.

---

[c]Not to be confused with the *identity matrix*.

Let's now compute the probability p_{ij} of finding the chain in state j given that it came from state i:

$$\begin{aligned} P(X_n = j|X_{n+1} = i) &= \frac{P(X_n = j \cap X_{n+1} = i)}{P(X_{n+1} = i)} \\ &= \frac{P(X_{n+1} = i|X_n = j)P(X_n = j)}{P(X_{n+1} = i)} \\ p_{ij}^n &= p_{ji}^n \frac{\pi_j^n}{\pi_i^{n+1}}. \end{aligned} \tag{3.89}$$

Therefore, the reverse transition is not homogeneous in time since it is a function of n. Nonetheless, π does not depend on n if it is the stationary distribution. The chain becomes homogeneous in time in this situation:

$$p_{ij} = \frac{p_{ji}\pi_j^*}{\pi_i^*} = \tilde{p}_{ji}. \tag{3.90}$$

In this case, the chain is reversible and we obtain a detailed balance equation:

$$\pi_i^* p_{ij} = \pi_j^* p_{ji}, \ \forall i, j, \tag{3.91}$$

Which leads to:

$$\begin{aligned} \pi_i^* p_{ij} &= \pi_j^* p_{ji} \\ \sum_i \pi_i^* p_{ij} &= \sum_i \pi_j^* p_{ji} = \pi_j^* \sum_i p_{ji} = \pi_j^* \\ \therefore \ \pi_j^* &= \sum_i \pi_i^* p_{ij}, \ \forall j. \end{aligned} \tag{3.92}$$

Example 3.2.3. A three-state market

Suppose that the chance the market depicted in Fig. 3.3 stays in the same state is zero. If it is in a *crab* state, it can move to a *bear* state with probability a or it can move to a state *bull* with probability $1-a$. The same applies to the other configurations. If it is in a *bear* state, it can move to a *bull* state with probability a and to a *crab* state with probability $1-a$. Finally, if it is in a *bull* state it can move to a *crab* state with probability a and to a *bear* state with probability $1-a$. Therefore, the transition matrix is given by:

$$P = \begin{bmatrix} 0 & a & 1-a \\ 1-a & 0 & a \\ a & 1-a & 0 \end{bmatrix}. \tag{3.93}$$

From the detailed balance equation (Eq: 3.92) we can write:

$$\begin{aligned} \pi_1^* p_{12} &= \pi_2^* p_{21} \rightarrow \pi_1^* a = \pi_2^*(1-a) \\ \pi_2^* p_{23} &= \pi_3^* p_{32} \rightarrow \pi_2^* a = \pi_3^*(1-a). \\ \pi_3^* p_{31} &= \pi_1^* p_{13} \rightarrow \pi_3^* a = \pi_1^*(1-a) \end{aligned} \tag{3.94}$$

This can be recast in matrix formulation:

$$\begin{bmatrix} a & a-1 & 0 \\ 0 & a & a-1 \\ a-1 & 0 & a \end{bmatrix} \begin{bmatrix} \pi_1^* \\ \pi_2^* \\ \pi_3^* \end{bmatrix} = 0. \tag{3.95}$$

From this we can compute the parameter a:

$$\begin{aligned} \begin{vmatrix} a & a-1 & 0 & a & a-1 \\ 0 & a & a-1 & 0 & a \\ a-1 & 0 & a & a-1 & 0 \end{vmatrix} &= 0 \\ a^3 + (a-1)^3 &= 0 \\ a &= 1/2. \end{aligned} \tag{3.96}$$

Let's now suppose that the stationary state is $\pi^* = (bull = 1/6, crab = 4/6, bear = 1/6)$ and we want to find the transition matrix. From the detailed balance equation we have:

$$\begin{aligned} \frac{P(bull \to crab)}{P(crab \to bull)} &= \frac{\pi_2^*}{\pi_1^*} = 4 \to P(bull \to crab) = 4P(crab \to bull), \\ \frac{P(bear \to crab)}{P(crab \to bear)} &= \frac{\pi_2^*}{\pi_3^*} = 4 \to P(bear \to crab) = 4P(crab \to bear), \\ \frac{P(bull \to bear)}{P(bear \to bull)} &= \frac{\pi_2^*}{\pi_1^*} = 1 \to P(bull \to bear) = P(bear \to bull). \end{aligned} \tag{3.97}$$

We know that the sum of the columns must be one. Therefore, we initially choose $P(crab \to bull) = P_{12} = 1$, $P(crab \to bear) = P_{32} = 1$ e $P(bear \to bull) = P_{13} = 1$. From this we have $P(bull \to crab) = P_{21} = 4$, $P(bear \to crab) = P_{23} = 4$ and $P(bull \to bear) = P_{31} = 1$.

$$P = \frac{1}{5}\begin{bmatrix} 0 & 1 & 1 \\ 4 & 3 & 4 \\ 1 & 1 & 0 \end{bmatrix}. \tag{3.98}$$

■

The detailed balance is the basis for the Markov chain Monte Carlo simulations (MCMC - see Sec. B).

3.2.3 FIRST PASSAGE TIME

The expected number of interactions needed to reach a state j departing from a state i is known as *first passage time* [29,31,73] μ_{ij}. In order to find it, let's start with the probability measurement of a random variable A:

$$\begin{aligned}P(A) &= P(A\cap\Omega) = P\left(A\cap\bigcup_n B_n\right)\\ &= \sum_n P(A\cap B_n)\\ &= \sum_n P(A|B_n)P(B_n).\end{aligned} \tag{3.99}$$

Therefore, the expected time of first passage is given by:

$$\begin{aligned}\mu_{ij} &= \langle\text{number of steps from } i \text{ to } j\rangle\\ &= \sum_k \langle\text{number of steps to } j|X_0 = i, X_1 = k\rangle\, P(X_0 = i, X_1 = k)\\ &= \sum_k \langle\text{number of steps to } j|X_0 = i, X_1 = k\rangle\, P_{ik}.\end{aligned} \tag{3.100}$$

If $k = j$, the number of steps is 1. On the other hand, if $k \neq j$, then the number of steps is μ_{kj} added by the number of steps from i to k. Thus:

$$\begin{aligned}\mu_{ij} &= 1\cdot P_{ij}\Big|_{j=k} + \sum_{k\neq j}\left(1+\mu_{kj}\right)\cdot P_{ik}\\ &= \sum_k P_{ik} + \sum_{k\neq j}\mu_{kj}P_{ik}\\ &= 1 + \sum_{k\neq j} P_{ik}\mu_{kj}\end{aligned} \tag{3.101}$$

In matrix notation:

$$\begin{aligned}\mu &= \mathbf{J} + \mathbf{P}(\mu - \mu_d)\\ (\mathbf{I}-\mathbf{P})\mu + \mathbf{P}\mu_d &= \mathbf{J},\end{aligned} \tag{3.102}$$

where μ_d is the diagonal component of μ.

Example 3.2.4. First passage time for the Markov chain in Fig. 3.3

Assigning crab=0, bull=1 and bear=2, Eq. 3.102 gives:

$$\begin{aligned}\mu_{00} &= 1 + P_{01}\mu_{10} + P_{02}\mu_{20}\\ \mu_{01} &= 1 + P_{00}\mu_{01} + P_{02}\mu_{21}\\ \mu_{10} &= 1 + P_{00}\mu_{02} + P_{01}\mu_{12}\\ \mu_{10} &= 1 + P_{11}\mu_{10} + P_{12}\mu_{20}\\ \mu_{11} &= 1 + P_{10}\mu_{01} + P_{12}\mu_{21}\\ \mu_{12} &= 1 + P_{10}\mu_{02} + P_{11}\mu_{12}\\ \mu_{20} &= 1 + P_{21}\mu_{10} + P_{22}\mu_{20}\\ \mu_{21} &= 1 + P_{20}\mu_{01} + P_{22}\mu_{21}\\ \mu_{22} &= 1 + P_{20}\mu_{02} + P_{21}\mu_{12}.\end{aligned} \tag{3.103}$$

After some calculation:

$$\mu = \begin{array}{c} i\backslash j \\ \text{crab} \\ \text{bull} \\ \text{bear} \end{array} \begin{array}{c} \begin{array}{ccc} \text{crab} & \text{bull} & \text{bear} \end{array} \\ \begin{pmatrix} 1.19 & 16.04 & 12.92 \\ 1.51 & 14.22 & 11.32 \\ 1.38 & 15.49 & 11.45 \end{pmatrix} \end{array}. \quad (3.104)$$

■

If it is possible to reach any state departing from any state, than the chain is said to be *irreducible*. If μ_{ii} is finite, then the state is known as *positive recurrent*. On the other hand, it is known as *null recurrent*. Also, we say that a state is *ergodic* if it is aperiodic and positive recurrent. If the probability of taking finite steps to reach a state after departing itself is less than one, then we say that the state is *transient*. Otherwise, we say that the state is *periodic*.

One interesting result shows up when we compute:

$$\begin{aligned} \mu &= \mathbf{J} + \mathbf{P}(\mu - \mu_d) \\ \pi\mu &= \pi\mathbf{J} + \pi\mathbf{P}(\mu - \mu_d) \\ \pi\mu &= \mathbf{e} + \pi(\mu - \mu_d) \\ \pi &= \mathbf{e}\mu_d^{-1} \\ &= \begin{bmatrix} 1/\mu_{00} & 1/\mu_{11} & \dots & 1/\mu_{NN} \end{bmatrix}. \end{aligned} \quad (3.105)$$

This implies that the faster the chain returns to a specific state, the more likely it is that this state is in the stationary distribution.

3.2.4 BIRTH-AND-DEATH PROCESS

The birth-and-death[15] process [30, 31] is a Markov chain in which the next state corresponds to a unitary increase or decrease in the number of objects under study. Its diagram is illustrated in Fig. 3.4 wherein λ and μ are birth and death rates per unit time respectively, and Δ_t is a short period. During this period, the population can increase by one individual, decrease by one individual, or it may remain with the same population.

The transition matrix for this process is given by:

$$\mathbf{P} = \begin{bmatrix} 1-(\lambda_0)\Delta_t & \lambda_0\Delta_t & & & \\ \mu_1\Delta_t & 1-(\lambda_1+\mu_1)\Delta_t & \lambda_1\Delta_t & \dots & \\ & & \vdots & \ddots & \lambda_{N-1}\Delta_t \\ & & & \mu_N\Delta_t & 1-(\lambda_N+\mu_N)\Delta\_t \end{bmatrix}. \quad (3.106)$$

For the first and following transitions we get for the probability X_n of finding the system in state n:

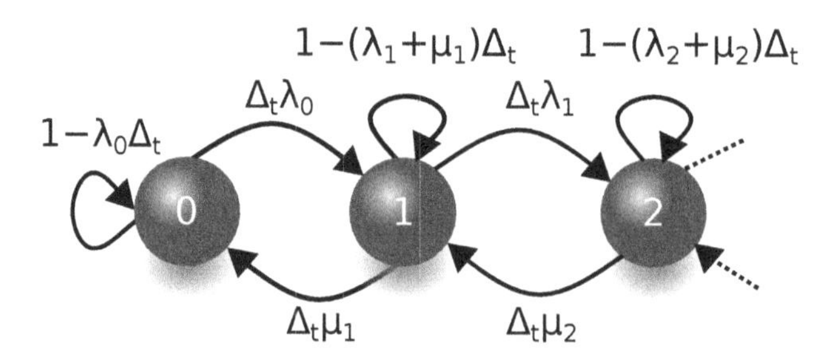

Figure 3.4: Diagram for the birth-and-death process. λ and μ are the birth and death rates per unit time and Δ_t is a short period

$$\begin{aligned} X_0(t+\Delta) &= X_0(t)(1-\lambda_0\Delta_t)+X_1(t)\mu_1\Delta_t \\ X_n(t+\Delta) &= X_n(t)(1-[\lambda_n+\mu_n]\Delta_t)+X_{n+1}(t)\mu_{n+1}\Delta_t+X_{n-1}(t)\lambda_{n-1}\Delta_t. \end{aligned} \tag{3.107}$$

In the limit of very short periods $\Delta \to 0$:

$$\begin{aligned} \frac{dX_0(t)}{dt} &= -\lambda_0 X_0(t)+X_1(t)\mu_1 \\ \frac{dX_n(t)}{dt} &= -(\lambda_n+\mu_n)X_n(t)+X_{n+1}(t)\mu_{n+1}+X_{n-1}(t)\lambda_{n-1}. \end{aligned} \tag{3.108}$$

3.2.4.1 Pure Birth Process

It is not simple to find a closed-form solution to generic birth-and-death processes. Nonetheless, it is possible to study a system with a negligible death rate. In this case the Markov chain consists of only transient states and the previous set of equations can be written as:

$$\begin{aligned} \frac{dX_0(t)}{dt} &= -\lambda_0 X_0(t) \\ \frac{dX_1(t)}{dt} &= -\lambda_1 X_1+X_0(t)\lambda_0. \end{aligned} \tag{3.109}$$

From the first equation we get for $X_0(0)=1$:

$$X_0(t)=e^{-\lambda_0 t}, \tag{3.110}$$

and from the second:

$$\begin{aligned} \frac{dX_1(t)}{dt} &= -\lambda_1 X_1(t)+X_0(t)\lambda_0 \\ &= -\lambda_1 X_1(t)+\lambda_0 e^{-\lambda_0 t}. \end{aligned} \tag{3.111}$$

We can proceed a bit further considering the situation where the birth rate does not depend on the population ($\lambda_n = \lambda, \forall n$):

$$X_1'(t) + \lambda X_1(t) = \lambda e^{-\lambda t}. \tag{3.112}$$

This is a non-homogeneous differential equation. In order to solve it, we create:

$$u = e^{\int \lambda dt} = e^{\lambda t}, \tag{3.113}$$

and multiply the whole equation by this integrating factor:

$$\begin{aligned} e^{\lambda t}X_1'(t) + e^{\lambda t}\lambda X_1(t) &= \lambda \\ \left(e^{\lambda t}X_1(t)\right)' &= \lambda \\ X_1(t)e^{\lambda t} &= \lambda t + \xi \\ X_1(t) &= (\lambda t + \xi)e^{\lambda t}. \end{aligned} \tag{3.114}$$

At the beginning of the process $X_n(t) = 0,\ n \neq 0$. Therefore, $\xi = 0$ and we get:

$$X_1(t) = \lambda t e^{\lambda t}. \tag{3.115}$$

Following the same procedure for the next terms we get:

$$X_n(t) = \frac{(\lambda t)^n}{n!} e^{-\lambda t}, \tag{3.116}$$

which is a Poisson[16] distribution, which is expected for a model of successive events. Therefore, $\langle X_n(t) \rangle = \text{VAR}(X_n(t)) = \lambda t$.

3.2.4.2 Yule-Furry Process

In a Yule-Furry[17] process [82] the birth rate is proportional to the population size: $\lambda_n = n\lambda$. In this situation, we have:

$$\begin{aligned} \frac{dX_n(t)}{dt} &= -n\lambda X_n(t) + (n-1)\lambda X_{n-1} \\ \frac{dX_0(t)}{dt} &= 0. \end{aligned} \tag{3.117}$$

As initial conditions, let's assume only one individual: $X_1(0) = 1$ and $X_i(0) = 0,\ i \neq 1$. For $n = 1$:

$$\frac{dX_1(t)}{dt} = -\lambda X_1 \rightarrow X_1(t) = e^{-\lambda t}. \tag{3.118}$$

For $n = 2$:

$$\begin{aligned}
\frac{dX_2(t)}{dt} &= -2\lambda X_2 + \lambda X_1 \\
X_2'(t) + 2\lambda X_2(t) &= \lambda e^{-\lambda t} \\
e^{2\lambda t} X_2'(t) + 2e^{2\lambda t}\lambda X_2(t) &= \lambda e^{\lambda t} \\
\left(e^{2\lambda t} X_2(t)\right)' &= \lambda e^{\lambda t} \\
X_2(t) &= e^{-2\lambda t}(e^{\lambda t} + \xi)
\end{aligned} \tag{3.119}$$

Since $X_2(0) = 0$ we get $\xi = -1$. Therefore:

$$X_2(t) = e^{-\lambda t}\left(1 - e^{-\lambda t}\right). \tag{3.120}$$

If we continue this procedure, we obtain:

$$X_n(t) = e^{-\lambda t}\left(1 - e^{-\lambda t}\right)^{n-1}, \; n \geq 1. \tag{3.121}$$

This is a geometric distribution with $p = e^{-\lambda t}$.

If the initial population consists of m individuals, we can say that each of them obeys an independent Yule process. Since each process has a geometric distribution, the sum will have a negative binomial distribution:

$$X_{n,m}(t) = \binom{n-1}{m-1} e^{-\lambda t}\left(1 - e^{-\lambda t}\right)^{n-m}, \; n > m. \tag{3.122}$$

One simulation of the Yule-Furry process is shown in Fig. 3.5. It is instructive to note that as time progresses, the tail of the distribution becomes longer.

3.2.4.3 Pure Death Process

In a pure death process we get:

$$\frac{dX_n(t)}{dt} = -\mu_n X_n(t) + \mu_{n+1} X_{n+1}(t). \tag{3.123}$$

Let's start from the end of the chain setting $X_{N+1}(t) = 0$, $X_N(0) = 1$, and $\mu_n = \mu$:

$$X_N'(t) = -N\mu X_N(t) \rightarrow X_N(t) = e^{-N\mu t}. \tag{3.124}$$

Going backwards in the chain:

$$\begin{aligned}
X_{N-1}'(t) + (N-1)\mu X_{N-1}(t) &= N\mu e^{-N\mu t} \\
e^{(N-1)\mu t} X_{N-1}'(t) + e^{(N-1)\mu t}(N-1)\mu X_{N-1}(t) &= N\mu e^{(N-1)\mu t} e^{-N\mu t} \\
\left(e^{(N-1)\mu t} X_{N-1}(t)\right)' &= N\mu e^{-\mu t} \\
e^{(N-1)\mu t} X_{N-1}(t) &= -Ne^{-\mu t} + \xi \\
X_{N-1}(t) &= \left(-Ne^{-\mu t} + \xi\right) e^{-(N-1)\mu t}.
\end{aligned} \tag{3.125}$$

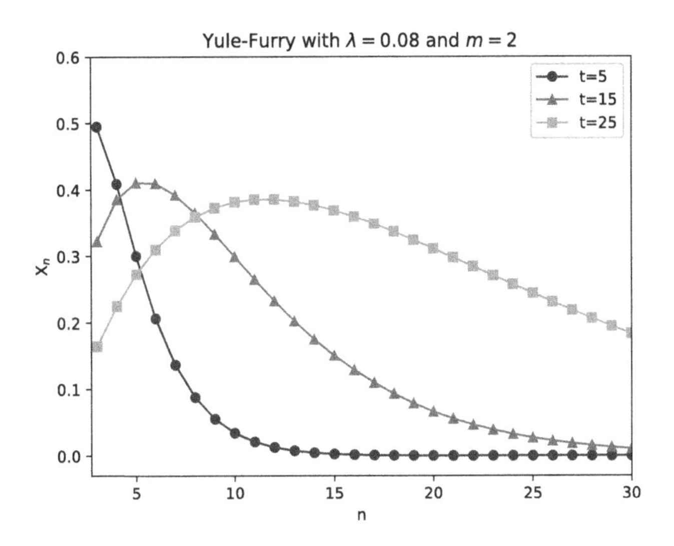

Figure 3.5: Probability X_n of finding the system in state n for a Yule-Furry process with $\lambda = 0.08$, $m = 2$

At the end of the chain, $X_{N-1}(0) = 0$. Therefore, $\xi = N$ and we get:

$$X_{N-1}(t) = N\left(1 - e^{-\mu t}\right) e^{-(N-1)\mu t}. \tag{3.126}$$

Proceeding recursively, one obtains:

$$X_n(t) = \binom{N}{n} e^{-n\mu t}\left(1 - e^{-\mu t}\right)^{N-n}, \tag{3.127}$$

which is a binomial distribution with a survival probability of $e^{-\mu t}$. From this equation, it is easy to infer that the probability of extinction increase exponentially fast to one:

$$X_0(t) = (1 - e^{-\mu t})^N. \tag{3.128}$$

3.2.4.4 Linear Birth and Death Process

In a linear birth-and-death process we set $\lambda_n = n\lambda$ and $\mu_n = n\mu$. Therefore,

$$\begin{aligned}
\frac{dX_0(t)}{dt} &= \mu X_1(t) \\
\frac{dX_n(t)}{dt} &= -(\lambda + \mu)nX_n(t) + (n+1)\mu X_{n+1}(t) + (n-1)\lambda X_{n-1}(t).
\end{aligned} \tag{3.129}$$

In steady state, the time derivatives are zero and we get $X_1(t \to \infty) = 0$. If $X_0(t \to \infty) = 1$, then the ultimate extinction is certain. On the other hand, we still have that

$X_n(t \to \infty) = 0$ even if $X_0(t \to \infty) = P < 1$. This implies that the population grows without bounds with probability $1 - P$. Either way, in steady state, the population either goes to zero or increases indefinitely.

The expected population is:

$$
\begin{aligned}
M(t) &= \sum_{n=1}^{\infty} n X_n(t) \\
\frac{dM(t)}{dt} &= \sum_{n=1}^{\infty} n \frac{dX_n(t)}{dt} \\
&= -(\lambda+\mu) \sum_{n=1}^{\infty} n^2 X_n(t) + \mu \sum_{n=1}^{\infty} n(n+1) X_{n+1}(t) + \lambda \sum_{n=1}^{\infty} n(n-1) X_{n-1}(t).
\end{aligned}
\tag{3.130}
$$

Changing variables $m = n+1$ and $k = n-1$:

$$
\begin{aligned}
\frac{dM(t)}{dt} &= -(\lambda+\mu) \sum_{n=0}^{\infty} n^2 X_n(t) + \mu \sum_{m=0}^{\infty} (m-1) m X_m(t) + \lambda \sum_{k=0}^{\infty} (k+1) k X_k(t) \\
&= \sum_{n=0}^{\infty} \left[\lambda \left(-n^2 + n(n+1)\right) + \mu \left(-n^2 + n(n-1)\right) \right] X_n(t) \\
&= (\lambda+\mu) \sum_{n=0}^{\infty} n X_n(t) = (\lambda - \mu) M(t).
\end{aligned}
\tag{3.131}
$$

Therefore,

$$
M(t) = M(0) e^{(\lambda-\mu)t},
\tag{3.132}
$$

which goes asymptotically to zero if $\mu > \lambda$ or to infinite otherwise.

3.3 ORNSTEIN-UHLENBECK PROCESSES

The Ornstein-Uhlenbeck[18] [28–31, 36, 55, 64, 73, 83] process (OU) is widely used in physics to model problems that show reversion to the mean. An OU process X is described by:

$$
dV = \gamma(V_0 - V)dt + \beta dW_t,
\tag{3.133}
$$

where dW_t is a Wiener process (see Sec. 2.2.5), V_0, γ and β are constants.

In this section we will study general properties of OU processes as well as two important processes that are used in finances: the Vašíček and CIR models.

3.3.1 LANGEVIN EQUATION

Let's begin this study with a simple example from physics. According to Newton's laws, the speed of a particle is given by the simple equation:

$$\frac{dx(t)}{dt} = v(t). \tag{3.134}$$

If we assume that the position of the particle is a random variable X, then it has to satisfy the Markov property since its current position depends only on its previous position. This also has to be a continuous process $\lim_{\Delta \to 0} X(t+\Delta) = X(t)$ and

$$V(t) = \lim_{\Delta \to 0} \frac{X(t+\Delta) - X(t)}{dt} \tag{3.135}$$

is the velocity of the particle.

We can rewrite Eq. 3.134 for the stochastic case as:

$$\lim_{\Delta \to 0} [X(t+\Delta) - X(t)] = dX = V(t)dt = G[X(t), dt], \tag{3.136}$$

where $G[X(t), dt]$ is a Markov propagator that generalizes classical dynamics to the stochastic domain. For instance, Einstein [84][19] in his work on diffusion considered a propagator:

$$G[X(t), dt] = \sqrt{\theta^2 dt} \mathcal{N}(0,1), \tag{3.137}$$

where $\mathcal{N}(0,1)$ indicates a zero centered normal distribution with unitary variance. In this situation we get:

$$\begin{aligned} dX &= \sqrt{\theta^2 dt} \mathcal{N}(0,1) = \theta dW_t \\ \langle dX \rangle &= \langle dW \rangle = 0 \\ \mathrm{VAR}(dX) &= \theta^2 dt. \end{aligned} \tag{3.138}$$

This, however, does not take Newton's second law into consideration. Langevin[20] [28, 32, 55, 73] fixed this problem doing:

$$\frac{dV}{dt} = A(t) = \frac{F(t)}{m}, \tag{3.139}$$

where F is the resulting force including the viscous drag created by the other particles. Given a linear drag $F(t) \propto -V(t)$, we get:

$$\lim_{\Delta \to 0} [V(t+\Delta) - V(t)] = dV = -\gamma V(t)dt + \sqrt{\beta^2 dt} \mathcal{N}(0,1). \tag{3.140}$$

Adding an additional force created by an external field, we get the OU process:

$$dV = \gamma(V_0 - V)dt + \beta dW. \tag{3.141}$$

3.3.1.1 Solution

In order to find an analytical solution for a general OU process, we make:

$$\begin{aligned} f(V,t) &= V(t)e^{\gamma t} \\ df &= e^{\gamma t}dV + V\gamma e^{\gamma t}dt. \end{aligned} \tag{3.142}$$

Placing this in the OU process:

$$\begin{aligned} df &= e^{\gamma t}\left[\gamma(V_0 - V)dt + \beta dW\right] + V\gamma e^{\gamma t}dt \\ &= e^{\gamma t}\gamma V_0 dt - V\gamma e^{\gamma t}dt + \beta e^{\gamma t}dW + V\gamma e^{\gamma t}dt \\ &= e^{\gamma t}V_0\gamma dt + \beta e^{\gamma t}dW. \end{aligned} \tag{3.143}$$

Integrating from 0 to t:

$$\begin{aligned} f(t) &= f(0) + V_0\gamma\int_0^t e^{\gamma s}ds + \beta\int_0^t e^{\gamma s}dW_s \\ Ve^{\gamma t} &= V(0) + V_0 e^{\gamma s}|_0^t + \beta\int_0^t e^{\gamma s}dW_s \\ V(t) &= V(0)e^{-\gamma t} + V_0\left(e^{\gamma t} - 1\right)e^{-\gamma t} + \beta\int_0^t e^{\gamma s}dW_s e^{-\gamma t} \\ V(t) &= V(0)e^{-\gamma t} + V_0\left(1 - e^{-\gamma t}\right) + \beta\int_0^t e^{-\gamma(t-s)}dW_s. \end{aligned} \tag{3.144}$$

We can now calculate the expected value of $V(t)$:

$$\begin{aligned} \langle V(t)\rangle &= \left\langle V(0)e^{-\gamma t}\right\rangle + \left\langle V_0\left(1 - e^{-\gamma t}\right)\right\rangle + \beta\int_0^t e^{-\gamma(t-s)}\langle dW_s\rangle \\ &= V(0)e^{-\gamma t} + V_0\left(1 - e^{-\gamma t}\right). \end{aligned} \tag{3.145}$$

For a long period:

$$\lim_{t\to\infty}\langle V\rangle = V_0. \tag{3.146}$$

This verifies the main property of OU processes. In the long run, the expected velocity converges to a specific value.

In order to calculate the variance of $V(t)$, we must use the Ito isometry[21].

Ito's Isometry

An *elementary process* adapted to a filtration $\mathscr{F}_t$ has the form:

$$F_t = \sum_j e_j \mathbf{1}_{(t_j,t_{j+1}]}(t), \tag{3.147}$$

where e_j is a random variable measurable with respect to the filtration.

The Ito integral [29, 36, 55, 64, 73, 85, 86] of an elementary process F is given by:

$$\begin{aligned} I_t[F] &= \int_0^t F_s dW_s = \sum_{i\geq 0} e_j \left[W_{t_{j+1}\wedge t} - W_{t_j \wedge t}\right], \\ &= \sum_{i\geq 0} e_j \Delta W_j, \end{aligned} \tag{3.148}$$

where W_t is a Wiener process adapted to the filtration such that $(W_t, \mathscr{F}_t)$ is a Markov process.

According to Fubini's theorem[22], the expected value of an integral has to be the same as the integral of the expected value. Since, the expected value of $W_t^2 = 1$, we have, considering independent Wiener processes ΔW_i and ΔW_j:

$$\begin{aligned} \left\langle (I_t[F])^2 \right\rangle &= \left\langle \sum_{ij} e_i e_j \Delta W_i \Delta W_j \right\rangle \\ &= \sum_{ij} \langle e_i^2 \rangle (t_{j+1} - t_j) \delta_{ij} \\ &= \left\langle \int_0^t F_s^2 ds \right\rangle. \end{aligned} \tag{3.149}$$

This is known as Ito's isometry [36, 73, 86].

Therefore, the variance of the OU process is given by:

$$\begin{aligned} \text{VAR}\{V(t)\} &= \langle (V(t) - \langle V(t) \rangle)^2 \rangle \\ &= \left\langle \left(\beta \int_0^t e^{-\gamma(t-s)} dW_s \right)^2 \right\rangle = \beta^2 \left\langle \int_0^t e^{-2\gamma(t-s)} ds \right\rangle \\ &= \beta^2 \int_0^t e^{-2\gamma(t-s)} dt = \frac{\beta^2}{2\gamma} \left(1 - e^{-2\gamma t}\right). \end{aligned} \tag{3.150}$$

Again, for a sufficiently long period:

$$\lim_{t\to\infty} \text{VAR}\{V(t)\} = \frac{\beta^2}{2\gamma}. \tag{3.151}$$

OU processes are used to model, for example, the interest rate. High interest rates hamper the economic activity as the propensity to borrow and invest is reduced. On the other hand, if the interest rate is low, financial institutions do not have incentives to land money. As a result, the interest rate typically stays within a band. Thus, interest rates may oscillate in the short term but tend to be stable in the long term. The Vašíček[23] model [87] captures this idea making V the *instantaneous interest rate*, γ the *speed of reversion*, V_0 the *long term level*, β the *instantaneous volatility*.

3.3.1.2 Dissipation-Fluctuation Theorem

In OU processes there is a competition between damping (γ parameter) and fluctuations (β^2 parameter). However, in the long term, the average kinetic energy of a particle is given by:

$$\mathscr{E}_K = \frac{m}{2}\lim_{t\to\infty} \text{VAR}\{V(t)\} = m\frac{\beta^2}{4\gamma}. \tag{3.152}$$

At this limit, the particle reaches equilibrium with the medium. Thus, according to the energy equipartition theorem [88], for one degree of freedom we have:

$$\mathscr{E}_K = \frac{k_B T}{2} = m\frac{\beta^2}{4\gamma}, \tag{3.153}$$

where T is the temperature of the medium. Therefore, there is a clear relationship between fluctuation and dissipation. Knowing one of the parameters, it is possible to obtain the other:

$$\beta^2 = 2\gamma k_B T/m. \tag{3.154}$$

In the limit where the inertial effects are minimized with respect to the randomized component, the OU process is reduced to a Wiener process (see Sec. 2.2.5). This is captured by considering a zero differential velocity:

$$0 = -\gamma V dt + \beta dW_t \to V dt = dX = \frac{\beta}{\gamma} dW_t. \tag{3.155}$$

The diffusion parameter, in this case, is given by $\delta^2 = \beta^2/\gamma^2$. This is known as the *Smoluchowski limit*[24].

3.3.2 LIMITED OU PROCESS

One big limit of the Vašíček model is that it allows for negative oscillations of the interest rate. One way to circumvent this problem is by making:

$$dV = \gamma(V_0 - V)dt + \beta\sqrt{V}dW. \tag{3.156}$$

With this modification, the closer the interest rate approaches zero, the smaller is the random component. A detailed analysis [89] indicates that if $2\gamma V_0 > \beta^2$ then the interest rate will never reach zero. This is known as the Cox-Ingersoll-Ross[25] (CIR) model [26, 29, 55, 90].

Following the same procedure, it is simple to verify that the expected value of the interest rate is the same as for the Vašíček model since $\sqrt{V}$ only appears in the random component. For calculating the variance, Eq. 3.150 becomes:

$$\text{VAR}\{V(t)\} = \left\langle \left(\beta \int_0^t \sqrt{V_s} e^{-\gamma(t-s)} dW_s \right)^2 \right\rangle. \tag{3.157}$$

Applying Ito's isometry:

$$\begin{aligned} \text{VAR}\{V(t)\} &= \beta^2 \left\langle \int_0^t V_s e^{2\gamma(s-t)} ds \right\rangle \\ &= \beta^2 \int_0^t \langle V_s \rangle e^{2\gamma(s-t)} ds. \end{aligned} \tag{3.158}$$

We can now use Eq. 3.145 to obtain the expected value of V_s. Therefore:

$$\begin{aligned} \text{VAR}\{V(t)\} &= \beta^2 \int_0^t \left(V(0)e^{-\gamma s} + V_0 \left(1 - e^{-\gamma s}\right) \right) e^{2\gamma(s-t)} ds \\ &= \beta^2 \int_0^t \left(V(0)e^{\gamma s} + V_0 \left(e^{2\gamma s} - e^{\gamma s}\right) \right) e^{-2\gamma t} ds \\ &= e^{-2\gamma t} \beta^2 \left(\frac{V(0)e^{\gamma s}}{\gamma} \bigg|_0^t + \frac{V_0 e^{2\gamma s}}{2\gamma} \bigg|_0^t - \frac{V_0 e^{\gamma s}}{\gamma} \bigg|_0^t \right) \\ &= e^{-2\gamma t} \frac{\beta^2}{\gamma} \left(V(0)(e^{\gamma t} - 1) + \frac{V_0}{2} \left(e^{2\gamma t} - 1\right) - V_0 \left(e^{\gamma t} - 1\right) \right) \\ &= \frac{\beta^2}{\gamma} \left(V(0)e^{-\gamma t} - V(0)e^{-2\gamma t} + \frac{V_0}{2} \left(1 - e^{-2\gamma t}\right) - V_0 \left(e^{-\gamma t} - e^{-2\gamma t}\right) \right) \\ &= \frac{V(0)\beta^2}{\gamma} \left(e^{-\gamma t} - e^{-2\gamma t}\right) + \frac{\beta^2 V_0}{2\gamma} \left(1 - e^{-\gamma t}\right)^2. \end{aligned} \tag{3.159}$$

For a long period, the variance becomes:

$$\lim_{t \to 0} \text{VAR}\{x\} = \frac{\beta^2 V_0}{2\gamma}, \tag{3.160}$$

which is the same long term variance obtained in the Vašíček model multiplied by V_0.

3.4 POINT PROCESSES

A point process [30, 32, 64, 91] is defined as an ordered sequence of non-negative random variables $\{t_i | t_i < t_{i+1},\ i \in \mathbb{N}^*\}$, such that t_i can indicate, for instance, the

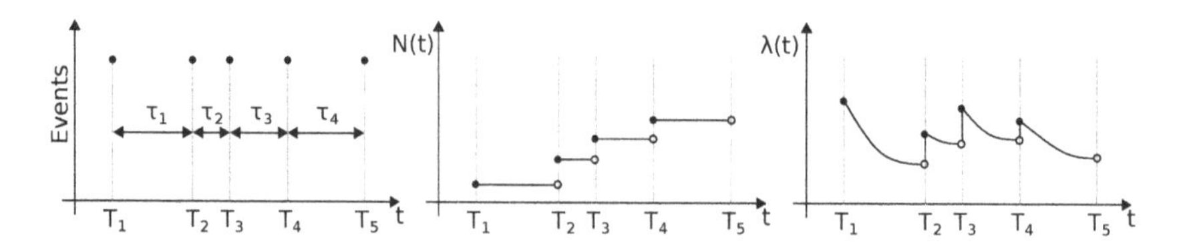

Figure 3.6: Left: a set of discrete events, Middle: a counter of these events, and Right: an intensity function related to these events

time it takes for an order to reach the stock broker. A pictorial representation of point processes is shown in Fig. 3.6.

One type of point process is the *duration process* that is described by:

$$\tau_i = t_i - t_{i-1}. \tag{3.161}$$

This process basically describes the time period between events. Related to these discrete events, it is also possible to define a *counting process* that counts the total number of them. We can formalize a counting process as a stochastic process $\{N_t\}_{t\geq 0}$ such that $N_t \in \mathbb{Z}^+$, t belongs to an ordered set T, the additions $N_{t+\Delta} - N_t \geq 0$, $\forall \Delta > 0$, and $N_0 = 0$. Thus, the one dimensional counting process is given by:

$$N(t) = \sum_{i \in \mathbb{N}^*} \mathbf{1}_{t_i \leq t}. \tag{3.162}$$

If this counting process is adapted to a filtration $\mathscr{F}_t$ with independent increments such that:

$$P(N(t) - N(s) = n | \mathscr{F}_s) = e^{-\lambda(t-s)} \frac{\lambda^n (t-s)^n}{n!} \quad n \in \mathbb{N}, \tag{3.163}$$

then we say that it is a *Poisson process*.

It is also possible to associate an *intensity jump process* to these events. This is defined, with respect to a filtration $\mathscr{F}_t$, as a measure of the change rate of the counting process:

$$\begin{aligned} \lambda(t|\mathscr{F}_t) &= \lim_{h \to 0} \left\langle \frac{N(t+h) - N(t)}{h} \middle| \mathscr{F}_t \right\rangle \\ &= \lim_{h \to 0} \frac{1}{h} Pr\left[N(t+h) > N(t) | \mathscr{F}_t\right]. \end{aligned} \tag{3.164}$$

Example 3.4.1. Intensity of a Poisson process

$$\lim_{\Delta\to 0}\left\langle \frac{1}{\Delta}[N(t+\Delta)-N(t)]\middle|\mathscr{F}_t\right\rangle =$$
$$\lim_{\Delta\to 0}\frac{1}{\Delta}\sum_{n=0} nP(N(t+\Delta)-N(t)=n|\mathscr{F}_t) = \lim_{\Delta\to 0}\frac{1}{\Delta}\sum_{n=1} ne^{-\lambda\Delta}\frac{\lambda^n\Delta^n}{n!}$$
$$= \lim_{\Delta\to 0}\frac{1}{\Delta}e^{-\lambda\Delta}(\lambda\Delta)\sum_{n=1}^{\infty}\frac{\lambda^{n-1}\Delta^{n-1}}{(n-1)!} = \lim_{\Delta\to 0} e^{-\lambda\Delta}\lambda\sum_{m=0}^{\infty}\frac{\lambda^m\Delta^m}{m!}$$
$$= \lim_{\Delta\to 0}\lambda e^{-\lambda\Delta}e^{\lambda\Delta} = \lambda \;\therefore\; \langle dN(t)|\mathscr{F}_t\rangle = \lambda dt. \tag{3.165}$$

■

A counting process $\{N(t), t\in[0,\infty)\}$ is a *non-homogeneous Poisson process* if $N(0)=0$, the increments are independent, and:

$$P(N(t+\Delta)-N(t)=n|\mathscr{F}(t)) = \begin{cases}\lambda(t)\Delta+o(\Delta) & \text{if } n=1\\ o(\Delta) & \text{if } n>1\\ 1-\lambda(t)\Delta & \text{if } n=0\end{cases}, \tag{3.166}$$

where $\lambda(t):[0,\infty)\mapsto[0,\infty)$ is an integrable function.

A *self-excited process* is an intensity jump process such that:

$$\lambda(t|\mathscr{F}_t) = \mu + \int_0^t \phi(t-u)dN(u)$$
$$= \mu + \sum_{t_u<t}\phi(t-t_u,) \tag{3.167}$$

where μ is the exogenous intensity and $\phi:\mathbb{R}_+\mapsto\mathbb{R}_+$ is a kernel that takes into account the positive influence of past events.

3.4.1 HAWKES PROCESS

A Hawkes[26] process is a self-exciting process where the kernel is given by [92]:

$$\phi(t) = \sum_{j=1}^{P}\alpha_j e^{-\beta_j t}\mathbf{1}_{t\in\mathbb{R}_+}. \tag{3.168}$$

Thus, the Hawkes process can be written as:

$$\lambda(t|\mathscr{F}_t) = \lambda_0(t) + \int_0^t\sum_{j=1}^{P}\alpha_j e^{-\beta_j(t-s)}dN_s$$
$$= \lambda_0(t) + \sum_{t_i<t}\sum_{j=1}^{P}\alpha_j e^{-\beta_j(t-t_i)}. \tag{3.169}$$

For $P=1$, the Hawkes process can be written as:

$$\lambda(t|\mathscr{F}_t) = \lambda_0(t) + \alpha \int_0^t e^{-\beta(t-u)} dN_u$$
$$= \lambda_0(t) + \alpha \sum_{t_u < t} e^{-\beta(t-t_u)}. \tag{3.170}$$

In the differential form, the Hawkes process is given by:

$$d\lambda_t = \beta(\lambda^* - \lambda_t)dt + \alpha dN_t. \tag{3.171}$$

3.4.1.1 Solution

It is possible to solve this equation using an approach similar to that we used for solving OU processes. Setting $x(t) = \lambda(t|\mathscr{F}_t)e^{\beta t}$ and using a simpler notation:

$$\begin{aligned} dx &= d\lambda_t e^{\beta t} + \lambda_t \beta e^{\beta t} dt \\ &= \beta(\lambda^* - \lambda_t)e^{\beta t} dt + \alpha dN_t e^{\beta t} + \lambda_t \beta e^{\beta t} dt \\ &= \beta \lambda^* e^{\beta t} dt + \alpha t e^{\beta t} dN \\ x &= x_0 + \beta\lambda^* \int_0^t e^{\beta s} ds + \alpha \int_0^t e^{\beta s} dN_s \\ &= x_0 + \lambda^* \left(e^{\beta t} - 1\right) + \alpha \int_0^t e^{\beta s} dN_s. \end{aligned} \tag{3.172}$$

Multiplying the whole equation by $e^{-\beta t}$, we get:

$$\begin{aligned} \lambda(t|\mathscr{F}_t) &= \lambda(0)e^{-\beta t} + \lambda^* e^{-\beta t}\left(e^{\beta t} - 1\right) + \alpha \int_0^t e^{\beta(s-t)} dN_s \\ &= \lambda_0(t) + \alpha \int_0^t e^{-\beta(t-s)} dN_s, \end{aligned} \tag{3.173}$$

where $\lambda_0(t) = \lambda(0)e^{-\beta t} + \lambda^*(1 - e^{-\beta t})$.

The expected value of a Hawkes process is given by:

$$\begin{aligned} \mu = \langle \lambda(t) \rangle &= \left\langle \lambda_0 + \int_0^t \phi(t-s) dN_s \right\rangle \\ &= \lambda_0 + \left\langle \int_0^t \phi(t-s)\lambda(s) ds \right\rangle \\ &= \lambda_0 + \int_0^t \phi(t-s)\mu ds \\ &= \lambda_0 + \mu \int_0^t \phi(x) dx \\ \mu &= \frac{\lambda_0}{1 - \int_0^t \phi(x) dx}, \end{aligned} \tag{3.174}$$

where $x = t - s$. For a Hawkes kernel:

$$\int_0^\infty \sum_{j=1}^{P} \alpha_j e^{-\beta_j t} dt = - \sum_{j=1}^{P} \frac{\alpha_j}{\beta_j} e^{-\beta_j t} \Bigg|_0^\infty = \sum_{j=1}^{P} \frac{\alpha_j}{\beta_j}. \tag{3.175}$$

Thus, Eq. 3.174 becomes:

$$\mu = \frac{\lambda_0}{1 - \sum_{j=1}^{P} \alpha_j / \beta_j}. \tag{3.176}$$

From this equation, we can see that for a stationary process, it is necessary that:

$$\sum_{j=1}^{P} \frac{\alpha_j}{\beta_j} < 1. \tag{3.177}$$

Hawkes processes have been used, for instance, to model the Epps[27]effect. This is an inverse relationship between the empirical correlation between two assets and the interval for which their prices are measured [93]. Coupled Hawkes processes can adequately capture the mean reversion of the microstructure of noise and reproduce the period depending correlation [94].

Notes

[1]Jean-André Ville (1910–1989) French mathematician.

[2]Jean le Rond dÁlembert (1717–1783) French philosopher, mathematician and physicist.

[3]Concept introduced by Joseph Leo Doob (1910–2003) American mathematician, advisee of Joseph Leonard Walsh and adviser of Paul Halmos among others.

[4]Joseph Leo Doob (1910–2004) American mathematician, advisee of Joseph Leonard Walsh and adviser of Paul Hamos, among others.

[5]Georege Pólya (1887–1985) Hungarian mathematician, adviser of Albert Einstein's son Hans Einstein among others.

[6]Jacob Bernoulli (1655–1705) Swiss mathematician, advisee of Gottfried Wilhelm Leibniz and adviser of his nephew Nikolaus I Bernoulli and his brother Johan Bernoulli among others.

[7]James Stirling (1692–1770) Scottish mathematician.

[8]Robert Gibrat (1904–1980) French engineer.

[9]Abraham Wald (1902–1950), Romanian mathematician advisee of Karl Menger, son of Carl Menger; adviser of Herman Chernoff among others.

[10]Developed by the Sir Francis Galton (1822–1911) English polymath, adviser of Karl Pearson, and Henry William Watson (1827–1903) English mathematician.

[11]Named after Kazuoki Azuma (1939–) Japanese mathematician, and Wassily Hoeffding (1914–1991) Finnish statistician.

[12]Rudolf Lipschitz (1832–1903) German mathematician, advisee of Gustav Dirichlet and Martin Ohm (younger brother of Georg Ohm), adviser of Christian Felix Klein.

[13]Andrei Andreyevich Markov (1856–1922) Russian mathematician, advisee of Pafnuty Chebyshev and adviser of Abram Beicovitch and Georgy Voronoy among others.

[14]Sydney Chapman (1888–1970) British mathematician and geophysicist, advisee of Godfrey Harold Hardy.

[15]Created by William Feller (1906–1970) Croatian mathematician.

[16]Siméon Denis Poisson (1781–1840) French mathematician and physicist, advisee of Joseph-Louis Lagrange and Pierro-Simon Laplace, adviser of Joseph Liouville among others.

[17]Wendel Hinkle Furry (1907–1984) American physicist.

[18]Leonard Salomon Ornstein (1880–1941) Dutch physicist, advisee of Hendrik Lorentz; George Eugene Uhlenbeck (1900–1988) Dutch physicist, advisee of paul Ehrenfest.

[19]Albert Einstein (1879–1955) German physicist, Nobel laureate in 1921.
[20]Paul Lagenvin (1872–1946) French physicsit, advisee of Pierre Curie, Joseph Thomson and Lippmann, adviser of Louis de Broglie and Léon Brillouin and Iréne Joliot-Curie.
[21]Kiyoshi Ito (1915–2008) Japanese mathematician.
[22]Guido Fubini (1879–1943) Italian mathematician.
[23]Oldřich Alfons Vašíček (1942–) Czech mathematician.
[24]Marian Smoluchowski (1872–1917) Polish physicist, advisee of Joseph Stefan.
[25]John Carrington Cox (1943–), Jonathan Edwards Ingersoll Jr. (1949–) advisee of Robert C. Merton, and Stephen Alan Ross (1944–2017) American economists.
[26]Alan Geoffrey Hawkes (1938–) British mathematician.
[27]Thomas Wake Epps, American economist.

4 Options Pricing

In the early 2000's, the Federal Reserve lowered interest rates in order to counter increasing government deficits and the financial effects of the dot-com bubble. This, however, flooded the market with funds, which in turn encouraged predatory lending to many who could not afford to pay it back. On the other hand, those debts were backed by houses used as collateral in a process called *securitization* in which mortgage-backed securities were bundled into *structured-investment vehicles*. The house prices, however, had been severely elevated due to the Community Reinvestment Act from 1977 that supported affordable housing to low-income citizens. When payments started to vanish, it was already too late. The housing bubble had popped, confidence in financial institutions had already disappeared, and panic spread worldwide. Banks lost more than $1 trillion on toxic assets and, by the end of 2007, the world experienced the biggest financial recession since the great depression of 1929.

Between October of 2007 and February of 2008, the S&P 500 index dropped from 1500 to about 700 points. Those who had investments in gold, however, saw its price move from $783 to $969 in the same period in the London bullion market. Also, many investors who were skilled in operating options made literally billions in profit.

When investing, it is usually a good idea to buy an insurance and make a hedge to avoid possible losses. As described in Chapter 1, this can be accomplished not only with gold, but with options. But how can one tell whether the price of the option is high, low or fair? In this chapter we will look at some strategies to value options starting with the binomial tree and then progressing to the famous Black-Scholes model. Both strategies are based on a Wiener process, so before describing them we will take a look at random walks and see how the Wiener process emerges as we consider very small steps in a random walk.

4.1 FROM A RANDOM WALK TO A WIENER PROCESS

Let's consider a random walk in which the agent can either move to the left or to the right after some small period. If it starts at an arbitrary position x, after a period Δ_t, it can move either to $x+\Delta_x$ with probability $1/2$ or to $x-\Delta_x$ with the same probability. This is clearly a martingale since $\langle x(t+\Delta_t)|\mathscr{F}\rangle = x(t)$.

If after a period $2\Delta_t$ we find the walker at an arbitrary position x, we can say that there is a probability $1/2$ that it had come from $x+\Delta_x$ and the same probability that it had come from $x-\Delta_x$ as depicted in Fig. 4.1. On the other hand, in order to find it at $x+\Delta_x$ after a period Δ_t, it must have come from x with probability $1/2$ or from $x+2\Delta_x$ with the same probability. The same happens if, after a period Δ_t, the walker is at $x-\Delta_x$. It must have come from x with probability $1/2$ or from $x-2\Delta_x$ with the same probability. Mathematically, we can write the probability of finding the walker at position x after a period $2\Delta_t$ as:

DOI: 10.1201/9781003127956-4

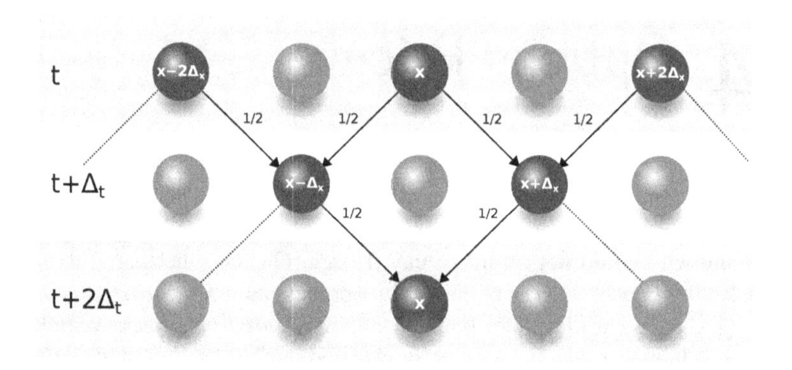

Figure 4.1: A random walker moving in constant finite steps

$$\begin{aligned} p(x,t+2\Delta_t) &= 1/4\cdot p(x+2\Delta_x,t) + 1/2\cdot p(x,t) + 1/4\cdot p(x-2\Delta_x,t) \\ p(x,t+2\Delta_t) - p(x,t) &= 1/4\cdot p(x+2\Delta_x,t) - 1/2\cdot p(x,t) + 1/4\cdot p(x-2\Delta_x,t) \\ \frac{p(x,t+2\Delta_t)-p(x,t)}{(2\Delta_t)}(2\Delta_t) &= 1/4\cdot\frac{[p(x+2\Delta_x,t)-2\cdot p(x,t)+p(x-2\Delta_x,t)]}{(2\Delta_x)^2}(2\Delta_x)^2. \end{aligned} \tag{4.1}$$

At the limit where $\Delta_x \to 0$ e $\Delta_t \to 0$:

$$\begin{aligned} \frac{\partial p(x,t)}{\partial t} &= \lim_{\substack{\Delta_x\to 0\\ \Delta_t\to 0}} \frac{1}{2}\frac{\Delta_x^2}{\Delta_t}\frac{\partial^2 p(x,t)}{\partial x^2} \\ \frac{\partial p(x,t)}{\partial t} &= D\frac{\partial^2 p(x,t)}{\partial x^2}, \end{aligned} \tag{4.2}$$

where D is the *diffusion coefficient* given by $D = \lim_{\substack{\Delta_x\to 0\\ \Delta_t\to 0}} 1/2\Delta_x^2/\Delta_t$.

The kernel of this diffusion equation (the impulse response, or Green's function[1]) can be found by taking its Fourier transform:

$$\begin{aligned} \mathscr{F}\left\{\frac{\partial p(x,t)}{\partial t}\right\} &= D\mathscr{F}\left\{\frac{\partial^2 p(x,t)}{\partial x^2}\right\} \\ \frac{\partial P(k,t)}{\partial t} &= -k^2 D P(k,t) \\ P(k,t) &= P(k,0)e^{-k^2Dt}. \end{aligned} \tag{4.3}$$

Since we want to find the response to a point source excitation, we set $p(x,0) = \delta(x) \to \mathscr{F}\{p(x,0)\} = P(k,0) = \mathscr{F}\{\delta(x)\} = 1$. Therefore:

$$P(k,t) = e^{-k^2Dt}. \tag{4.4}$$

Taking the inverse Fourier transform:

$$
\begin{aligned}
p(x,t) &= \mathscr{F}^{-1}\left\{e^{-k^2Dt}\right\} \\
&= \frac{1}{2\pi}\sqrt{\frac{\pi}{Dt}}\exp\left\{-\frac{x^2}{4Dt}\right\} \\
&= \frac{1}{\sqrt{2Dt}\sqrt{2\pi}}\exp\left\{-\frac{1}{2}\left(\frac{x}{\sqrt{2Dt}}\right)^2\right\}.
\end{aligned}
\tag{4.5}
$$

This is a zero-centered Gaussian distribution with variance $2Dt$. Assuming a scaling limit $\lim_{\substack{\Delta_x \to 0 \\ \Delta_t \to 0}} \Delta_x^2/\Delta_t = 1$, we obtain a Wiener process (W_t) with variance $\text{VAR}(W_t) = t$. This limit is also known as *Brownian motion*[2] and is formalized by *Donsker's*[3] *theorem* [95].

Thus, a Wiener process has an equivalent Fokker-Planck equation (Appendix C):

$$
\frac{\partial p(x,t)}{\partial t} = \frac{1}{2}\frac{\partial^2 p(x,t)}{\partial x^2},
\tag{4.6}
$$

that leads to a probability density function:

$$
p(x,t) = \frac{1}{\sqrt{2\pi t}}\exp\left\{-\frac{x^2}{t}\right\}.
\tag{4.7}
$$

4.2 BINOMIAL TREES

Let's consider a situation where we hold Δ shares of a specific stock that is currently being traded for \$100 and buy one call option for that stock with a strike of \$99 for a price of f. The current value of our portfolio is then $100\Delta - f$. We expect that the price of the shares after one month will either increase or decrease by 5 %. If the former happens, then the price of the option will be \$1 and the payoff of the operation is $105\Delta - 1$. On the other hand, if the latter happens, then the option has no value and the payoff of the operation is 95Δ. This tree, neglecting dividends, is illustrated in Fig. 4.2.

In order to achieve the same result in any situation, it is necessary to have $105\Delta - 1 = 95\Delta \rightarrow \Delta = 1/10$. This strategy of creating a portfolio free of uncertainty is called *delta hedging*.

Now, if we consider a risk-free investment, it is possible to go backwards and find the value of f using the compound rate (Eq. 2.6). In this example, the future value of our portfolio is, in either case, $P_T = \$9.5$. Discounting the value in the period with a risk-free rate of 5%, the present value of the portfolio is $P_0 = P_T e^{-rT} = 9.5e^{-5/100\times 1/12} \approx \$9.46$. Equating this value to $100\Delta - f$, we find that the price of the option has to be $f \approx \$0.54$ so that no arbitrage opportunities exist.

It is possible to generalize this process. The stock price can go up by a percentage $u > 1$ and the value of the option can be described by f_u. This creates a payoff $S_0 u\Delta - f_u$. On the other hand, the stock price can move down by a percentage $0 \leq$

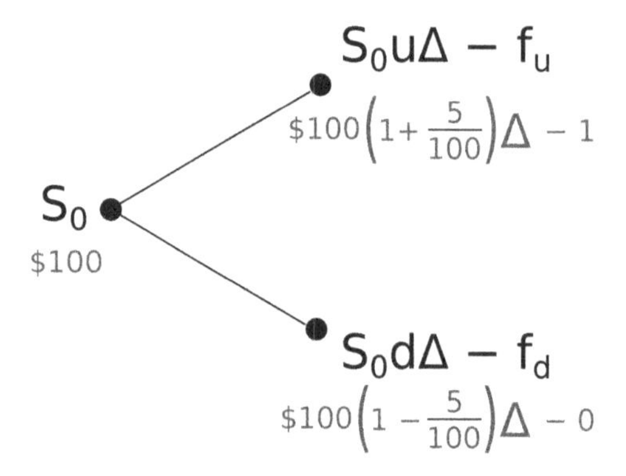

Figure 4.2: An one-step binomial tree

$d < 1$ and the value of the option is described by f_d. This creates a payoff $S_0 d\Delta - f_d$. In order to obtain a risk-free portfolio we equal both payoffs and get:

$$\Delta = \frac{f_u - f_d}{S_0(u-d)} = \frac{1-0}{100(1.05-0.95)} = 0.1. \tag{4.8}$$

The delta for a one step binomial tree [96, 97] is given by the ratio between the difference of payoffs and the difference of stock prices at time T.

In this scenario, the value of the portfolio at the end of period T is given by:

$$\begin{aligned} P_T &= S_0 u\Delta - f_u = S_0 u \frac{(f_u - f_d)}{S_0(u-d)} - f_u \\ &= \frac{df_u - uf_d}{u-d} = \frac{1.05 \times 1 - 0.95 \times 0}{1.05 - 0.95} = \$9.5. \end{aligned} \tag{4.9}$$

Again, we can discount this portfolio to the present:

$$\begin{aligned} P_0 &= e^{-rT} P_T = S_0\Delta - f \\ e^{-rT} \frac{df_u - uf_d}{u-d} &= S_0\Delta - f \\ f &= \frac{f_u - f_d}{u-d} - e^{-rT} \frac{(df_u - uf_d)}{u-d} \\ &= e^{-rT} \left(\frac{e^{rT} - d}{u-d} f_u - \frac{e^{rT} + u}{u-d} f_d \right) \\ &= e^{-rT} \left[pf_u + (1-p) f_d \right], \end{aligned} \tag{4.10}$$

where

$$p = \frac{e^{rT} - d}{u-d}. \tag{4.11}$$

In the example we are considering, for a period of 1 month, that we have $T = 1/12$, $r = 5/100$, $u = 1.05$, $d = 0.95$, and $f_u = 1$, $f_d = 0$. Therefore:

$$
\begin{aligned}
p &= \frac{e^{5/100\times 1/12} - 0.95}{1.05 - 0.95} \approx 0.54 \\
f &= e^{-5/100\times 1/12}[0.54 \times 1 + 0] \approx \$0.54,
\end{aligned}
\tag{4.12}
$$

which is exactly what we obtained before. It is interesting to note that the valuation of the option does not depend on the probability that the stock moves up or down. This happens because the valuation of the option is calculated with respect to the underlying stock, and this carries together the probability of price movement.

Another interesting point is the following. If $e^{rT} > u$, it might be interesting to short the stock, invest the whole amount in a risk-free investment, buy the stock back at the end of the period for a smaller value, and cash in the difference. In order to avoid this arbitrage opportunity, we must impose that $u \leq e^{rT} \leq d$. This, however, implies that:

$$
\begin{aligned}
d \leq e^{rT} &\leq u \\
0 \leq e^{rT} - d &\leq u - d \\
0 \leq \frac{e^{rT} - d}{u - d} &\leq 1 \\
0 \leq p &\leq 1.
\end{aligned}
\tag{4.13}
$$

Thus, p is a probability measure corresponding to an upward price movement of the stock price. Therefore, Eq. 4.10 states that the price of the option is its expected future price discounted at the risk-free rate.

We considered a one-step binomial tree. But this process can continue to many more steps as shown in Fig. 4.3. For each point in the binomial tree there are two possibilities, the stock price moving up with probability p or down with probability $1 - p$, which corresponds to a Bernoulli trial. Therefore, we can assign to every step of the tree a random variable X_i which assumes only two values 0 and 1 corresponding, respectively, to a downwards and an upwards price movement. For each individual (and independent) trial there is a state space $\Omega_2 = \{up, down\}$ such that the state space for a process with N trials is given by $\Omega = \Omega_2^N$. The process starts with $X(0) = 0$ and the probability measure for each step n is given by $P(X(n) = x) = p^x(1-p)^{n-x}$.

A Bernoulli counting process is defined as an additive random walk $N_n = \sum_{i=1}^{n} X_i$ with an associated probability measure $P(N_n = s) = \binom{n}{s} p^s (1-p)^{n-s}$. Inspecting Fig. 4.3 we see that the stock price in this model is given by $S_t = S_0 u^{N_n} d^{t-N_n}$, which is a multiplicative random walk.

Example 4.2.1. Two period stock market

In a two period stock market, the sample space is given by $\Omega = \{(d,d),(d,u),(u,d),(u,u)\}$. Therefore, the σ-algebras are given by $\Sigma_0 = \{\emptyset, \Omega\}$, $\Sigma_1 = \{\Sigma_0, \Omega_d, \Omega_u\}$, where $\Omega_d = \{(d,d),(d,u)\}$ and $\Omega_u = \{(u,u),(u,d)\}$. Σ_2 is the power set of Ω.

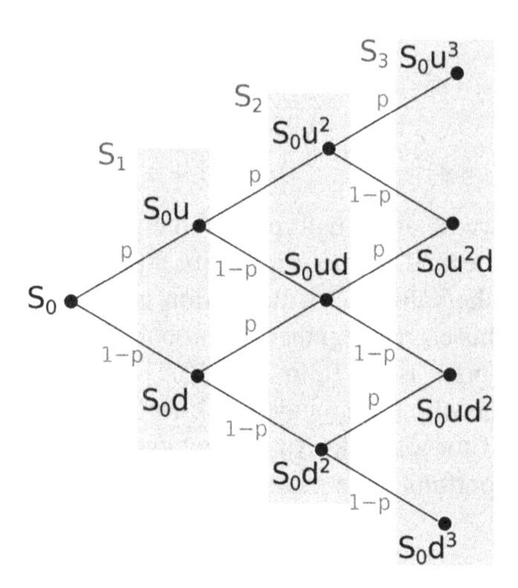

Figure 4.3: Multistep binomial tree

Note, however, that $\{S_2 = S_0ud\} = \{(u,d),(d,u)\}$ and this set is not in Σ_2. Therefore, the lattice representation of the Bernoulli process is not the information tree of the partitions. ■

4.2.1 COX-ROSS-RUBINSTEIN MODEL

What determines u and d? Certainly, they must account for the mean and variance of the stocks. The Cox-Ross-Rubinstein[4] (CRR) model [96–98] proposes a link between discrete binomial trees and geometric random walks and a strategy to value options.

Bachelier argued that, in the continuous-time limit, prices followed a random walk. However, an investor is interested in returns despite the underlying price of the asset. Therefore, if we consider a compound interest rate:

$$\begin{aligned}\lim_{\Delta_t \to 0} S(t+\Delta_t) &= \lim_{\Delta_t \to 0} S(t)e^{r\Delta_t} = \lim_{\Delta_t \to 0} S(t)(1+r\Delta_t)\\ \lim_{\Delta_t \to 0} [S(t+\Delta_t) - S(t)] &= \lim_{\Delta_t \to 0} rS(t)\Delta_t \\ dS(t) &= rS(t)dt.\end{aligned} \tag{4.14}$$

The same is expected for the volatility of the asset. Therefore, one should expect a stochastic component $\propto S(t)dW$. Thus, the price variation can be written as *geometric random walk*:

$$dS = \mu S dt + \sigma S dW. \tag{4.15}$$

In order to solve this equation, we must use Ito's lemma.

Ito's Lemma

For a doubly differentiable function f, its Taylor's expansion is:

$$df = \frac{\partial f}{\partial t}dt + \frac{\partial f}{\partial S}dS + \frac{1}{2}\frac{\partial^2 f}{\partial S^2}dS^2 + \ldots \quad (4.16)$$

Substituting the differential dS by a stochastic differential equation of the form $dS = a(S,t)dt + b(S,t)dW$, we get:

$$\begin{aligned} df =& \frac{\partial f}{\partial t}dt + \frac{\partial f}{\partial S}(a(S,t)dt + b(S,t)dW) + \\ &+ \frac{1}{2}\frac{\partial^2 f}{\partial S^2}(a^2(S,t)dt^2 + b^2(S,t)dW^2 + 2a(S,t)b(S,t)dtdW) + \ldots \end{aligned} \quad (4.17)$$

Eliminating high order terms we obtain Ito's lemma:

$$\begin{aligned} df &\approx \frac{\partial f}{\partial t}dt + \frac{\partial f}{\partial S}(a(S,t)dt + b(S,t)dW) + \frac{1}{2}b^2(S,t)\frac{\partial^2 f}{\partial S^2}dt \\ &= \left(\frac{\partial f}{\partial t} + a(S,t)\frac{\partial f}{\partial S} + \frac{1}{2}b^2(S,t)\frac{\partial^2 f}{\partial S^2}\right)dt + b(S,t)\frac{\partial f}{\partial S}dW. \end{aligned} \quad (4.18)$$

The term inside the parenthesis:

$$\mathscr{A}f = a(S,t)\frac{\partial f}{\partial S} + \frac{1}{2}b^2(S,t)\frac{\partial^2 f}{\partial S^2} \quad (4.19)$$

is known as the *infinitesimal generator* for the process S.

For a geometric random walk, $a(S,t) = \mu S$ and $b(S,t) = \sigma S$. For $f = \log(S)$, we have that $\partial f/\partial t = 0$, $\partial f/\partial S = 1/S$, and $\partial^2 f/\partial S^2 = -1/S^2$. Therefore:

$$d\log(S) = \left(\mu - \frac{1}{2}\sigma^2\right)dt + \sigma dW. \quad (4.20)$$

Thus, the variance of the infinitesimal log-return is simply $\sigma^2 dt$.

Let's now match the variance of the binomial tree to the variance of the geometric random walk:

$$\begin{aligned} \sigma^2\Delta_t &= \langle X^2\rangle - \langle X\rangle^2 \\ &= pu^2 + (1-p)d^2 - (pu + (1-p)d)^2 \\ &= p\left(u^2 - d^2 - pu^2 - pd^2 + 2d^2 - 2ud + 2pud\right) \\ &= p\left((u-d)^2 - p(u-d)^2\right) \\ &= (u-d)^2 p(1-p) \end{aligned} \quad (4.21)$$

Using the probability given in Eq. 4.11:

$$\sigma^2\Delta_t = (u-d)^2 \frac{e^{rt}-d}{u-d}\frac{u-e^{rt}}{u-d}. \tag{4.22}$$

Let's assume that $u = 1/d = e^x$. Therefore:

$$\begin{aligned}\sigma^2\Delta_t &= \left(e^{rt}-e^{-x}\right)\left(e^x-e^{rt}\right) \\ &= e^{rt+x}-e^{2rt}-1+e^{rt-x}.\end{aligned} \tag{4.23}$$

Doing a Taylor expansion up to second order and keeping only terms up to $o(t)$:

$$\begin{aligned}\sigma^2\Delta_t &\approx 1+rt+x+{}^1\!/\!_2(rt+x)^2-1-2rt-1+1+rt-x+{}^1\!/\!_2(rt-x)^2 \\ &= 2rtx+{}^1\!/\!_2x^2-2rtx+{}^1\!/\!_2x^2 \\ &= x^2.\end{aligned} \tag{4.24}$$

Thus:

$$\begin{aligned}u &= e^{\sigma\sqrt{\Delta_t}} \\ d &= e^{-\sigma\sqrt{\Delta_t}}.\end{aligned} \tag{4.25}$$

In the example we are considering in this section, indeed $u = 1.05 \approx 1/0.95 = 1/d$. For the period of one month, we get:

$$1.05 = e^{\sigma} \rightarrow \sigma \approx 0.05\$. \tag{4.26}$$

4.3 BLACK-SCHOLES-MERTON MODEL

Let's consider i) an European option that ii) pays no dividends, iii) operates in an efficient market, and iv) has no associated transactions costs to be bought or sold (also known as a *frictionless market*). Let's also assume that v) the underlying asset has normally distributed homoscedastic returns, and that vi) there exists a known risk-free investment.

It is important to point out that these assumptions are difficult to be met in real trading scenarios. It is known that assets can have fat tail distributions, transactions are hardly costless, trading does not occur in a continuous time and markets are not perfectly liquid.

Under these assumptions, nonetheless, the price of the underlying asset follows a geometric Brownian motion:

$$dS = \mu S dt + \sigma S dW_t. \tag{4.27}$$

Since the price of the option G depends on the price of the underlying asset, we can apply Ito's lemma:

$$dG = \left(\frac{1}{S}\frac{\partial G}{\partial t} + \mu\frac{\partial F}{\partial G} + \frac{\sigma^2}{2}S\frac{\partial^2 G}{\partial S^2}\right) S dt + \sigma\frac{\partial G}{\partial S} S dW. \tag{4.28}$$

Let's make a delta hedging and build a portfolio with Δ units of the underlying asset and on call option. Its value is:

$$P = \Delta S - G. \tag{4.29}$$

We must make the assumption of a *self-financing portfolio*. This means that there is no exogenous net flux of funds and the purchase of an asset can only be done with the sale of another asset. Thus, we can write:

$$dP = \Delta dS - dG. \tag{4.30}$$

Combining the previous equations:

$$\begin{aligned} dP &= \Delta\mu S dt + \Delta\sigma S dW_t - \left(\frac{1}{S}\frac{\partial G}{\partial t} + \mu\frac{\partial F}{\partial G} + \frac{\sigma^2}{2}S\frac{\partial^2 G}{\partial S^2}\right) S dt + \sigma\frac{\partial G}{\partial S} S dW \\ &= \left(\Delta - \frac{\partial G}{\partial S}\right)\sigma S dW_t - \left(\frac{1}{S}\frac{\partial G}{\partial t} + \mu\left[\frac{\partial G}{\partial S} - \Delta\right] + \frac{\sigma^2}{2}S\frac{\partial^2 G}{\partial S^2}\right) S dt. \end{aligned} \tag{4.31}$$

We can now eliminate the stochastic component making:

$$\Delta = \frac{\partial G}{\partial S}. \tag{4.32}$$

The value of the portfolio becomes:

$$dP = -\left(\frac{\partial G}{\partial t} + \frac{\sigma^2}{2}S^2\frac{\partial^2 G}{\partial S^2}\right) dt. \tag{4.33}$$

The compound interest rate for a risk-free investment, though, is given by (see Sec. 4.2.1):

$$dP = rPdt, \tag{4.34}$$

where r is the risk-free interest rate.

Combining equations 4.29, 4.32, 4.33, and 4.34 we get an equation for the price of the option in a manner corresponding to a risk-free investment:

$$-\left(\frac{\partial G}{\partial t} + \frac{\sigma^2}{2}S^2\frac{\partial^2 G}{\partial S^2}\right) dt = r\left(\frac{\partial G}{\partial S}S - G\right) \tag{4.35}$$

Rearranging terms we obtain the famous Black-Scholes[5] equation [99] (BSE):

$$\frac{\partial G}{\partial t} + rS\frac{\partial G}{\partial S} + \frac{\sigma^2}{2}S^2\frac{\partial^2 G}{\partial S^2} - rG = 0. \tag{4.36}$$

4.3.1 DIFFUSION-ADVECTION

Let's create a risk-neutral BSE by making a change of variables:

$$S = e^X \to X = \ln(S),\ dX = \frac{1}{S}dS. \tag{4.37}$$

Immediately we verify that:

$$\begin{aligned}
\Delta &= \frac{\partial G}{\partial S} = \frac{\partial G}{\partial X}\frac{\partial X}{\partial S} = \frac{1}{S}\frac{\partial G}{\partial X} \\
\Gamma &= \frac{\partial^2 G}{\partial S^2} = \frac{\partial}{\partial S}\left(\frac{\partial G}{\partial X}\frac{\partial X}{\partial S}\right) = \frac{\partial^2 G}{\partial S \partial X}\frac{\partial X}{\partial S} + \frac{\partial G}{\partial X}\frac{\partial^2 X}{\partial S^2} \\
&= \frac{\partial^2 G}{\partial X^2}\left(\frac{\partial X}{\partial S}\right)^2 - \frac{1}{S^2}\frac{\partial G}{\partial X} = \frac{1}{S^2}\left(\frac{\partial^2 G}{\partial X^2} - \frac{\partial G}{\partial X}\right).
\end{aligned} \tag{4.38}$$

Δ and Γ are some of *the Greeks* and represent risk measures. Other Greeks are $\Theta = \partial G/\partial t$, $\rho = \partial G/\partial r$ and[a] $\nu = \partial G/\partial \sigma$.

Applying these results in the BSE:

$$\frac{\partial G}{\partial t} + \left(r - \frac{\sigma^2}{2}\right)\frac{\partial G}{\partial X} + \frac{\sigma^2}{2}\frac{\partial^2 G}{\partial X^2} - rG = 0, \tag{4.39}$$

which is an advection-diffusion equation. To see this, let's consider Fick's[6] law for some concentration C:

$$J_D = -D\frac{\partial C}{\partial x}, \tag{4.40}$$

where D is a diffusion constant and J_D is the flux generated by the gradient of the concentration.

In advective transport, the flux is proportional to the concentration since it carries matter. A good example would be a stream of water carrying grains of sand. If Δ_x is the distance traveled by the particles in a period Δ_t, then the number of particles Q is given by $C\Delta_x A$, where A is the cross sectional area of the flow. Doing a little algebra:

$$\begin{aligned}
\lim_{\Delta_t \to 0} \frac{Q}{\Delta_t A} &= C \lim_{\substack{\Delta_t \to 0 \\ \Delta_x \to 0}} \frac{\Delta_x}{\Delta_t} \\
J_A &= C\frac{\partial x}{\partial t}.
\end{aligned} \tag{4.41}$$

The total flow is then given by:

$$J = J_A + J_D = C\frac{\partial x}{\partial t} - D\frac{\partial C}{\partial x}. \tag{4.42}$$

Applying the continuity equation:

---

[a]Although, 'nu' is used, it is read as 'vega', which is not really a Greek letter.

$$
\begin{aligned}
\frac{\partial C}{\partial t} &= -\frac{\partial J}{\partial x} \\
&= -\frac{\partial}{\partial x}\left(C\frac{\partial x}{\partial t} - D\frac{\partial C}{\partial x}\right) \\
&= -\frac{\partial}{\partial x}\left(C\frac{\partial x}{\partial t}\right) + \frac{\partial}{\partial x}\left(D\frac{\partial C}{\partial x}\right) \\
&= -u\frac{\partial C}{\partial x} + D\frac{\partial^2 C}{\partial x^2},
\end{aligned}
\tag{4.43}
$$

where u is the velocity $\partial x/\partial t$. The time derivative corresponds to an *accumulation*, whereas the first spatial derivative corresponds to *advection* and the second one to *diffusion*.

If chemical reactions also occur, then it is also necessary to include a kinetic term directly proportional do the concentration kC and another term related to sources and drains F. Therefore we finally obtain:

$$
\frac{\partial C}{\partial t} = -u\frac{\partial C}{\partial x} + D\frac{\partial^2 C}{\partial x^2} + kC + F. \tag{4.44}
$$

4.3.2 SOLUTION

There are many ways to solve the BSE. We will, however, show the elegant martingale measure approach [26, 55].

Let's discount the price of the underlying asset for the risk-free rate using $dP = rPdt$ (Eq. 4.34) so that $S^* = S/P$:

$$
dS^* = \frac{PdS - SdP}{P^2} = \frac{dS}{P} - S^* r dt. \tag{4.45}
$$

Considering that the price of the underlying asset follows a geometric Brownian motion, we get:

$$
\begin{aligned}
dS^* &= \frac{\mu S dt + \sigma S dW_t}{P} - S^* r dt \\
&= \mu S^* dt + \sigma S^* dW_t - S^* r dt \\
&= (\mu - r) S^* dt + \sigma S^* dW_t.
\end{aligned}
\tag{4.46}
$$

Let's now change the probability measure using Girsanov's theorem (see Appendix D) so that we use a Brownian motion with drift. To make it explicit that W_t is a Brownian motion under the probability measure $\mathbb{P}$ let's change the notation to $W_t^{\mathbb{P}}$. Thus:

$$
W_t^{\mathbb{Q}} = W_t^{\mathbb{P}} + \frac{\mu - r}{\sigma} t \rightarrow dW_t^{\mathbb{P}} = dW_t^{\mathbb{Q}} - \frac{\mu - r}{\sigma} dt. \tag{4.47}
$$

The term $(\mu - r)/\sigma$ is known as *Sharpe ratio* [100][7] or the *market price of risk*. Substituting this last expression in the discounted Brownian motion:

$$dS^* = \sigma S^* dW_t^{\mathbb{Q}}. \tag{4.48}$$

$W_t^{\mathbb{P}}$, in the interval $0 \leq t \leq T$, is a standard Wiener process on the probability space $(\Omega, \Sigma_T, \mathbb{P})$, whereas $W_t^{\mathbb{Q}}$ is a standard Wiener process on $(\Omega, \Sigma_T, \mathbb{Q})$. Under the risk-neutral measure $\mathbb{Q}$, the discounted price is clearly a martingale under $(\Omega, \Sigma_T, \mathbb{Q})$.

Since it is a martingale, we can write:

$$\langle S_T^* | \mathscr{F}_t \rangle_{\mathbb{Q}} = S_t^*, \tag{4.49}$$

where $\mathscr{F}_t$ is a Brownian filtration generated by the Wiener process. For a discount of e^{rt}, we get:

$$S_t = e^{-r(T-t)} \langle S_T | \mathscr{F}_t \rangle_{\mathbb{Q}}, \tag{4.50}$$

which is basically the Feynman-Kac formula (see Appendix E).

At the exercise date T, the price of the call option c is given by the difference between the price of the underlying asset $S(T)$ and the strike k. If this difference is negative, then the price of the option is zero. Therefore, we can write the exercise price of the option as:

$$c(S,T) = \max(S_T - k, 0). \tag{4.51}$$

The call price is then given by:

$$\begin{aligned} c(S,t) &= e^{-r(T-t)} \langle \max(S_T - k, 0) \rangle_{\mathbb{Q}} \\ &= e^{-r(T-t)} \langle S_T \mathbf{1}_{S_T \geq k} \rangle_{\mathbb{Q}} - k e^{-r(T-t)} \langle \mathbf{1}_{S_t \geq k} \rangle_{\mathbb{Q}} \\ &= e^{-r(T-t)} \langle S_T \mathbf{1}_{S_T \geq k} \rangle_{\mathbb{Q}} - k e^{-r(T-t)} P_{\mathbb{Q}}(S_T \geq k). \end{aligned} \tag{4.52}$$

In order to continue, we must know the distribution of S_T. We can find it adapting the geometric Brownian for the measure $\mathbb{Q}$ using the result of Eq. 4.47:

$$\begin{aligned} \frac{dS}{S} &= \mu dt + \sigma dW_t^{\mathbb{P}} \\ &= \mu dt + \sigma dW_t^{\mathbb{Q}} - (\mu - r) dt \\ &= r dt + \sigma dW_t^{\mathbb{Q}}. \end{aligned} \tag{4.53}$$

We can solve this equation applying Ito's lemma to $f = \ln(S)$. For this, we use $a(S,t) = rS$, $b(S,t) = \sigma S$, $\partial f / \partial t = 0$, $\partial f / \partial S = 1/S$, and $\partial^2 f / \partial S^2 = -1/S^2$. Then:

$$\begin{aligned} d\ln(S) &= \left(0 + rS\frac{1}{S} + \frac{1}{2}\sigma^2 S^2 \left(-\frac{1}{S^2}\right)\right) dt + \sigma S \frac{1}{S} dW_t^{\mathbb{Q}} \\ &= \left(r - \frac{1}{2}\sigma^2\right) dt + \sigma dW_t^{\mathbb{Q}} \\ \ln(S_T) - \ln(S_t) &= \left(r - \frac{1}{2}\sigma^2\right)(T - t) + \sigma \left(W_T^{\mathbb{Q}} - W_t^{\mathbb{Q}}\right). \end{aligned} \tag{4.54}$$

Its expected value is:

$$\tilde{\mu} = \langle \ln(S_T) \rangle = \ln(S_t) + \left(r - \frac{1}{2}\sigma^2 \right) \tau \tag{4.55}$$

and the variance is:

$$\text{VAR}\left(\ln(S_T)\right)_{\mathbb{Q}} = \sigma^2 \tau, \tag{4.56}$$

where $\tau = (T - t)$. The PDF of the log-return is therefore:

$$f(S_T) = \frac{1}{S_T \sigma \sqrt{2\pi\tau}} \exp\left\{ -\frac{1}{2} \frac{\left(\ln\left(\frac{S_T}{S_t}\right) - \left(r - \frac{1}{2}\sigma^2\right)\tau \right)^2}{\sigma^2 \tau} \right\}. \tag{4.57}$$

We can now resume Eq. 4.52. Considering some properties of the log-normal distribution and the fact that the log is a monotonic function, we get:

$$\begin{aligned} c(S,t) &= e^{-r\tau} \langle S_T \mathbf{1}_{\ln(S_T) \geq \ln(k)} \rangle_{\mathbb{Q}} - k e^{-r\tau} P_{\mathbb{Q}} \left(\ln(S_T) \geq \ln(k) \right) \\ &= e^{-r\tau} \exp\left\{ \tilde{\mu} + \frac{1}{2}\sigma^2 \tau \right\} \Phi\left(\frac{\tilde{\mu} - \ln(k) + \sigma^2 \tau}{\sigma \sqrt{\tau}} \right) \\ &\quad - k e^{-r\tau} \Phi\left(\frac{\tilde{\mu} - \ln(k)}{\sigma \sqrt{\tau}} \right) \\ &= S_t \Phi(d_1) - k e^{-r\tau} \Phi(d_2), \end{aligned} \tag{4.58}$$

where:

$$\begin{aligned} d_1 &= \frac{\tilde{\mu} - \ln(k) + \sigma^2 \tau}{\sigma \sqrt{\tau}} \\ d_2 &= d_1 - \sigma \sqrt{\tau}, \end{aligned} \tag{4.59}$$

and Φ is the CDF (see Sec. 2.2.4) of the standard normal distribution.

As described in Sec. 1.3.1.1 and in Fig. 1.3, it is expected that the option payoff will vary linearly with the price of the underlying asset after it reaches the strike price. It is fair to say that the option price given by the Black-Scholes model takes into account the premium related to the risk of the operation. Eq. 4.58 is composed of two terms. The first one is given by the price of the underlying asset weighted by its probability. Therefore, this product gives the expected value of the option if the exercise was right now. The second term gives the expected value of the stock being above the strike price discounted for the time value of money. Thus, the difference between both components indicates the price of the option as of today. Figure 4.4 shows the option price for a stock with a strike price of \$125 at different expiry times. A snippet used to generate Fig. 4.4 is given below.

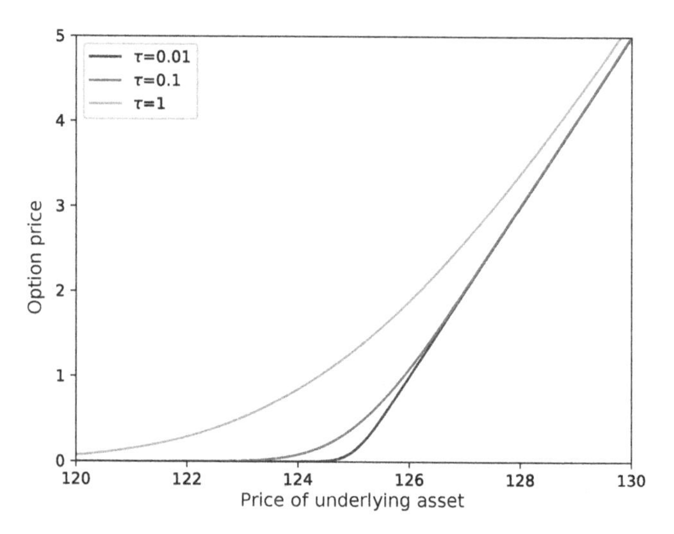

Figure 4.4: Call option payoff as a function of the price of the underlying asset for different expiry times

```
for tau in tau_space:
    price = []
    for S in S_space:
        mu = np.log(S) + (r-0.5*sigma2)*tau
        d1 = (mu-np.log(k) + sigma2*tau)/(sigma*np.sqrt(tau))
        d2 = d1 - sigma*np.sqrt(tau)

        c = S*phi(d1)-k*np.exp(-r*tau)*phi(d2)
        price.append(c)

    pl.plot(S_space,price)
```

Despite providing a limited description of reality, one may assign aesthetic beauty to the mathematical simplicity of the Black-Schole formula. Also, many efforts have been undertaken to improve the model [101–103]. Indiscriminate use of the model, however, has even been attributed as one of the causes of the financial crisis of 2008 [104, 105].

Notes

[1]George Green (1793–1841) British mathematical physicist.

[2]In reference to Robert Brown (1773–1858) Stottish botanist who observed the random motion of pollen in water in 1827.

[3]Monroe David Donsker (1924–1991) American mathematician.

[4]Mark Edward Rubinstein (1944–2019) American economist.

[5]Fischer Sheffey Black (1938–1995) American economist; Myron Samuel Scholes (1941–) Canadian economist, advisee of Eugene Fama, Nobel laureate in 1997; The model was expanded by: Robert Cox Merton (1944–) American economist, son of Robert King Merton. Merton was the advisee of Paul Samuelson and adviser of Jonathan Edward Ingersoll. He received the Nobel prize of economics in 1997 together with Scholes. Black did not receive the Nobel price since it is not given posthumously.

[6]Adolf Eugen Fick (1829–1901) German physician.

[7]William Forsyth Sharpe (1934–) American economist, advisee of Harry Markowitz. Sharpe won the Nobel prize in economics in 1990.

5 Portfolio Theory

"[Economics] is the philosophy of human life..." L. von Mises[1]

Right after entering the Eurozone in 2001, the expectations for the economy of Greece were high. Attraction of private investors, backing by the European central bank, and a lowering of transaction costs were only a few of the beliefs at the time. Nonetheless, the government of Greece defaulted 1.6 billions of euros to the International Monetary Fund (IMF) in 2015 after openly admitting that it had distorted some of its budget figures in order to facilitate its acceptance in the Eurozone. That resulted in the bankruptcy of hundreds of thousands of Greek companies, a fall of 26 % in gross domestic product (GDP), and an explosion in the unemployment rate.[a]

Risk! Given the intrinsic stochastic nature of the financial activity, economists have been devising sophisticated methods to reduce the exposition of investors to risk. Options are used for hedging and the Black-Scholes model studied in the previous chapter is a tool that can be used (under a lot of restrictions) to assess the correct price of options. In this chapter we will study a different strategy to mitigate risk using portfolios of assets. We will begin studying the modern portfolio theory (MPT) and then move to more elaborated analyses such as the random matrix theory.

5.1 MARKOWITZ MODEL

Let's begin our study creating a portfolio consisting of N assets. If each asset has a return $R_n(t)$ at time t, then a vector of returns is given by:

$$\mathbf{R(t)} = \begin{bmatrix} R_1(t) & R_2(t) & \dots & R_N(t) \end{bmatrix}^T \tag{5.1}$$

and their expected returns are given by:

$$\langle \mathbf{R} \rangle = \alpha = \begin{bmatrix} \alpha_1 & \alpha_2 & \dots & \alpha_N \end{bmatrix}^T. \tag{5.2}$$

One may think that assets are completely uncorrelated, but in practice prices often show some level of correlation. In some markets, assets can be strongly correlated as for example in the market of digital goods. Thus, we create a covariance matrix:

$$\text{COV}_{\mathbf{R}} = \mathbf{S} = \left\langle (\mathbf{R} - \alpha)(\mathbf{R} - \alpha)^T \right\rangle \tag{5.3}$$

Let's now create a portfolio by adding W_n fractions of each asset:

$$\mathbf{W} = \begin{bmatrix} W_1 & W_2 & \dots & W_N \end{bmatrix}^T, \tag{5.4}$$

[a]See [106] and [107] for some ways a central authority may afflict the economy.

DOI: 10.1201/9781003127956-5

such that

$$\mathbf{W} \cdot \mathbf{e} = 1, \tag{5.5}$$

since it is a distribution. This restriction is called *budget equation.*

The return of this portfolio is given by:

$$R_W = \mathbf{W} \cdot \mathbf{R}, \tag{5.6}$$

which is known as *reward equation.* The expected return of this portfolio, on the other hand, is given by:

$$\alpha_W = \langle R_W \rangle = \mathbf{W} \cdot \alpha. \tag{5.7}$$

The variance of this portfolio is given by:

$$\begin{aligned}
\mathrm{VAR}(R_W) &= \mathrm{VAR}\left(\mathbf{W} \cdot \mathbf{R}\right) \\
&= \left\langle \left(\mathbf{W}^T(\mathbf{R} - \alpha)\right)\left(\mathbf{W}^T(\mathbf{W} - \alpha)^T\right)^T \right\rangle \\
&= \left\langle \mathbf{W}^T(\mathbf{R} - \alpha)(\mathbf{R} - \alpha)^T\mathbf{W} \right\rangle \\
&= \mathbf{W}^T\left\langle (\mathbf{R} - \alpha)(\mathbf{R} - \alpha)^T \right\rangle \mathbf{W} \\
&= \mathbf{W}^T\mathbf{S}\mathbf{W}.
\end{aligned} \tag{5.8}$$

Markowitz[2] proposed a strategy to find a locus of points ($\alpha_W \times \sigma_W^2$) that produce an *efficient frontier* [108]. This is a set of points where each return is maximized for a given level of risk. It can be found creating a Lagrangian[3] function for minimizing the variance given the restrictions given by Eqs. 5.5 and 5.7:

$$\Lambda = 1/2\mathbf{W}^{\mathrm{T}}\mathbf{S}\mathbf{W} + \lambda\left(1 - \mathbf{W} \cdot \mathbf{e}\right) + \mu\left(\alpha_0 - \mathbf{W} \cdot \alpha\right), \tag{5.9}$$

where λ and μ are Lagrange multipliers.

From matrix calculus:

$$\frac{\partial \Lambda}{\partial \mathbf{W}} = 1/2\mathbf{W}^T(\mathbf{S} + \mathbf{S}^T) - \lambda\mathbf{e}^T - \mu\alpha^T. \tag{5.10}$$

Since, the covariance matrix is symmetric, we get:

$$\begin{aligned}
\frac{\partial \Lambda}{\partial \mathbf{W}} &= \mathbf{W}_0^T\mathbf{S}^T - \lambda\mathbf{e}^T - \mu\alpha^T = 0 \\
\mathbf{W}_0^T &= \lambda\mathbf{e}^T\mathbf{S}^{T^{-1}} + \mu\alpha^T\mathbf{S}^{T^{-1}} \\
\mathbf{W}_0 &= \lambda\mathbf{S}^{-1}\mathbf{e} + \mu\mathbf{S}^{-1}\alpha,
\end{aligned} \tag{5.11}$$

where $\mathbf{S}^{-1}$ is known as the *precision matrix.*

Minimizing the Lagrangian function with respect to the multipliers, we get:

$$\begin{aligned}
\alpha_0 &= \mathbf{W}_0 \cdot \alpha \\
&= \left[\lambda\mathbf{S}^{-1}\mathbf{e} + \mu\mathbf{S}^{-1}\alpha\right] \cdot \alpha \\
&= \lambda\left[(\mathbf{S}^{-1}\mathbf{e}) \cdot \alpha\right] + \mu\left[(\mathbf{S}^{-1}\alpha) \cdot \alpha\right]
\end{aligned} \tag{5.12}$$

and

$$\begin{aligned}1 &= \mathbf{W}_0 \cdot \mathbf{e}\\ &= \left[\lambda \mathbf{S}^{-1}\mathbf{e} + \mu \mathbf{S}^{-1}\alpha\right] \cdot \mathbf{e}\\ &= \lambda\left[(\mathbf{S}^{-1}\mathbf{e}) \cdot \mathbf{e}\right] + \mu\left[(\mathbf{S}^{-1}\alpha) \cdot \mathbf{e}\right].\end{aligned} \tag{5.13}$$

We can put these two equations in a matrix form:

$$\begin{aligned}\begin{bmatrix}(\mathbf{S}^{-1}\mathbf{e}) \cdot \mathbf{e} & (\mathbf{S}^{-1}\alpha) \cdot \mathbf{e}\\ (\mathbf{S}^{-1}\mathbf{e}) \cdot \alpha & (\mathbf{S}^{-1}\alpha) \cdot \alpha\end{bmatrix}\begin{bmatrix}\lambda\\ \mu\end{bmatrix} &= \begin{bmatrix}1\\ \alpha_0\end{bmatrix}\\ \begin{bmatrix}\lambda\\ \mu\end{bmatrix} &= \mathbf{M}^{-1}\begin{bmatrix}1\\ \alpha_0\end{bmatrix},\end{aligned} \tag{5.14}$$

where

$$\mathbf{M} = \begin{bmatrix}\mathbf{e}^T\mathbf{S}^{-1}\mathbf{e} & \alpha^T\mathbf{S}^{-1}\mathbf{e}\\ \mathbf{e}^T\mathbf{S}^{-1}\alpha & \alpha^T\mathbf{S}^{-1}\alpha\end{bmatrix}. \tag{5.15}$$

Note that $\left(\mathbf{e}^T\mathbf{S}^{-1}\alpha\right)^T = \alpha^T\mathbf{S}^{-1}\mathbf{e}$, and consequently the matrix M is symmetric.

Once the Lagrange multipliers are found, the Markowitz portfolio can be constructed with Eq. 5.11. The weights, however, can assume negative values. Some investors use this information as a signal to sell shares. But what if the investor does not have these shares? In order to obtain only positive weights, the following normalization [109] can be used:

$$W_n^* = \frac{\max(0, W_n)}{\sum_m \max(0, W_m)}. \tag{5.16}$$

5.1.1 RISK ESTIMATION

In order to obtain the risk associated to the Markowitz portfolio, we first rewrite Eq. 5.11 as:

$$\mathbf{W}_0 = \left[\mathbf{S}^{-1}\mathbf{e} \quad \mathbf{S}^{-1}\alpha\right]\begin{bmatrix}\lambda\\ \mu\end{bmatrix} \tag{5.17}$$

In the Markowitz model, the risk is given by the variance of the portfolio (Eq. 5.8):

$$\begin{aligned}\sigma_w^2 &= \mathbf{W}_0^T\mathbf{S}\mathbf{W}_0\\ &= \left(\left[\mathbf{S}^{-1}\mathbf{e} \quad \mathbf{S}^{-1}\alpha\right]\begin{bmatrix}\lambda\\ \mu\end{bmatrix}\right)^T\mathbf{S}\left(\left[\mathbf{S}^{-1}\mathbf{e} \quad \mathbf{S}^{-1}\alpha\right]\begin{bmatrix}\lambda\\ \mu\end{bmatrix}\right)\\ &= \left[\lambda \quad \mu\right]\begin{bmatrix}\mathbf{e}^T\mathbf{S}^{-1}\\ \alpha^T\mathbf{S}^{-1}\end{bmatrix}\left[\mathbf{e} \quad \alpha\right]\begin{bmatrix}\lambda\\ \mu\end{bmatrix}\\ &= \left[\lambda \quad \mu\right]\begin{bmatrix}\mathbf{e}^T\mathbf{S}^{-1}\mathbf{e} & \mathbf{e}^T\mathbf{S}^{-1}\alpha\\ \alpha^T\mathbf{S}^{-1}\mathbf{e} & \alpha^T\mathbf{S}^{-1}\alpha\end{bmatrix}\begin{bmatrix}\lambda\\ \mu\end{bmatrix}\\ &= \left[\lambda \quad \mu\right]\mathbf{M}\begin{bmatrix}\lambda\\ \mu\end{bmatrix}.\end{aligned} \tag{5.18}$$

Using the result of Eq. 5.14:

$$\begin{aligned}\sigma_w^2 &= \begin{bmatrix}1 & \alpha_0\end{bmatrix} \mathbf{M}^{-1}\mathbf{M}\mathbf{M}^{-1} \begin{bmatrix}1 \\ \alpha_0\end{bmatrix} \\ &= \begin{bmatrix}1 & \alpha_0\end{bmatrix} \mathbf{M}^{-1} \begin{bmatrix}1 \\ \alpha_0\end{bmatrix}.\end{aligned} \tag{5.19}$$

It is possible to use this last equation to plot the locus of the standard deviation-expected return pairs. This is known as *Markowitz bullet* or *efficient frontier*. Figure 5.1 shows the efficient frontier calculated for daily returns of IBM, AMD and Intel.

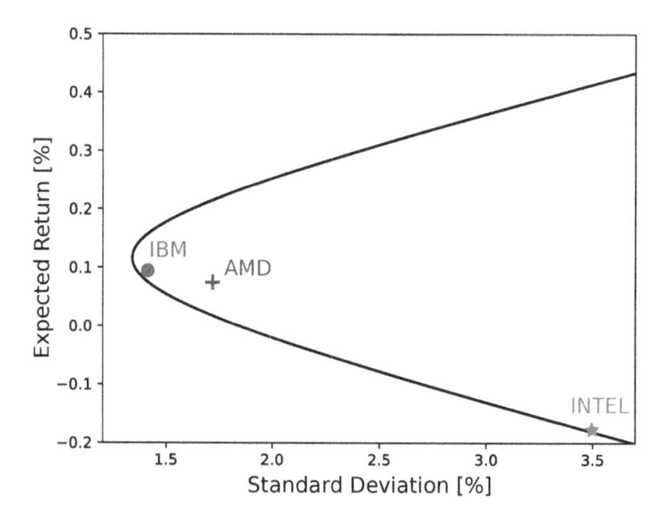

Figure 5.1: Standard deviation-expected return space with the Markowitz bullet calculated for daily returns of IBM, AMD, and Intel

For a code that calculates the Markowitz bullet, we will use NumPy for array handling, MatPlotLib for plotting, Pandas for obtaining the time series and DateTime for handling datetime data:

```
import numpy as np
import matplotlib.pyplot as pl
import pandas_datareader as pdr

from datetime import datetime
```

Next, we must define functions for the matrix multiplication of three elements (tri) and the covariance of two series (cv):

```
def tri(a,b,c):
        return np.dot(a,np.dot(b,c))

def cv(x,y):
        return np.cov(x,y)[0][1]
```

Time series can be obtained with Pandas:

```
ibm = pdr.get_data_yahoo(symbols='IBM',start=datetime(2011,1,1),
                                    end=datetime(2012,1,1))
amd = pdr.get_data_yahoo(symbols='AMD',start=datetime(2011,1,1),
                                    end=datetime(2012,1,1))
itl = pdr.get_data_yahoo(symbols='INTC',start=datetime(2011,1,1),
                                    end=datetime(2012,1,1))

# Select Close Prices
pibm = ibm['Adj Close']
pamd = amd['Adj Close']
pitl = itl['Adj Close']

# Get Returns
ribm = ret(pibm)
ramd = ret(pamd)
ritl = ret(pitl)
```

The alpha vector, **M** and covariance matrices are computed with:

```
a = np.array([[np.average(ribm)],[np.average(ramd)],[np.average(ritl)]])
aT = np.transpose(a)

# Covariance Matrix
cii = cv(ribm,ribm)
caa = cv(ramd,ramd)
ctt = cv(ritl,ritl)

S = np.array([[cii, cv(ribm,ramd), cv(ribm,ritl)],
              [cv(ramd,ribm), caa, cv(ramd,ritl)],
              [cv(ritl,ribm), cv(ritl,ramd), ctt]])
Si = np.linalg.inv(S)

# M Matrix
e = np.array([[1.0] for i in range(3)])
eT = np.transpose(e)

M = np.array([[ tri(eT,Si,e)[0][0], tri(aT,Si,e)[0][0]],
              [ tri(eT,Si,a)[0][0], tri(aT,Si,a)[0][0]]])
Mi = np.linalg.inv(M)
```

Finally, the bullet is computed with:

```
a_space = np.linspace(-0.002,0.005,100)
s_space = []
for ao in a_space:
    m = np.array([[1],[ao]])
    mT = np.transpose(m)

    s = tri(mT,Mi,m)[0][0]
    s_space.append(s)

s_space = [np.sqrt(s)*100 for s in s_space]
a_space = [a*100 for a in a_space]

pl.plot(s_space,a_space)
```

5.1.2 RISK-FREE ASSET

What happens if we include one risk-free asset in our portfolio? Although such asset does not really exist, one can use the safest assets in the market, such as a certificate of deposit (CD) or savings bonds[b]. Let's, however, consider that its variance is negligible and its expected value is r_0.

Our portfolio is then constructed with M risky assets such that $\mathbf{W}\cdot\mathbf{e}$ is the fraction of such assets and $1-\mathbf{W}\cdot\mathbf{e}$ is the fraction allocated in risk-free assets. The expected value of the portfolio is then:

$$\begin{aligned}\alpha_0 &= \mathbf{W}\cdot\alpha+(1-\mathbf{W}\cdot\mathbf{e})\,r_0\\ \alpha_0-r_0 &= \mathbf{W}\cdot(\alpha-\mathbf{e}r_0).\end{aligned} \tag{5.20}$$

Let's once again minimize the risk associated with this portfolio with a Langrangian function:

$$\Lambda = 1/2\mathbf{W}^{\mathrm{T}}\mathbf{S}\mathbf{W}+\lambda_P\left(\alpha_0-r_0+\mathbf{W}\cdot(\mathbf{e}r_0-\alpha)\right). \tag{5.21}$$

Minimizing this function with respect to $\mathbf{W}$ we get:

$$\begin{aligned}\frac{d\Lambda}{d\mathbf{W}} &= \mathbf{W}_P^T\mathbf{S}^T+\lambda_P\left(\mathbf{e}^T r_0-\alpha^T\right)=0\\ \mathbf{W}_P^T\mathbf{S}^T &= \lambda_P\left(\alpha^T-\mathbf{e}^T r_0\right)\\ \mathbf{W}_P &= \lambda_P\mathbf{S}^{-1}(\alpha-\mathbf{e}r_0).\end{aligned} \tag{5.22}$$

From Eq. 5.20 we get:

$$\begin{aligned}\alpha_0-r_0 &= \lambda_P\left[\mathbf{S}^{-1}(\alpha-\mathbf{e}r_0)\right]\cdot(\alpha-\mathbf{e}r_0)\\ \lambda_P &= \frac{\alpha_0-r_0}{(\alpha-\mathbf{e}r_0)\cdot\left[\mathbf{S}^{-1}(\alpha-\mathbf{e}r_0)\right]}.\end{aligned} \tag{5.23}$$

The risk associated with this portfolio is given by:

$$\begin{aligned}\sigma_P^2 &= \mathbf{W}_P^T\mathbf{S}\mathbf{W}_P\\ &= \lambda_P^2(\alpha-\mathbf{e}r_0)^T\mathbf{S}^{-1}(\alpha-\mathbf{e}r_0)\\ &= (\alpha_0-r_0)^2\frac{(\alpha-\mathbf{e}r_0)^T\mathbf{S}^{-1}(\alpha-\mathbf{e}r_0)}{\left[(\alpha-\mathbf{e}r_0)^T\mathbf{S}^{-1}(\alpha-\mathbf{e}r_0)\right]^2}\\ &= \frac{(\alpha_0-r_0)^2}{(\alpha-\mathbf{e}r_0)^T\mathbf{S}^{-1}(\alpha-\mathbf{e}r_0)}.\end{aligned} \tag{5.24}$$

[b]This class of asset is usually protected by insurance corporations such as the Federal Deposit Insurance Corporation (FDIC) in the United States.

5.1.2.1 Market Portfolio

Let's define a *market portfolio* as the optimal portfolio (which lies on the efficient frontier) fully invested in risky assets. According to this definition, we must take the portfolio from Eq. 5.22 and apply $\mathbf{W}\cdot\mathbf{e}=1$. Thus, we get:

$$\begin{aligned}\left[\lambda_M \mathbf{S}^{-1}(\alpha-\mathbf{e}r_0)\right]\cdot\mathbf{e}&=1\\ \lambda_M&=\frac{1}{\left[\mathbf{S}^{-1}(\alpha-\mathbf{e}r_0)\right]\cdot\mathbf{e}}.\end{aligned} \tag{5.25}$$

The expected value of this portfolio, according to Eq. 5.22, is:

$$\begin{aligned}\langle R_M\rangle&=\mathbf{W}_M\cdot\alpha=\lambda_M\mathbf{S}^{-1}(\alpha-\mathbf{e}r_0)\cdot\alpha\\ &=r_0-r_0+\frac{\left[\mathbf{S}^{-1}(\alpha-\mathbf{e}r_0)\right]\cdot\alpha}{\left[\mathbf{S}^{-1}(\alpha-\mathbf{e}r_0)\right]\cdot\mathbf{e}}\\ &=r_0+\frac{\left[\mathbf{S}^{-1}(\alpha-\mathbf{e}r_0)\right]\cdot\alpha-\left[\mathbf{S}^{-1}(\alpha-\mathbf{e}r_0)\right]\cdot\mathbf{e}r_0}{\left[\mathbf{S}^{-1}(\alpha-\mathbf{e}r_0)\right]\cdot\mathbf{e}}\\ &=r_0+\frac{(\alpha-\mathbf{e}r_0)^T\mathbf{S}^{-1}(\alpha-\mathbf{e}r_0)}{\left[\mathbf{S}^{-1}(\alpha-\mathbf{e}r_0)\right]\cdot\mathbf{e}}.\end{aligned} \tag{5.26}$$

The risk associated with this portfolio is:

$$\begin{aligned}\sigma_M^2&=\mathbf{W}_M^T\mathbf{S}\mathbf{W}_M\\ &=\lambda_M^2\left[\mathbf{S}^{-1}(\alpha-\mathbf{e}r_0)\right]^T\mathbf{S}\left[\mathbf{S}^{-1}(\alpha-\mathbf{e}r_0)\right]\\ &=\frac{(\alpha-\mathbf{e}r_0)^T\mathbf{S}^{-1}(\alpha-\mathbf{e}r_0)}{\left(\left[\mathbf{S}^{-1}(\alpha-\mathbf{e}r_0)\right]\cdot\mathbf{e}\right)^2}\\ &=\frac{(\langle R_M\rangle-r_0)^2}{(\alpha-\mathbf{e}r_0)^T\mathbf{S}^{-1}(\alpha-\mathbf{e}r_0)}.\end{aligned} \tag{5.27}$$

5.1.3 TOBIN'S SEPARATION THEOREM

According to Eqs. 5.22 and 5.26, we can write:

$$\mathbf{W}_P=\frac{\lambda_P}{\lambda_M}\mathbf{W}_M. \tag{5.28}$$

The ratio between lambdas can be computed as:

$$
\begin{aligned}
\gamma &= \frac{\lambda_P}{\lambda_M} \\
&= (\alpha_0 - r_0)\frac{\left[\mathbf{S}^{-1}(\alpha - \mathbf{e}r_0)\right]\cdot\mathbf{e}}{(\alpha - \mathbf{e}r_0)\cdot\left[\mathbf{S}^{-1}(\alpha - \mathbf{e}r_0)\right]} \\
&= (\alpha_0 - r_0)\left[\frac{\left[\mathbf{S}^{-1}(\alpha - \mathbf{e}r_0)\right]\cdot(\alpha - \mathbf{e}r_0)}{\left[\mathbf{S}^{-1}(\alpha - \mathbf{e}r_0)\right]\cdot\mathbf{e}}\right]^{-1} \\
&= \frac{\alpha_0 - r_0}{\langle R_M\rangle - r_0}.
\end{aligned}
\tag{5.29}
$$

On the other hand, according to the relation above it is also possible to write:

$$
\begin{aligned}
\mathrm{VAR}(\mathbf{W}_P) &= \gamma^2\,\mathrm{VAR}(\mathbf{W}_M) \\
\gamma &= \frac{\sigma_p}{\sigma_M}.
\end{aligned}
\tag{5.30}
$$

Matching both expressions and considering that α_0 is the expected return of the optimal portfolio:

$$
\begin{aligned}
\frac{\sigma_P}{\sigma_M} &= \frac{\alpha_0 - r_0}{\langle R_M\rangle - r_0} \\
\langle R_P\rangle &= r_0 + \sigma_P\left(\frac{\langle R_M\rangle - r_0}{\sigma_M}\right).
\end{aligned}
\tag{5.31}
$$

This is known as the *capital market line* (CML). The slope within parenthesis (known as Sharpe[4] ratio [110], [100], [111]) describes a normalized risk premium and is exactly the slope corresponding to a *tangent portfolio*. We can see this by working with Eq. 5.27:

$$
\begin{aligned}
\sigma_M = \frac{\langle R_M\rangle - r_0}{K(R)} &\rightarrow K(R)\sigma_M = \langle R_M\rangle - r_0 \\
dK(R)\sigma_M &+ K(R)d\sigma_M = d\langle R_M\rangle \\
\frac{d\langle R_M\rangle}{d\sigma_M} &= K(R) + \sigma_M\frac{dK(R)}{d\sigma_M} \\
&= K(R) = \frac{\langle R_M\rangle - r_0}{\sigma_M}.
\end{aligned}
\tag{5.32}
$$

Therefore, we can say that these efficient portfolios can be written as a linear combination of risk-free assets and a tangent portfolio and this is known as *Tobin's*[5] *separation theorem* [112].

It is possible to add the Markowitz bullet as shown in Fig. 5.2 with the following snippet:

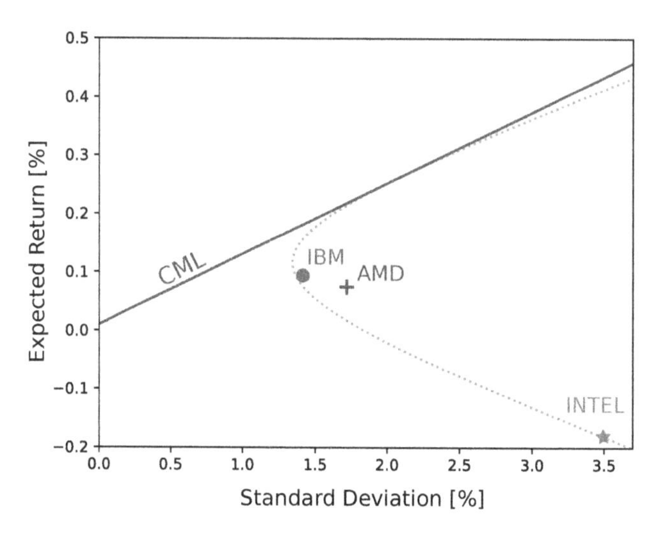

Figure 5.2: Markowitz bullet (dotted curve) and the capital market line (CML) for a risk-free asset with an expected return of 0.01%

```
ro = 0.01/100
d = (tri(np.transpose(a-e*ro),Si,a-e*ro))[0][0]
Rm = (ro + d/tri(np.transpose(a-e*ro),Si,e))[0]
sigma_m = (np.sqrt(((Rm-ro)**2)/d))[0]

R_space = []
n_space = np.linspace(0,np.sqrt(max(s_space)),100)
for n in n_space:
        Rp = ro + n*(Rm-ro)/sigma_m
        R_space.append(Rp)

n_space = [n*100 for n in n_space]
R_space = [r*100 for r in R_space]
pl.plot(n_space,R_space)
```

The modern portfolio theory is usually criticized with respect to the assumption of risk. Indeed, variance can be a poor estimator of risk and some distributions may not even have a well-defined variance. We have seen, for instance, that many returns have fat tail distributions. Furthermore, there can also be non-quantifiable risks[c] [113]. On the other hand, the modern portfolio theory has some results that are very appealing such as composing a diversified portfolio do reduce risk.

[c] Also known as Knightian[6] uncertainty.

5.1.3.1 CAPM

Equation 5.31 can also be written as:

$$\frac{\langle R_M \rangle - r_0}{\beta_M} = \langle R_P \rangle - r_0. \tag{5.33}$$

The term on the left hand side of the equation is known as the *Treynor*[7] *index* [114] and quantifies a risk-adjusted excess return of the investment. This equation can be slightly modified to value assets in the *capital asset pricing model* [110, 114–116] (CAPM):

$$\langle R_i \rangle = r_0 + \beta_i \left(\langle R_m \rangle - r_0 \right), \tag{5.34}$$

where R_m is the expected return of the market and R_i is the expected return of a new asset. The locus of the expected return of an investment as a function of β_i in this equation is known as *security market line* (SML). Here, the angular coefficient β_i quantifies the *systematic risk* of such investment. The latter can be given by the linear regression:

$$\beta_i = \frac{\text{COV}(\langle R_i \rangle, \langle R_m \rangle)}{\text{VAR}(\langle R_i \rangle)}. \tag{5.35}$$

Equation 5.33 assumes that the ratio between excess returns is given by β. However, it is not uncommon to find an extra *abnormal return* given by α added to the expression. Therefore, we can generalize this equation to:

$$\left(\langle R_i \rangle_t - r_0 \right) = \alpha_i + \beta_i \left(\langle R_m \rangle_t - r_0 \right) + \varepsilon_{it}, \tag{5.36}$$

where ε_{it} is an *idiosyncratic risk*. The locus of the excess return of the asset as a function of the excess return of the portfolio is known as *security characteristic line* (SCL). If the abnormal return is positive, then the investment is paying an extra price for the assumed risk.

Based on this model, it is possible to check whether an asset is under- or overvalued. In order to do this we find its β and plot the pair $(\beta_i, \langle R_i \rangle)$ together with the SML. If the point is below the SML, then the asset is paying less than the market for an assumed risk. On the other hand, if the point is above the SML, then the asset pays more than the market for the assumed risk. This, however, is a violation of the EMH (see Sec. 1.3) that states that it should not be possible to beat the market. An example, nonetheless, is given in Fig. 5.3 for the daily returns of stocks of some technology companies between 2013/1/1 and 2014/1/1. The NASDAQ-100 Technology Sector (NDXT) is used as an estimator for the market.

5.2 RANDOM MATRIX THEORY

Let's take a set of assets and organize the time series of their normalized returns in rows:

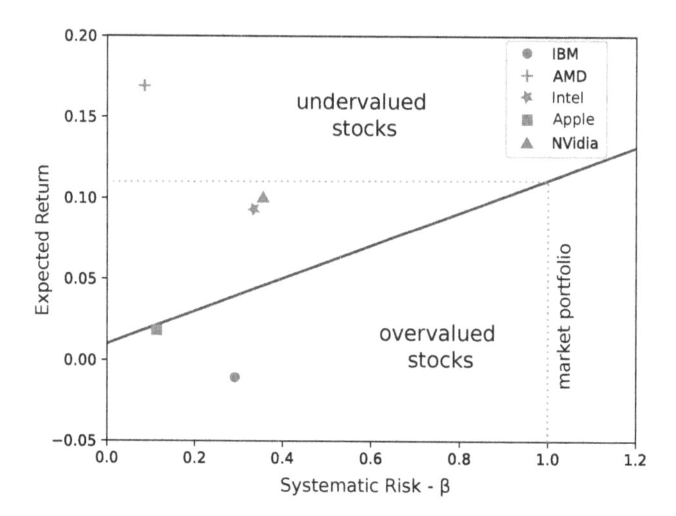

Figure 5.3: Pricing of stocks under CAPM. The solid line is the SML estimated from the NDXT index, whereas the market portfolio has $\beta = 1$ by definition

$$\mathbf{A} = \begin{bmatrix} \tilde{r}_1(t_1) & \tilde{r}_2(t_1) & \dots & \tilde{r}_N(t_1) \\ \tilde{r}_1(t_2) & \tilde{r}_2(t_2) & \dots & \tilde{r}_N(t_2) \\ \vdots & & \ddots & \vdots \\ \tilde{r}_1(t_N) & \tilde{r}_2(t_N) & \dots & \tilde{r}_N(t_N) \end{bmatrix} \tag{5.37}$$

A random matrix [117–119] is a matrix whose all elements are random variables. If matrix $\mathbf{A}$ is unitary and its elements are i.i.d. and uniformly distributed, then we say that this matrix belongs to the *Circular Ensemble*. If, on the other hand, we lift the restriction of $\mathbf{A}$ being unitary but require that its elements are i.i.d. and follow a Gaussian distribution, then we say that this matrix belongs to a *Gaussian ensemble*. Considering that $\mathbf{A}$ belongs to the Gaussian ensemble, symmetric matrices created as $1/2\left(\mathbf{A}+\mathbf{A}^T\right)$ belong to the *Orthogonal Gaussian Ensemble* (GOE). In this case, the diagonal elements of such matrix are $\mathcal{N}(0,1)$ distributed, whereas non-diagonal elements are $\mathcal{N}(0,1/2)$ distributed.

If a new matrix is created as $1/2\left(\mathbf{A}+\mathbf{A}^H\right)$, then it belongs to the *Gaussian Unitary Ensemble* (GUE). Furthermore, if the matrix is created as $1/2\left(\mathbf{A}+\mathbf{A}^D\right)$, where the subscript D indicates the quaternion dual transpose, then this new matrix belongs to the *Gaussian Sympletic Ensemble* (GSE).

It is also possible to create a new matrix $\mathbf{A}'\mathbf{A}$, where $'$ is T, H or D depending if the matrix A is real, complex or quaternion. In this case, the new matrix belongs to the *Wishart Ensemble*[8]. The covariance matrix, for instance, follows this ensemble.

Matrices that belong to these ensembles have a set of properties and have been used to study, for instance, quantum chaos [120]. Here, we are particularly interested in the covariance matrix and that can follow the Wishart ensemble. In order to explore some of its properties, let's use Green's functions to obtain its density of eigenvalues.

5.2.1 DENSITY OF EIGENVALUES

For a square matrix **H**, its resolvent (or, equivalently, the retarded Green's function) can be written as:

$$\mathbf{G}^R(\varepsilon) = [(\varepsilon + i\eta)\mathbf{I} - \mathbf{H}]^{-1}. \tag{5.38}$$

The resolvent can be decomposed in the base of the eigenstates of **H**:

$$\mathbf{G}^R(\varepsilon) = \sum_n |n\rangle\, \mathbf{C}_n. \tag{5.39}$$

Substituting this decomposition in the definition of the resolvent:

$$\begin{aligned} [(\varepsilon + i\eta)\mathbf{I} - \mathbf{H}]\, \mathbf{G}(\varepsilon) &= \mathbf{I} \\ [(\varepsilon + i\eta)\mathbf{I} - \mathbf{H}] \sum_n |n\rangle\, \mathbf{C}_n &= \mathbf{I} \\ \sum_n [(\varepsilon + i\eta) - \lambda_n]\, |n\rangle\, \mathbf{C}_n &= \mathbf{I}. \end{aligned} \tag{5.40}$$

Applying an eigenstate on the left side:

$$\begin{aligned} \langle m| \sum_n [(\varepsilon + i\eta) - \lambda_n]\, |n\rangle\, \mathbf{C}_n &= \langle m| \\ [(\varepsilon + i\eta) - \lambda_m]\, \mathbf{C}_m &= \langle m| \\ \mathbf{C}_m &= \frac{\langle m|}{(\varepsilon + i\eta) - \lambda_m}. \end{aligned} \tag{5.41}$$

Using this result in Eq. 5.39:

$$\mathbf{G}^R(\varepsilon) = \sum_n |n\rangle\, C_n = \sum_n \frac{|n\rangle\langle n|}{(\varepsilon + i\eta) - \lambda_n}. \tag{5.42}$$

Let's now use this eigenexpansion of the resolvent to obtain the density of eigenvalues of **H**:

$$\rho(\lambda) = \frac{1}{N} \sum_{\alpha=1}^{N} \delta(\lambda - \lambda_\alpha). \tag{5.43}$$

Therefore, using the eigenexpansion of the Green's function:

$$\begin{aligned} \mathbf{G}^R(\varepsilon) &= \sum_\alpha \frac{|\alpha\rangle\langle\alpha|}{\varepsilon + i\eta - \varepsilon_\alpha} \\ \langle m|\, \mathbf{G}^R(\varepsilon)\, |m\rangle &= \frac{1}{(\varepsilon - \varepsilon_m) + i\eta} \\ &= \frac{(\varepsilon - \varepsilon_m) - i\eta}{(\varepsilon - \varepsilon_m)^2 + \eta^2} \\ \mathrm{Im}\left\{\mathbf{G}^R_{mm}(\varepsilon)\right\} &= -\frac{\eta}{(\varepsilon - \varepsilon_m)^2 + \eta^2} \\ \lim_{\eta \to 0} \mathrm{Im}\left\{\mathbf{G}^R_{mm}(\varepsilon)\right\} &= -\pi\delta(\varepsilon - \varepsilon_m). \end{aligned} \tag{5.44}$$

Thus, from Eq. 5.43, the spectral density of $\mathbf{H}$ is:

$$\rho(\varepsilon) = -\frac{1}{N\pi} \lim_{\eta \to 0} \mathrm{Im}\left\{\mathrm{Tr}\left\{\mathbf{G}^R(\varepsilon)\right\}\right\}. \tag{5.45}$$

This is sometimes stated as the Plemelj-Sokhotski[9] formula.

5.2.2 WIGNER'S SEMI-CIRCLE LAW

In the light of Dyson's equation, we can write matrix $\mathbf{H}$ as:

$$\mathbf{H} = \begin{bmatrix} h_{11} & \tau_1^\dagger \\ \tau_2 & \mathbf{H}_0 \end{bmatrix}. \tag{5.46}$$

Dyson's Equation

It is possible to simplify many problems related to the resolvent by using Dyson's[10] equation. Let's begin with an unperturbed matrix $\mathbf{H}_0$, apply a perturbation $\mathbf{V}$ and compute its resolvent:

$$\begin{aligned} \mathbf{G}^R(\varepsilon) &= [(\varepsilon + i\eta)\mathbf{I} - \mathbf{H}_0 - \mathbf{V}]^{-1} \\ &= \left[\mathbf{G}_0^{-1} - \mathbf{V}\right]^{-1} \\ &= \left[\mathbf{G}_0^{-1}(\mathbf{I} - \mathbf{G}_0\mathbf{V})\right]^{-1} \\ &= (\mathbf{I} - \mathbf{G}_0\mathbf{V})^{-1}\mathbf{G}_0 \\ (\mathbf{I} - \mathbf{G}_0\mathbf{V})\,\mathbf{G}^R(\varepsilon) &= \mathbf{G}_0 \\ \mathbf{G}^R(\varepsilon) &= \mathbf{G}_0 + \mathbf{G}_0\mathbf{V}\mathbf{G}^R(\varepsilon), \end{aligned} \tag{5.47}$$

where $\mathbf{G}_0$ is the resolvent for the unperturbed matrix $\mathbf{H}_0$. Using the completness relation, we can find an expression for its elements:

$$\langle a|\,\mathbf{G}^R(\varepsilon)\,|b\rangle = \sum_{m,n} \langle a|\,\mathbf{G}_0\,|m\rangle \langle m|\,\mathbf{V}\,|n\rangle \langle n|\,\mathbf{G}^R(\varepsilon)\,|b\rangle \tag{5.48}$$

or simply:

$$G^R_{ab}(\varepsilon) = G_{0ab} + \sum_{n,m} G_{0am}V_{nm}G^R_{nb}(\varepsilon). \tag{5.49}$$

Applying Dyson's equation[d], we get:

$$\begin{aligned} G_{11} &= G^0{}_{11} + G^0{}_{11}\tau_1^\dagger\mathbf{G}_{21} \\ \mathbf{G}_{21} &= \mathbf{G}^0{}_{22}\tau_2 G_{11} \end{aligned} \tag{5.50}$$

[d] We changed the notation for the bare resolvent from a subscript to a superscript 0.

Solving this system:

$$\begin{aligned} G_{11} &= G_{11}^0 + G_{11}^0 \tau_1^\dagger \mathbf{G}_{22}^0 \tau_2 G_{11} \\ &= \left[\mathbf{I} - G_{11}^0 \Sigma_2\right]^{-1} G_{11}^0 \\ &= \left[(\varepsilon + i\eta) - h_{11} - \Sigma_2\right]^{-1}, \end{aligned} \quad (5.51)$$

where $\Sigma_2 = \tau_1^\dagger \mathbf{G}_{22}^0 \tau_2$ is known as the *self-energy*. Let's inspect the expected value of the self-energy:

$$\begin{aligned} \langle b | \tau^\dagger \mathbf{G}_{22}^0 \tau | b \rangle &= \sum_{n,m} \langle b | \tau_1^\dagger | n \rangle \langle n | \mathbf{G}_{22}^0 | m \rangle \langle m | \tau_1 | b \rangle \\ &= \sum_{n,m} \tau_{bn}^\dagger G_{22nm}^0 \tau_{mb}. \end{aligned} \quad (5.52)$$

Let's now consider that the elements of $\tau_1^\dagger$ and τ_2 are random variables with zero mean and variance σ^2. Thus, $\langle \tau_{bn}^\dagger \tau_{mb} \rangle = (\sigma^2/N)\delta_{n,m}$ and:

$$\begin{aligned} \langle b | \Sigma_2 | b \rangle &= \sum_{n,m} \frac{\sigma^2}{N} \delta_{n,m} G_{22nm}^0 \\ &= \frac{\sigma^2}{N} \mathrm{Tr}\left\{\mathbf{G}_{22}^0\right\}. \end{aligned} \quad (5.53)$$

Therefore, from Eq. 5.51:

$$\frac{1}{N}\mathrm{Tr}\{G_{11}\} = \frac{1}{\varepsilon + i\eta - h_{11} - \frac{\sigma^2}{N}\mathrm{Tr}\left\{\mathbf{G}_{22}^0\right\}}. \quad (5.54)$$

Given that the eigenvalues of a matrix and its minor interlace, their traces are similar. Therefore:

$$\begin{aligned} t &= \frac{1}{\varepsilon + i\eta - h_{11} - \sigma^2 t} \rightarrow \sigma^2 t^2 - at + 1 = 0 \\ t &= \frac{a \pm \sqrt{a^2 - 4\sigma^2}}{2\sigma^2}, \end{aligned} \quad (5.55)$$

where $t = (1/N)\mathrm{Tr}(G)$ and $a = \varepsilon + i\eta - h_{11}$.

The spectral density is obtained from Eq. 5.45:

$$\begin{aligned} \rho(\varepsilon) &= -\frac{1}{\pi} \lim_{\eta \to 0} \mathrm{Im}(t) \\ &= \frac{1}{2\pi\sigma^2} \sqrt{4\sigma^2 - \varepsilon^2}. \end{aligned} \quad (5.56)$$

Therefore, the locus of the spectral density of the Gaussian ensemble is a semi-circle. This result is indeed known as *Wigner's semi-circle law*[11].

In order to simulate this law we can follow the snippet below. We need NumPy for numerical calculations and Matplotlib for plotting. We also define a few constants:

```
import numpy as np
import matplotlib.pyplot as pl

eta = 0.005
N = 100
e_space = np.linspace(-2,2,100)
DOS = []
```

We then sample Wigner matrices from the normal distribution, calculate the resolvent and take averages:

```
for sample in range(300):
        H = [[np.random.normal(0,1.0/np.sqrt(N)) for i in range(N)]
                                                      for j in range(N)]
        H = 0.5*np.sqrt(2)*(H + np.transpose(H))

        g_space = []
        for e in e_space:
                G = np.linalg.inv((e+1j*eta)*np.eye(N)-H)
                r = -(1.0/(N*np.pi))*np.trace(G.imag)
                g_space.append(r)

        DOS.append(g_space)

DOS = np.average(DOS,axis=0)
```

Finally we plot the result together with the Wigner semi-circle law:

```
W = [(1.0/(2*np.pi))*np.sqrt(4-z**2) for z in e_space]

pl.plot(e_space,DOS,'.',color='gray')
pl.plot(e_space,W,'k')
pl.show()
```

The result of a simulation for 300 100×100 square matrices from the Gaussian orthogonal ensemble is shown in Fig. 5.4.

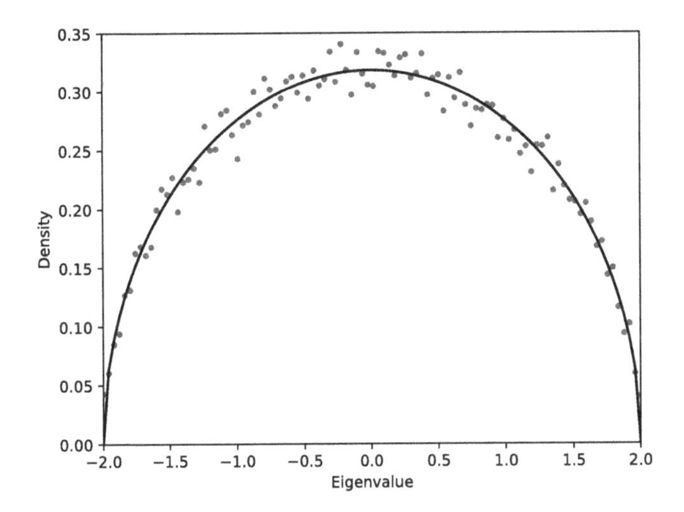

Figure 5.4: Spectral density for 300 100×100 square matrices from the Gaussian orthogonal ensemble. Gray dots are obtained from the resolvent, whereas the solid line is the Wigner semi-circle law

5.2.3 MARČENKO-PASTUR LAW

The density of eigenvalues for the Wishart ensemble is greatly modified if compared to that of Gaussian ensembles. In order to see that, we must use Marčenko-Pastur[12] theorem.

Marčenko-Pastur Theorem

Let's take a matrix of normalized returns $\mathbf{X}$ as a Wigner $n \times N$ matrix where $\langle X_{ij} \rangle = 0$ and $\langle X_{ij}^2 \rangle = 1$. Also, let's take $\mathbf{T}_n$ as an $n \times n$ diagonal matrix with elements τ_i such that the distribution of eigenvalues converge to some distribution $H(\tau)$. Finally, let's consider an $N \times N$ matrix $\mathbf{B}_N = \mathbf{A}_N + (1/N)\mathbf{X}^\dagger \mathbf{T}_n \mathbf{X}$, where $\mathbf{A}_N$ is an $N \times N$ matrix. If these three matrices are independent, then Marčenko-Pastur theorem [121] states that:

$$G_B(z) = G_A\left(z - c\int \frac{\tau dH(\tau)}{1 + \tau G_B(z)}\right), \tag{5.57}$$

where $c = n/N$ and G_Y is the Stieltjes[13] transformation of matrix $\mathbf{Y}$.

The Stieltjes transformation of a matrix $\mathbf{A}_{N\times N}$ is defined as:

$$G_A(\varepsilon) = \int \frac{\rho(t)}{t - \varepsilon} dt = \int \frac{1}{t - \varepsilon} dF^A(t), \tag{5.58}$$

where the measure $F^A(\varepsilon)$ of matrix $\mathbf{A}_{N\times N}$ is the density of eigenvalues described intuitively by the ratio between the number of eigenvalues smaller than ε and N.

The density of eigenvalues for a null matrix $\mathbf{A} = 0$ is given by $\delta(t)$. Therefore, its Stieltjes transformation is given by:

$$G_A(\varepsilon) = \int \frac{\delta(t)}{t-\varepsilon}dt = -\frac{1}{\varepsilon}. \tag{5.59}$$

In this picture, matrix $\mathbf{B}$ is given by $1/N\mathbf{X}^*\mathbf{T}\mathbf{X}$ and its Stieltjes transformation is given by:

$$\begin{aligned} G_B(z) &= -\frac{1}{z - c\int \frac{\tau dH(\tau)}{1+\tau G_B(z)}} \\ z &= c\int \frac{\tau dH(\tau)}{1+\tau G_B(z)} - \frac{1}{G_B(z)}. \end{aligned} \tag{5.60}$$

If we use the distribution $\mathbf{T} = \mathbf{I}$ then $dH(\tau) = \delta(\tau-1)d\tau$, and we get:

$$\begin{aligned} z &= c\int \frac{\tau\delta(\tau-1)d\tau}{1+\tau G_B(z)} - \frac{1}{G_B(z)} \\ &= \frac{c}{1+G_B(z)} - \frac{1}{G_B(z)} \\ zG_B^2(z) &+ (z+1-c)G_B(z) + 1 = 0. \end{aligned} \tag{5.61}$$

To find the density of eigenvalues, we must make $z = \varepsilon + i\eta$:

$$\begin{aligned} (\varepsilon + i\eta)G_B^2(z) + (\varepsilon + i\eta + 1 - c)G_B(z) + 1 &= 0 \\ \left[\varepsilon G_B^2(z) + (\varepsilon + 1 - c)G_B(z) + 1\right] + i\eta\left[G_B^2(z) + G_B(z)\right] &= 0 \\ G_B(z) = -\frac{\varepsilon + 1 - c}{2\varepsilon} \pm i\frac{\sqrt{4\varepsilon - (\varepsilon + 1 - c)^2}}{2\varepsilon} & \\ -\frac{1}{\pi}\mathrm{Im}\left\{G_B(\varepsilon + i.\eta)\right\} = \frac{\sqrt{4\varepsilon - (\varepsilon + 1 - c)^2}}{2\pi\varepsilon}. & \end{aligned} \tag{5.62}$$

The element in the square root can be factored:

$$\begin{aligned} 4\varepsilon - \left(\varepsilon^2 + (1-c)^2 + 2\varepsilon(1-c)\right) &= \varepsilon^2 + 2\varepsilon(1+c) + (1-c)^2 \\ \varepsilon &= \frac{2(1+c)^2 \pm \sqrt{4(1+c)^2 - 4(1-c)^2}}{2} \\ &= (1+c)^2 \pm 2\sqrt{c} \\ &= (1 \pm \sqrt{c})^2. \end{aligned} \tag{5.63}$$

Therefore, the density of eigenvalues can be written as:

$$\rho_B(\varepsilon) = \frac{\sqrt{(\varepsilon - \varepsilon_a)(\varepsilon_b - \varepsilon)}}{2\pi\varepsilon}, \tag{5.64}$$

where $\varepsilon_{ab} = (1 \pm \sqrt{c})^2$.

5.2.3.1 Correlation Matrix

Matrix $\mathbf{B}_N = 1/N\mathbf{X}^\dagger\mathbf{T}\mathbf{X}$ is not a correlation matrix. We can make it a correlation matrix doing:

$$\begin{aligned}\mathbf{C}_n &= \frac{1}{N}\mathbf{T}_n^{1/2}\mathbf{X}\mathbf{X}^\dagger\mathbf{T}_n^{1/2} \\ &= \frac{1}{N}\mathbf{F}\mathbf{G},\end{aligned} \tag{5.65}$$

where $\mathbf{F} = \mathbf{T}_n^{1/2}\mathbf{X}_n$ and $\mathbf{G} = \mathbf{X}_n^\dagger\mathbf{T}_n^{1/2}$. Since $\mathbf{T}_n$ is an $n\times n$ matrix and $\mathbf{X}_n$ is an $n\times N$ matrix, we have that $\mathbf{F}$ is an $n\times N$ matrix, and $\mathbf{G}$ is an $N\times n$ matrix. Thus, $\mathbf{C}_n$ is an $n\times n$ matrix.

On the other hand, we can write:

$$\begin{aligned}\frac{1}{N}\mathbf{G}\mathbf{F} &= \frac{1}{N}\mathbf{X}_n^\dagger\mathbf{T}_n^{1/2}\mathbf{T}_n^{1/2}\mathbf{X}_n \\ &= \frac{1}{N}\mathbf{X}_n^\dagger\mathbf{T}_n\mathbf{X}_n \\ &= \mathbf{B}_N.\end{aligned} \tag{5.66}$$

Since $\mathbf{C}_n = (1/N)\mathbf{F}\mathbf{G}$ and $\mathbf{B}_N = (1/N)\mathbf{G}\mathbf{H}$, we can state that $\mathbf{C}$ has all eigenvalues of $\mathbf{B}$ and its remaining $n-N$ eigenvalues are zero. Since we are assuming that $n > N$, we have the measure:

$$\begin{aligned}F^C(t) &= \left(\frac{n-N}{n}\right)I_{(0,\infty]}(t) + \frac{N}{n}F^B(t) \\ &= \left(1-\frac{1}{c}\right)I_{(0,\infty]}(t) + \frac{1}{c}F^B(t) \\ dF^C(t) &= \left(1-\frac{1}{c}\right)\delta(t)dt + \frac{1}{c}dF^B(t) \\ \int\frac{1}{z-t}dF^C(t) &= \left(1-\frac{1}{c}\right)\int\frac{1}{z-t}\delta(t)dt + \frac{1}{c}\int\frac{1}{z-t}dF^B(t) \\ G_C(z) &= \frac{\left(1-\frac{1}{c}\right)}{z} + \frac{1}{c}G_B(z).\end{aligned} \tag{5.67}$$

The density of eigenvalues of the correlation matrix is then:

$$
\begin{aligned}
-\frac{1}{\pi}\mathrm{Im}\{G_C(\varepsilon+i\eta)\} &= -\frac{1}{\pi}\mathrm{Im}\left\{\frac{1-1/c}{\varepsilon+i\eta}\right\} - \frac{1}{c}\cdot\frac{1}{\pi}\mathrm{Im}\{G_B(\varepsilon+i\eta)\} \\
\rho_C(\varepsilon) &= -\frac{1-\frac{1}{c}}{\pi}\mathrm{Im}\left\{\frac{\varepsilon-i\eta}{\varepsilon^2+\eta^2}\right\} + \frac{1}{c}\cdot\rho_B(\varepsilon) \\
&= \left(1-\frac{1}{c}\right)\delta(\varepsilon) + \frac{1}{c}\cdot\rho_B(\varepsilon) \\
&= \max\left(0, 1-\frac{1}{c}\right)\delta(\varepsilon) + \frac{c}{2\pi\varepsilon}\sqrt{(\varepsilon-\varepsilon_a)(\varepsilon_b-\varepsilon)}.
\end{aligned} \tag{5.68}
$$

Here we have used the maximum function because negative densities do not exist.

Generalizing for a variance σ^2, the Marčenko-Pastur law becomes [118]:

$$
\rho(\varepsilon) = \frac{\sqrt{(\varepsilon-\varepsilon_{\text{min}})(\varepsilon_{\text{max}}-\varepsilon)}}{2\pi\varepsilon\sigma^2}, \tag{5.69}
$$

where

$$
\varepsilon_{\text{max,min}} = \sigma^2\left(1 \pm \sqrt{c}\right)^2. \tag{5.70}
$$

The Marčenko-Pastur distribution for different values of c is shown in Fig. 5.5.

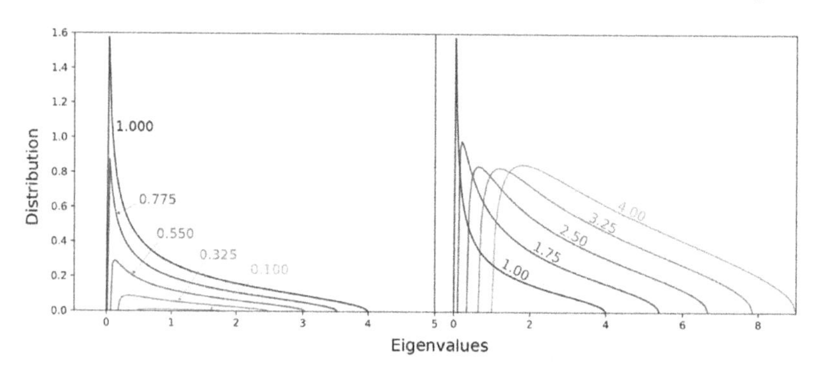

Figure 5.5: Theoretical Marčenko-Pastur distributions for different values of c

To simulate the Marčenko-Pastur law in Python, we must include the NumPy and the Matplotlib libraries. Also, we define a few constants:

```
import numpy as np
import matplotlib.pyplot as pl

n = 77
M = 100
eta = 1E-3
c = n/M
```

The theoretical curve can be calculated with:

```
ea = (1+np.sqrt(c))**2
eb = (1-np.sqrt(c))**2

esp = np.linspace(eb,ea,100)
rsp = []

for e in esp:
        r = (float(c)/(2*np.pi*e))*np.sqrt((e-ea)*(eb-e))
        rsp.append(r)
```

The experimental density can be calculated with:

```
DOS = []
for it in range(300):
        R = np.array([[np.random.normal(0,1) for cn in range(n)]
                                                        for cm in range(M)])
        H = np.dot(R,np.transpose(R)); N = len(H)
        H = np.dot(1.0/N,H)

        gsp = []
        for e in esp:
                G = np.linalg.inv((e+1j*eta)*np.eye(N)-H)
                gsp.append(-(1.0/(N*np.pi))*np.trace(G.imag))
        DOS.append(gsp)

DOS = np.average(DOS,axis=0)
```

The result of this simulation is shown in Fig. 5.6.

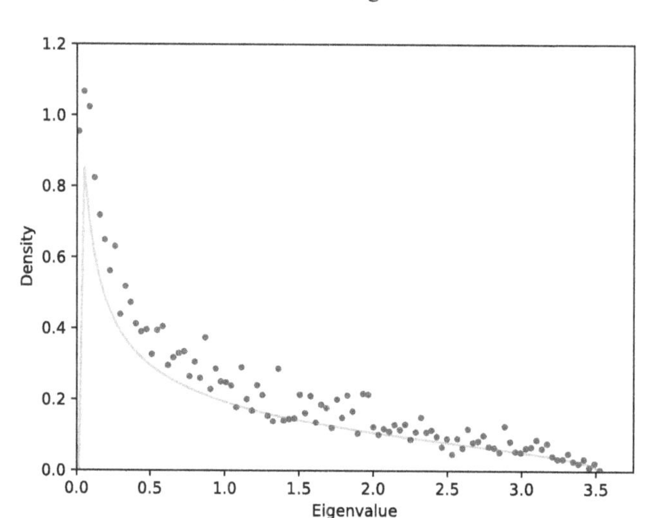

Figure 5.6: Marčenko-Pastur distribution for a random $n \times N$ correlation matrix. The solid line is the theoretical prediction and the dots are obtained with an ensemble of 300 random matrices

5.2.4 WIGNER'S SURMISE

In 1957 during a conference on neutron physics [122], a question about the distribution of eigenvalue spacings of heavy nuclei appeared. Wigner, who was attending the conference stepped up to the blackboard and guessed the following result:

Consider a 2×2 matrix:

$$H = \begin{bmatrix} x_1 & x_3 \\ x_3 & x_2 \end{bmatrix}, \tag{5.71}$$

in which elements x_1 and x_2 are $\mathcal{N}(0,1)$-distributed random variables and x_3 is a $\mathcal{N}(0,1/2)$-distributed random variable.

The eigenvalues of this matrix are given by:

$$\begin{aligned} \begin{vmatrix} x_1-\lambda & x_3 \\ x_3 & x_2-\lambda \end{vmatrix} &= x_1x_2 - x_1\lambda - x_2\lambda + \lambda^2 - x_3^2 = 0 \\ \lambda^2 - (x_1+x_2)\lambda + x_1x_2 - x_3^2 &= 0 \\ 2\lambda = x_1 + x_2 \pm \sqrt{(x_1+x_2)^2 - 4x_1x_2 + 4x_3^2}&. \end{aligned} \tag{5.72}$$

Therefore, the spacings between eigenvalues are given by:

$$s = \lambda_1 - \lambda_2 = \sqrt{(x_1 - x_2)^2 + 4x_3^2}. \tag{5.73}$$

The probability measure s is given by:

$$\begin{aligned} p(s) &= \int_{-\infty}^{\infty} \frac{1}{\sqrt{2\pi}} e^{-1/2x_1^2} \frac{1}{\sqrt{2\pi}} e^{-1/2x_2^2} \frac{1}{\sqrt{\pi}} e^{-x_3^2} \delta\left(s - \sqrt{(x_1 - x_2)^2 + 4x_3^2}\right) d\mathbf{x} \\ &= \frac{1}{2\pi^{3/2}} \int_{-\infty}^{\infty} \exp\left\{-(1/2x_1^2 + 1/2x_2^2 + x_3^2)\right\} \delta\left(s - \sqrt{(x_1 - x_2)^2 + 4x_3^2}\right) d\mathbf{x} \end{aligned} \tag{5.74}$$

Making the following change of variables:

$$\begin{aligned} x_1 - x_2 &= r\cos(\theta) \\ 2x^3 &= r\sin(\theta) \\ x_1 + x_2 &= \phi, \end{aligned} \tag{5.75}$$

we get:

$$\begin{bmatrix} 1 & -1 & 0 \\ 0 & 0 & 2 \\ 1 & 1 & 0 \end{bmatrix} \begin{bmatrix} x_1 \\ x_2 \\ x_3 \end{bmatrix} = \begin{bmatrix} r\cos(\theta) \\ r\sin(\theta) \\ \phi \end{bmatrix} \tag{5.76}$$

and

$$\begin{bmatrix} x_1 \\ x_2 \\ x_3 \end{bmatrix} = \frac{1}{2} \begin{bmatrix} 1 & 0 & 1 \\ -1 & 0 & 1 \\ 0 & 1 & 0 \end{bmatrix} \begin{bmatrix} r\cos(\theta) \\ r\sin(\theta) \\ \phi \end{bmatrix} = \begin{bmatrix} 1/2\,(r\cos(\theta) + \phi) \\ 1/2\,(\phi - r\cos(\theta)) \\ 1/2 r\sin(\theta) \end{bmatrix}. \tag{5.77}$$

We must also change the differential volume of integration. Therefore, we make $d\mathbf{x} = |J| d\mathbf{y}$, where J is the Jacobian[14] matrix given by:

$$J = \begin{bmatrix} \frac{dx_1}{dr} & \frac{dx_1}{d\theta} & \frac{dx_1}{d\phi} \\ \frac{dx_2}{dr} & \frac{dx_2}{d\theta} & \frac{dx_2}{d\phi} \\ \frac{dx_3}{dr} & \frac{dx_3}{d\theta} & \frac{dx_3}{d\phi} \end{bmatrix} = \begin{bmatrix} 1/2\cos(\theta) & -1/2 r\sin(\theta) & 1/2 \\ -1/2\cos(\theta) & 1/2 r\sin(\theta) & 1/2 \\ 1/2\sin(\theta) & 1/2 r\cos(\theta) & 0 \end{bmatrix} \tag{5.78}$$

Its determinant is $-r/4$. Therefore, the probability measure becomes:

$$p(s) = \frac{1}{8\pi^{3/2}} \int_{-\infty}^{\infty} dr \delta(s - r) \int_0^{2\pi} d\theta \int_{-\infty}^{\infty} d\phi e^{\beta(r,\theta,\phi)}, \tag{5.79}$$

where

$$\begin{aligned}\beta(r,\theta,\phi) &= -(1/2)(1/4)\left(r\cos(\theta)+\phi\right)^2 - (1/2)(1/4)\left(\phi - r\cos(\phi)\right)^2 - 1/4r^2\sin^2(\theta)\\ &= -1/4\left(\frac{r^2}{2}\cos^2(\theta)+\frac{\phi^2}{2}+r\phi\cos(\theta)+\frac{\phi^2}{2}+\frac{r^2}{2}\cos^2(\theta)-r\phi\cos(\theta)\right.\\ &\qquad\left.+r^2\sin^2(\theta)\right)\\ &= -1/4\left(r^2\cos^2(\theta)+r^2\sin^2(\theta)+\phi^2\right)\\ &= -1/4r^2 - \frac{\phi^2}{2\cdot 2}.\end{aligned} \tag{5.80}$$

The probability measure finally becomes:

$$\begin{aligned}p(s) &= \frac{1}{8\pi^{3/2}}\sqrt{2\pi 2}\int_{-\infty}^{\infty}d\phi\frac{e^{-\frac{\phi^2}{2\cdot 2}}}{\sqrt{2\pi 2}}\int_{-\infty}^{\infty}dr\cdot r\delta(s-r)\int_0^{2\pi}d\theta e^{-1/4r^2}\\ &= \frac{2}{8\pi}2\pi s\cdot e^{-1/4s^2}\\ &= \frac{1}{2}s\cdot e^{-1/4s^2}.\end{aligned} \tag{5.81}$$

It is convenient, though, to express it as $\bar{p}(s) = \langle s\rangle p(\langle s\rangle s)$, where:

$$\begin{aligned}\langle s\rangle &= \int_0^{\infty} sp(s)ds\\ &= \frac{1}{2}\int_0^{\infty} s^2 e^{-1/4s^2}ds\\ &= \frac{1}{2}\frac{\sqrt{\pi}2^3(2-1)!!}{2^2}\\ &= \sqrt{\pi}.\end{aligned} \tag{5.82}$$

Consequently,

$$\begin{aligned}\bar{p}(s) &= \langle s\rangle p(\langle s\rangle s)\\ &= \sqrt{\pi}p(\sqrt{\pi}s)\\ &= \frac{\pi}{2}se^{-1/4\pi s^2}.\end{aligned} \tag{5.83}$$

The spacing distribution shown in Fig. 5.7 implies that the eigenvalues "repel" each other since the distribution goes to zero for small values.

More generally, Gaussian ensembles have spacing distributions given by:

$$p_\beta(s) = a_\beta s^\beta \exp\left\{-b_\beta s^2\right\}, \tag{5.84}$$

where $\beta = 1,2$ and 4 for GOE , GUE and GSE respectively, and [117–119]:

$$
\begin{aligned}
a_\beta &= \frac{2\left[\Gamma\left(\frac{2+\beta}{2}\right)\right]^{\beta+1}}{\left[\Gamma\left(\frac{1+\beta}{2}\right)\right]^{\beta+2}} = \frac{\pi}{2}, \frac{32}{\pi^2}, \frac{2^{18}}{3^6\pi^3}, \dots \\
b_\beta &= \left(\frac{\Gamma\left(\frac{2+\beta}{2}\right)}{\Gamma\left(\frac{1+\beta}{2}\right)}\right)^2 = \frac{\pi}{4}, \frac{4}{\pi}, \frac{64}{9\pi}, \dots
\end{aligned}
\tag{5.85}
$$

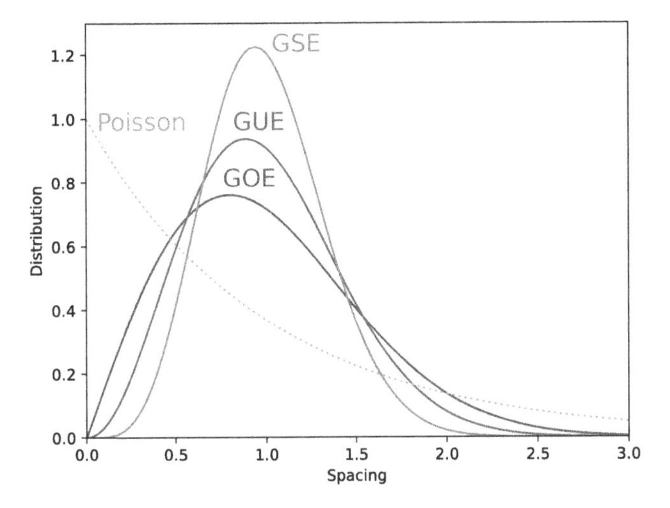

Figure 5.7: Spacing distribution for the GOE, GUE and GSE ensembles. The Poisson distribution $\exp(-s)$ is shown for comparison

5.3 OTHER PORTFOLIO MEASURES

Modern portfolio theory has some deficiencies. For instance, taking the variance of the portfolio as a measure of risk penalizes undesired high losses the same way as small losses. In order to compensate for this and other problems, it is possible to diversify the variance instead of the securities. This is known as *risk parity portfolio* [123, 124].

Let's take a volatility defined as:

$$
\sigma = \sqrt{\mathbf{W}^{\mathrm{T}}\mathbf{S}\mathbf{W}}. \tag{5.86}
$$

According to Euler's homogeneous functions theorem [125], we can also write this standard variation as:

$$
\sigma = \sum_{i=1}^{N} W_i \left(\frac{\partial \sigma}{\partial \mathbf{W}}\right)_i = \sum_{i=1}^{N} \sigma_i, \tag{5.87}
$$

where the summand is the *partial volatility*. The derivative is given by:

$$\begin{aligned}\frac{\partial\sigma}{\partial\mathbf{W}} &= \frac{\partial}{\partial\mathbf{W}}\sqrt{\mathbf{W}^T\mathbf{S}\mathbf{W}}\\ &= \frac{1}{2}\frac{1}{\sqrt{\mathbf{W}^T\mathbf{W}}}2\mathbf{S}\mathbf{W}\\ &= \frac{\mathbf{S}\mathbf{W}}{\sigma}.\end{aligned} \tag{5.88}$$

Therefore, the individual volatilities are given by:

$$\sigma_i = W_i\left(\frac{\partial\sigma}{\partial\mathbf{W}}\right)_i = W_i\frac{(\mathbf{S}\mathbf{W})_i}{\sigma}. \tag{5.89}$$

The strategy is now to compose a *risk budget* $\mathbf{b} = [b_1, b_2, \ldots, b_N]$ such that $\mathbf{b}\cdot\mathbf{1} = 1$. Thus, we get a portfolio where each security contributes to a specific proportion of the total risk σ:

$$\sigma_i = b_i\sigma \rightarrow W_i\frac{(\mathbf{S}\mathbf{W})_i}{\sigma} = b_i\sigma. \tag{5.90}$$

Example 5.3.1. Uncorrelated Portfolio

In the simplest case where the covariance matrix is diagonal and the assets are uncorrelated, we get:

$$\begin{aligned}W_iS_{ii}W_i &= b_i\sigma^2\\ W_i &= \frac{\sqrt{b_i}}{\sqrt{S_{ii}}}\sigma^2.\end{aligned} \tag{5.91}$$

Normalizing it such that $\sum_n W_n = 1$, we obtain:

$$\begin{aligned}w_i &= \frac{\frac{\sqrt{b_i}}{\sqrt{S_{ii}}}\sigma^2}{\sum_n \frac{\sqrt{b_n}}{\sqrt{S_{nn}}}\sigma^2}\\ &= \left(\sum_n\sqrt{\frac{b_n}{b_i}\frac{S_{ii}}{S_{nn}}}\right)^{-1}.\end{aligned} \tag{5.92}$$

■

For assets that show some level of correlation we can write the risk budget equation as [126]:

$$\frac{(\mathbf{S}\mathbf{W})_i}{\sigma} = \frac{b_i}{W_i}\sigma. \tag{5.93}$$

Or in matrix notation[e]:

[e] $\mathbf{a}\odot\mathbf{b}$ is the Hadamard[15] product between vectors $\mathbf{a}$ and $\mathbf{b}$.

$$\frac{\mathbf{SW}}{\sigma} = \mathbf{b} \odot \mathbf{M}\sigma, \tag{5.94}$$

where:

$$\begin{aligned} \mathbf{M} &= \begin{bmatrix} W_1^{-1} & W_2^{-1} & \dots & W_N^{-1} \end{bmatrix}^T, \\ \mathbf{b} &= \begin{bmatrix} b_1 & b_2 & \dots & b_N \end{bmatrix}^T. \end{aligned} \tag{5.95}$$

Making $\mathbf{X} = \mathbf{W}/\sigma$ e $\mathbf{Y} = \mathbf{M}\sigma$, it is possible to write:

$$\mathbf{SX} = \mathbf{b} \odot \mathbf{Y}. \tag{5.96}$$

This, however, is the result of:

$$\begin{aligned} \nabla_X \left[1/2\mathbf{X}^\mathrm{T}\mathbf{SX} - \mathbf{b} \cdot \log(\mathbf{X}) \right] &= 0 \\ (\mathbf{SX})_n - \frac{b_n}{X_n} &= 0. \ \square \end{aligned} \tag{5.97}$$

Therefore, the weights X_n can be found minimizing the function:

$$f(\mathbf{X}) = 1/2\mathbf{X}^\mathrm{T}\mathbf{SX} - \mathbf{b} \cdot \log(\mathbf{X}), \tag{5.98}$$

where the logarithmic term can be understood as a 'weighted entropy' [127] correction to the portfolio [128].

More recently, the hierarchical structure of portfolios has also been studied. This is performed calculating the ultrametric ordering, a concept that is borrowed from the theory of frustrated disordered systems [129, 130]. In short, an ultrametric space is a metric space (see Sec. 8.1.1) where the triangle inequality is restricted to $d(x,y) \leq \max\{d(x,y), d(y,z)\}$ for any two points x and y, where $d(x,z)$ is the distance function between x and z. This implies that there is no intermediate points between x and z, and a random walk in an ultrametric space is non-ergodic [131]. This is an appropriate way to describe hierarchical structures since the concept of ultrametricity is tightly related to the concept of hierarchy.

Hierarchical trees can be used to represent ultrametric sets by calculating the distances between nodes. Computing the distances between all assets creates a complete graph. This carries redundant information that can be eliminated if one uses a subdominant ultrametric structure such as the *minimum spanning tree* (MST) [132–134]. The distance between assets is typically computed as the Euclidean distance between their time series and the MST is computed using Prim's algorithm [135].

Finally, agglomerative hierarchical clustering algorithms are used to create hierarchical clusters (dendrograms) [136]. In this algorithm, two clusters with the smallest distance are permanently joined together forming a new one. This new cluster k has a distance to the remaining clusters m given by:

$$d_{km} = \alpha_i d_{im} + \alpha_j d_{jm} + \beta d_{ij} + \gamma |d_{im} - d_{jm}|. \tag{5.99}$$

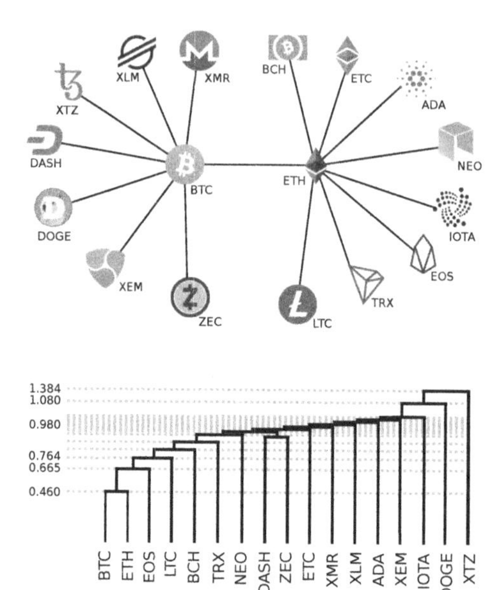

Figure 5.8: Minimum spanning tree (top) and dendrogram (bottom) calculated for 17 different digital goods for a period of 10^5 minutes ending on 2019-08-12

The MST and the equivalent dendrogram calculated for a set of different digital goods are shown in Fig. 5.8. We can see in this case that the MST separates the digital goods in two major groups, one dominated by BTC and another dominated by ETH.

Notes

[1]Ludwig Heinrich Edler von Mises (1881–1973) Austrian economist, advisee of Eugen Böhm von Bawerk, adviser of Oskar Morgenstern and Israel Kirzner among others.

[2]Harry Max Markowitz (1927–) American economist, advisee of Milton Friedman and Nobel laureate in 1990.

[3]Giuseppe Ludovico de la Grange Tournier (1736–1813) Italian mathematician, advisee of Leonhard Euler and adviser of Joseph Fourier and Siméon Poisson, among others.

[4]William Forsyth Sharpe (1934–) American economist, unofficially advised by Harry Markowitz, awarded the Nobel prize in 1990.

[5]James Tobin (1918–2002) American economist, advisee of Joseph Schumpeter, adviser of Edmund Phelps among others. He was awarded the Nobel prize in economics in 1981.

[6]Frank Hyneman Knight (1885–1972) American economist, adviser of Milton Friedman, George Stigler and James Buchanan among others.

[7]Jack Lawrence Treynor (1930–2016) American economist.

[8]John Wishart (1898–1956) Scottish mathematician, advisee of Karl Pearson.

[9]Julian Karol Sochocki (1842–1927) Russian mathematician, and Josip Plemelj (1873–1967) Slovene mathematician.

[10]Freeman John Dyson (1923–2020) British mathematical physicist, advisee of Hans Bethe. Dyson was awarded many prizes such as the Henri Pincaré Prize in 2012.

[11]Eugene Paul Wigner (1902–1995) Hungarian physicist, advisee of Michael Polanyi and adviser of John Bardeen among others. Wigner was awarded many prizes, including the Nobel in physics in 1963.

[12]Vladimir Alexandrovich Marčenko (1922–) and Leonid Andreevich Pastur (1937–) Ukranian mathematicians.

[13]Thomas Joannes Stieltjes (1856–1894) Dutch mathematician, advisee of Charles Hermite.

[14]Carl Gustav Jacob Jacobi (1804–1851) German mathematician, adviser of Ludwig Otto Hesse among others.

[15]Jaques Salomon Hadamard (1865–1963) French mathematician, adviser of Maurice Fréchet and Paul Lévy among others.

6 Criticality

By the end of the XVIII century, the French monarchy lead by Louis XIV was essentially insolvent. His mandate was characterized by a concentration of power that resulted in lengthy wars and financial misconducts. The Palace of Versailles, for instance, costed billions to be built and that helped to deepen the country's debt. In order to finance these debts, Louis XIV endorsed the monetary ideas of John Law[1]. Law proposed to substitute gold, which was used as currency, for the creation of paper money as a method to stimulate the economy and create revenue to cover the monarchy's debt. Although, his ideas were rejected in Scotland, Louis XIV embraced them in France and created a central bank (*Banque Générale Privée*) to issue paper money. Moreover, Law established the Mississippi company (the Company of the West and the Company of the Indies) to monopolistically trade minerals and then sold shares of this company in France to back the central bank.

The price of the shares rose quickly under the promise of high profits in America and this was working towards reducing the monarchy's debt. However, investors eventually perceived that the exploration of minerals in Mississippi was not so simple and many tried to convert their shares in gold. Law adopted many restrictive policies such as suspending the operation of banks for a few days and criminalized the sale of gold. This only increased concerns that eventually created panic, bank run, stampedes and riots. The end result was a massive crisis that made Law leave Paris in the cover of the night leaving all his properties behind to pay creditors [137].

This is known today as the *Mississippi bubble* and is one of the first examples of an economic bubble. One may ask, though, if all crises show any universality, or if it is possible to predict when an economic collapse is approaching. We will try to answer some of these questions in the following sections.

6.1 CRISES AND CYCLES

In order to study crises, first we must turn our attention to the *production-possibility frontier* (PPF). This is a tool that allows us to see the relationship between the production of two goods or the relationship between consumption and investment as shown in Fig. 6.1.

The line shown in the figure indicates the efficient possibility. For example, consider that production is at point X and a company wants to increase the production of a good B. The production of a good A will have to be reduced because resources will have to be taken and allocated to the production of good B. This illustrates the definition of *opportunity cost*[a][2]. Moreover, capital will be taken from the production of an item to another according to a *marginal rate of transformation*.

Any point outside the efficient frontier (such as point q) is impossible to be reached since it requires the overproduction of both items. Points inside the

[a]The total potential benefit of all alternatives that are discarded when one chooses some option.

DOI: 10.1201/9781003127956-6

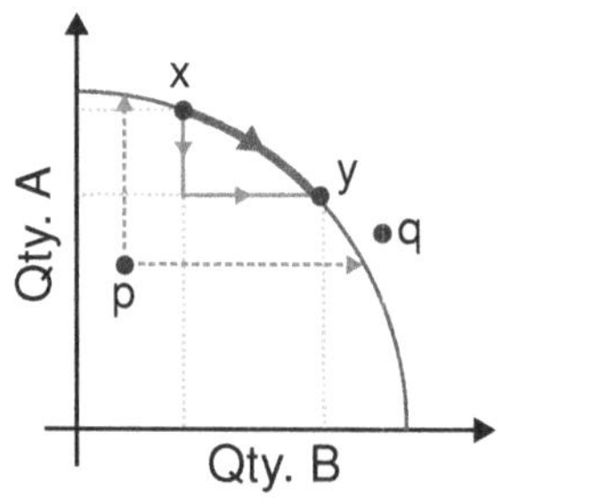

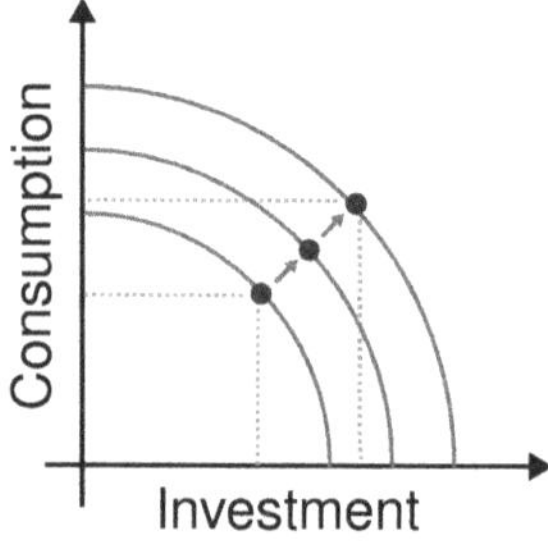

Figure 6.1: Production-possibility frontier for the (left) production of two goods *A* and *B*, and (right) relationship between consumption and investment

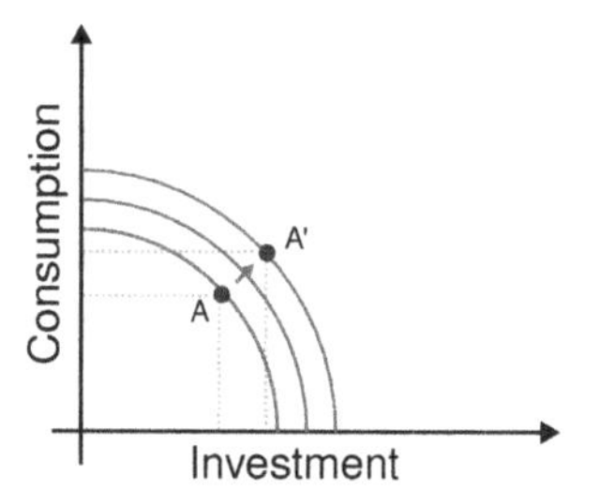

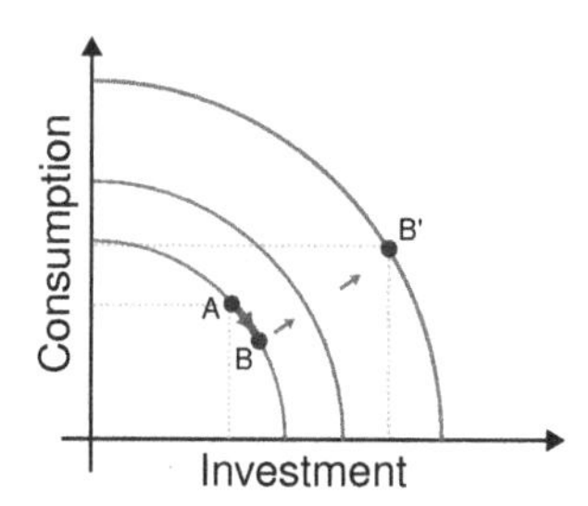

Figure 6.2: Production-possibility frontier for consumption × investment comparing the growth of an economy where the agents (left) do not increase their savings *vs.* (right) one where the agents increase their savings

possibility region (such as point *p*) are inefficient because it would be possible to produce more of a good *B* for the production of the same quantity of a good *A* and vice-versa.

The relationship between investment[b] and consumption is also shown in Fig. 6.1. The curves indicate an efficient frontier that expands due to economic growth. The curve on the right hand side of Fig. 6.2 shows the dynamics of an economy where the agents choose to consume less and increase their savings. When this happens, the economy moves from point *A* to point *B*. The immediate consequence is that the net savings available to investment increase and the economy grows at progressive higher rates. This contrasts with the curve on the left hand side of the figure where the economic growth is smaller due to less investment capital available.

The *loanable funds model* is a complementary tool that helps us understand the dynamics of economic growth. This model is illustrated by a plot of interest rate as a function of the volume of investment as shown in Fig. 6.3. The demand for funds with high interest rates must be low, therefore the curve must decay as the volume increases. On the other hand, the supply of funds with high interest rates must have the opposite behavior.

[b]gross investment = depreciation + net investment.

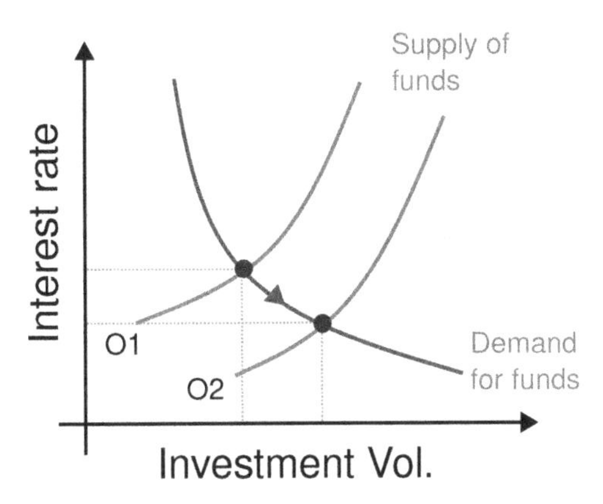

Figure 6.3: Loanable funds market: the shift from *O*1 towards *O*2 happens when the agents adjust their temporal preference towards the future

Let's see what happens when the agents adjust their temporal preference towards the future. In order for the agents to be able to sell funds, they must lower interest rates. This lowering of interest rates sends a signal to companies saying that funds are affordable and consumers have savings. Investment is thus stimulated and the effect in the model is a shift of the supply curve from *O*1 towards *O*2 because the interest rate has to be lower for the same volume of funds. As it can be seen in the figure, the quantity of loanable funds available for investment increases in agreement with the previous analysis using the PPF.

Hayek's triangles [138] is one last tool that we will use to explain this dynamics. This is a graphical representation of the intertemporal production structure of an economy. In order to understand the triangle shown in Fig. 6.4, consider a productive system segmented in four stages: i) mining, ii) ore refining, iii) manufacturing of the ore in a metallic part, and iv) selling of the metallic parts. The horizontal axis of the triangle indicates the production stage of the productive arrangement, whereas the vertical axis indicates the quantity of products produced at each stage.

When consumers adjust their temporal preference towards the future, the present consumption drops and investment increases. We see a corresponding drop in the price of money and an increase in the quantity of loanable funds in the loanable funds model. Accordingly, this implies in the Hayek's triangle that consumption has lowered. Consequently sellings also drop and the capital is transferred to the initial stages of production that are more immediate. Hence, the triangle shrinks vertically and expands horizontally.

More funds available for investment signals the companies that consumers have savings. Hence, investments are made and the economy grows. The increase in investment leads to an increase in the demand for funds, forcing the interest rate to go back to a position near its original value. The overall effect is higher production and consumption expanding the triangle both horizontally and vertically. This echoes *Say's*[3] *market law*, which states that the value of goods that individuals can purchase

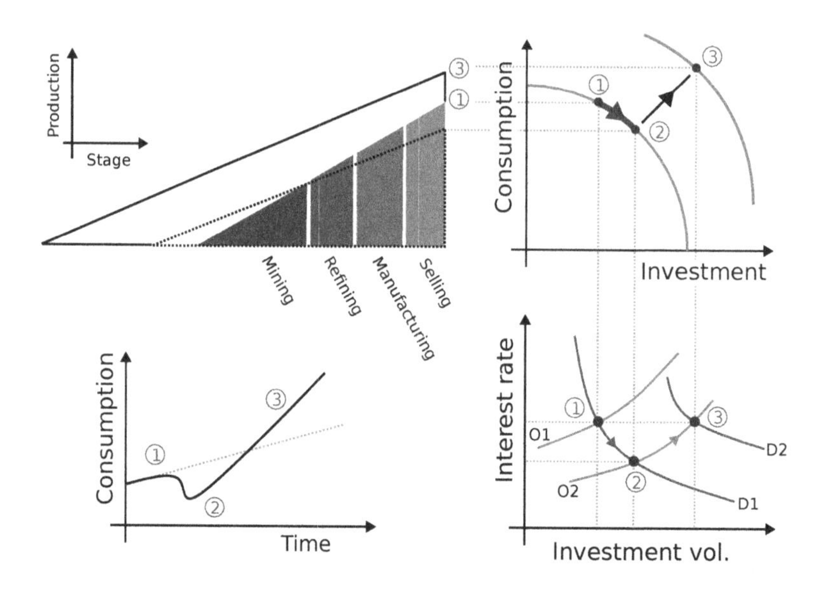

Figure 6.4: Hayek's triangle relating to the production-possibility frontier and the loanable funds model

equals the market value of those goods that these individuals can previously offer, or in simpler words, in order to consume, people must produce first.

What happens if a government injects money in the economy through a *monetary stimulus*? Initially, two things happen [139] as depicted in Fig. 6.5. Firstly, even if there is no actual savings, the supply curve shifts towards the right because of the credit expansion. Having more affordable credit, companies invest and expand their businesses. This produces a shift in the demand curve. Consumers, for their part, stay in the unaltered savings curve moving to a point of lower consumption.

This produces two competing effects in the production-possibility model. There is a tendency of a clockwise movement because of the increase in production by the companies that took credit. However, there is also a tendency for a counterclockwise movement because of consumers that have less savings. The result is a point outside the region of production possibility, which makes the economy unstable.

In Hayek's triangle we see an expansion in the initial production stages because of lower interest rates. On the other hand, higher consumption increases the height of the triangle while keeping its base unaltered. This discrepancy causes the capital not being allocated in the final production stages.

This situation becomes unsustainable since many enterprises do not find capital to complete their productions and many expanded businesses do not find buyers once they complete their productions. What was initially a boom becomes a bust, the economy becomes uncoordinated and goes into recession. This is known as the *Austrian business cycle theory* (ABCT) [140]. Although there is some empirical evidence (see, for example [141]) supporting this theory, it is also the object of many

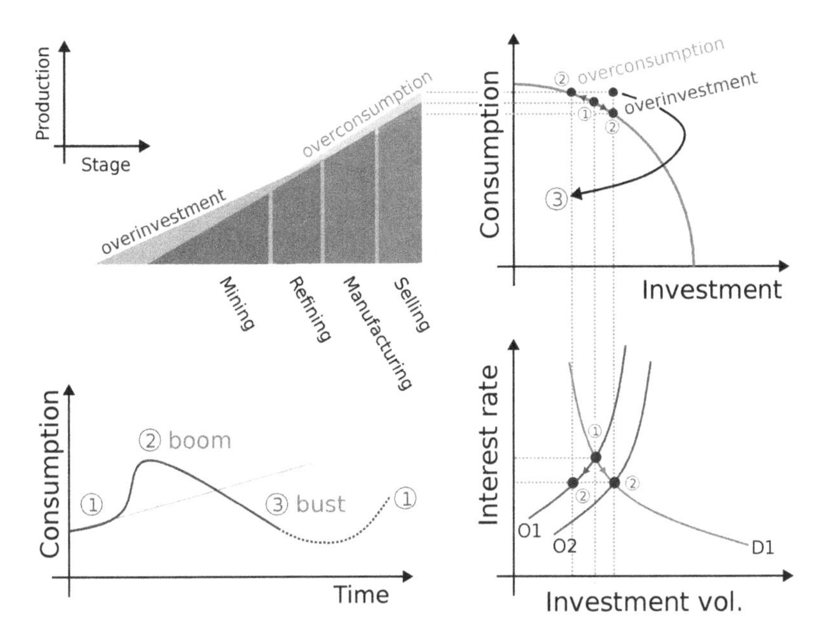

Figure 6.5: Austrian business cycle theory illustrated with Hayek's triangle, the production-possibility frontier and the loanable funds model

criticism. For instance, are crises always the result of malinvestment? Some early agent-based models, for instance state that even portfolio insurance strategies have the potential to cause crises [142]. Let's see what physics has to tell us by analyzing two topics: catastrophe theory and self-organized criticality.

6.2 CATASTROPHE THEORY

Most of the models we have seen so far are based on a normal distribution of returns, but how can we model abrupt changes that happen in real situations? Here we will discuss the *catastrophe theory* based on the works of Stewart[4] [143] but developed by Thom[5] [144] and Zeeman[6]. This theory gives a broad mathematical background for physical theories that explain phase transitions such as that of Landau[7].

Consider a mapping $f : M \to \mathbb{R}$, where M is a smooth manifold[c]. Consider also a family of mappings $F : M \times N \to \mathbb{R}$, where the parametrization N is a k-dimensional smooth manifold. If $F(x,0) = f(x)$, $\forall x$, then we say that f is *unfolded* by the family F.

A *fold catastrophe* is given by the unfolding:

$$V_f(x) = x^3 - ax, \tag{6.1}$$

[c]An infinitely differentiable manifold, where *manifold* is a topological space (a set of points and their neighborhoods that satisfy the axioms of emptiness, union and intersection) that is locally homeomorphic to an Euclidean space.

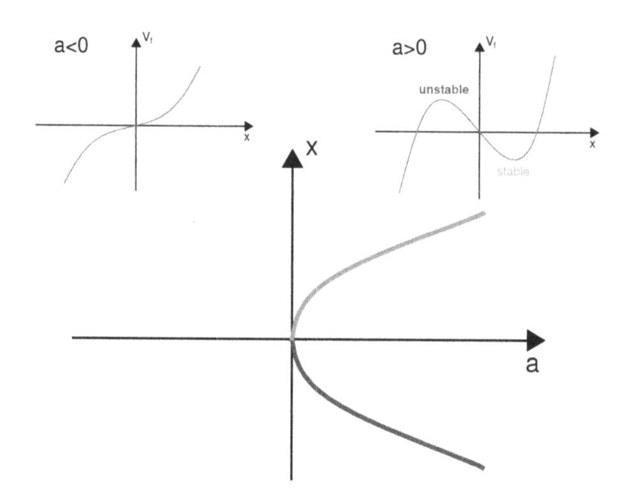

Figure 6.6: Equilibrium points of the fold catastrophe as a function of the control space. The two insets show the potential function for both $a < 0$ and $a > 0$

where $V : C \times X \to \mathbb{R}$ is a *potential function* (*Lyapunov function*[8]), $C = \{a : a \in \mathbb{R}\}$ is a *control space*, and $X = \{x : x \in \mathbb{R}\}$ is a *state space*. Its *catastrophe manifold* is the gradient of the potential function:

$$\nabla_x V_f(x) = 3x^2 - a. \tag{6.2}$$

The equilibrium points of the cusp catastrophe can be found by making its catastrophe manifold equal to zero:

$$\nabla_x V_f(x) = 3x^2 - a = 0 \to x = \pm\sqrt{\frac{a}{3}}. \tag{6.3}$$

For positive values of the control variable a there are two solutions, one stable and another unstable. When $a = 0$ both solutions collapse into a single point called *bifurcation point* or *tipping point*. For negative values of the control variable a, there is no equilibrium point for the fold catastrophe and a physical system would exhibit unpredictable behavior. This behavior is shown in Fig. 6.6.

6.2.1 CUSP CATASTROPHE

Let's now consider the *cusp catastrophe* that is given by the unfolding:

$$V_c(x) = x^4 - bx^2 + ax, \tag{6.4}$$

where the control space is given by $C = \{(a,b) : a, b \in \mathbb{R}\}$. This is a stable potential, since it is topologically invariant for any perturbation ε:

$$V_c(x; a, b, \varepsilon) = x^4 - bx^2 + ax + \varepsilon x = V_c(x; a + \varepsilon, b, 0). \tag{6.5}$$

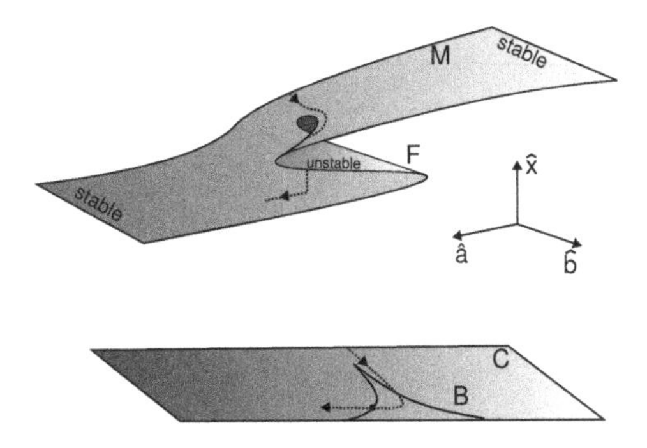

Figure 6.7: Catastrophe manifold (top) and the projection of the fold curve B in the control space C (bottom); the dotted line shows a discontinuous path that jumps between two stable states

Its catastrophe manifold, shown in Fig. 6.7, is given by:

$$\nabla_x V_c(x) = 4x^3 - 2bx + a, \tag{6.6}$$

which is a *depressed cubic* function. By moving on this manifold it is possible to move between two stable states through a hysteresis loop.

The *fold curve* is the edge of a surface when the latter is observed from a given direction. Mathematically, the *fold curve* for some direction z is the set of points on a surface such that the tangent plane to the surface at some point contains the direction z. The tangent plane to a point p, in turn, is the set of vectors tangent to this point for curves on the surface that contain this point. Since, the catastrophe manifold is a depressed cubic function, we can find the fold curve using Cardano's[9] discriminant:

$$D = 8b^3 - 27a^2. \tag{6.7}$$

The equilibrium frontier (*bifurcation set*) is found when $D = 0$. Also, when $D < 0$ there are three equilibrium points (two stable points and one unstable), whereas when $D > 0$, there is only one stable solution. This behavior is illustrated in Fig. 6.8. Path #2 shows a spontaneous symmetry breaking behavior where the system moves from one to two stable solutions.

Since parameter a controls the obliquity of the potential, it is known as *asymmetry factor*. Parameter b determines the bifurcation, hence, it is known as *bifurcation factor*.

6.2.2 CATASTROPHIC DYNAMICS

The cusp catastrophe has been used to model many socioeconomic phenomena, such as the failure of tax systems [145] and economic crises [146, 147]. Consider, for instance, the closing values of the S&P500 index shown in Fig. 6.9. Until approximately the end of February/2019, the movement of the closing price seemed to follow

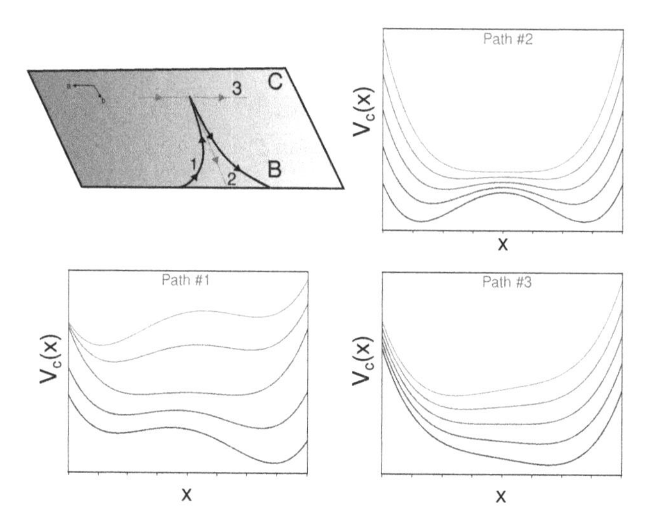

Figure 6.8: Solid black line B: the projection of the fold line on the control space C. There are three paths indicated on this surface. The first one follows exactly the projection of the fold curve. The second path starts at $(0,0)$ and moves towards $\hat{b}$ always within the bifurcation set. The third path moves in the direction from $-\hat{a}$ to $\hat{a}$. Five potentials are stacked vertically from bottom to top for each path

a geometric Brownian motion, but suddenly it dropped more than a thousand points in less than a month. It is intuitive then to try to model this movement with:

$$\begin{aligned} dX &= -\nabla_X V_c(X)dt + \sigma dW \\ &= -(4X^3 - 2bX + a)dt + \sigma dW, \end{aligned} \tag{6.8}$$

where V_c is the cusp potential function and σ^2 is the variance per time unit.

The asymmetry and bifurcation factors can be written as a function of exogenous parameters Y_n:

$$\begin{aligned} a &= a_0 + \sum_{n=1}^{N} a_n Y_n \\ b &= b_0 + \sum_{n=1}^{N} b_N Y_n, \end{aligned} \tag{6.9}$$

where a_n and b_n are parameters that can be estimated using the method of moments [148]. Observe that Eq. 6.8 is a gradient flow subject to white noise. We can use C.23 to find the equivalent Fokker-Plank representation:

$$\partial_t P(X,t) = \nabla_X \left(P(X,t)\nabla_X V_c(X)\right) + 1/2\nabla_X^2 \left(P(X,t)\sigma^2\right) \tag{6.10}$$

In order to find the steady state solution, let's try the ansatz:

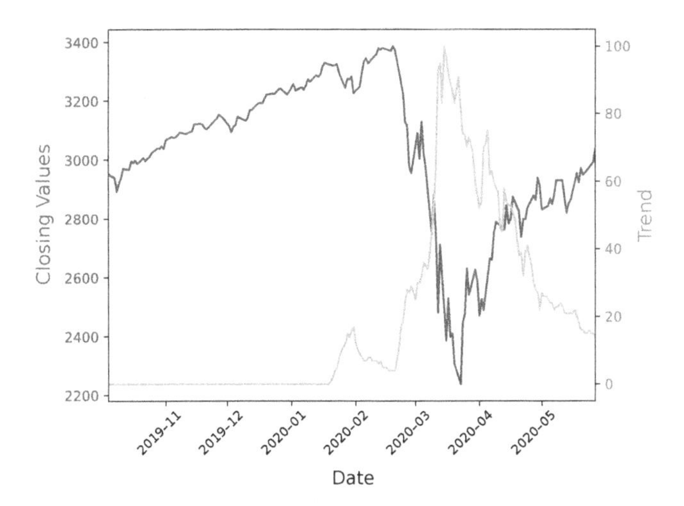

Figure 6.9: Dark curve: Closing values of the S&P500 index between 2019/10/10 and 2020/05/17, light curve: Trends obtained from Google Trends for the keyword 'coronavirus' for the same period

$$P(X) = e^{kV_C(X)}. \tag{6.11}$$

For this density we have:

$$\begin{aligned}
\frac{\partial P(X)}{\partial X} &= ke^{kV_C(X)}\frac{\partial V_C(X)}{\partial X} \\
&= kP(X)\frac{\partial V_C(X)}{\partial X} \\
\frac{\partial^2 P(X)}{\partial X^2} &= k^2 e^{kV(X)}\frac{\partial V_C(X)}{\partial X}\frac{\partial V(X)}{\partial X} + ke^{kV(X)}\frac{\partial^2 V_C(X)}{\partial X^2} \\
&= k^2 P(X)\frac{\partial V_C(X)}{\partial X}\frac{\partial V_C(X)}{\partial X} + kP(X)\frac{\partial^2 V_C(X)}{\partial X^2}.
\end{aligned} \tag{6.12}$$

The time-independent Fokker-Planck equation is:

$$\begin{aligned}
\frac{\partial}{\partial X}\left(P(X)\frac{\partial V_C(X)}{\partial X}\right) &= -\frac{\sigma^2}{2}\frac{\partial^2 P(X)}{\partial X^2} \\
\frac{\partial P(X)}{\partial X}\frac{\partial V_C(X)}{\partial X} + P(X)\frac{\partial^2 V_C(X)}{\partial X^2} &= -\frac{\sigma^2}{2}\frac{\partial^2 P(X)}{\partial X^2}.
\end{aligned} \tag{6.13}$$

Substituting the results from Eq. 6.12 in Eq. 6.13:

$$kP(X)\frac{\partial V_C(X)}{\partial X}\frac{\partial V_C(X)}{\partial X}+P(X)\frac{\partial^2 V_C(X)}{\partial X^2}= \\ -\frac{\sigma^2}{2}\left[k^2P(X)\frac{\partial V_C(X)}{\partial X}\frac{\partial V_C(X)}{\partial X}+kP(X)\frac{\partial^2 V_C(X)}{\partial X^2}\right]$$

$$k\frac{\partial V_C(X)}{\partial X}\frac{\partial V_C(X)}{\partial X}+\frac{\partial^2 V_C(X)}{\partial X^2}=-\frac{\sigma^2}{2}\left[k^2\frac{\partial V_C(X)}{\partial X}\frac{\partial V_C(X)}{\partial X}+k\frac{\partial^2 V_C(X)}{\partial X^2}\right] \tag{6.14}$$

Therefore, by comparing both sides we see that $k=-2/\sigma^2$ and obtain:

$$P(X)=\frac{1}{Z}\exp\left\{-\frac{2V_C(X)}{\sigma^2}\right\}=\frac{1}{Z}\exp\left\{-\frac{2}{\sigma^2}\int\nabla_X V_C(X)dX\right\}, \tag{6.15}$$

where Z is the partition function. This steady state solution is the known canonical Gibbs[10] distribution [73]:

Consider now a polynomial function $g(x,y)$. We can write:

$$\begin{aligned}\langle X^iY^jg(X,Y)\rangle &= \iint x^iy^jg(x,y)f_{X,Y}(x,y)dxdy \\ &= \iint x^iy^jg(x,y)f_{Y|X}(y|x)f_X(x)dxdy,\end{aligned} \tag{6.16}$$

where f_A is the density of variable A.

Let's also consider that the conditional density can be written as:

$$f_{Y|X}(y|x)=\varphi(x)\exp\left\{-\int g(x,y)dy\right\}, \tag{6.17}$$

where $\varphi(x)$ is a normalizing function. Note that this is exactly the situation that we have if we make $g(x,y)=2\nabla_X V_C(X)/\sigma^2$.

It is possible to rewrite Eq. 6.17 as:

$$g(x,y)=-\nabla_Y\log\left(f_{Y|X}(y|x)\right)=\frac{\nabla_Y f_{Y|X}(y|x)}{f_{Y|X}(y|x)}. \tag{6.18}$$

Therefore, we can rewrite Eq. 6.16 using Fubini's theorem as:

$$\langle X^iY^jg(X,Y)\rangle=\int x^if_X(x)\left[\int y^j\partial f_{Y|X}(y|x)\right]dx. \tag{6.19}$$

The integral within brackets can be solved by parts:

$$\int y^j\partial f_{Y|X}(y|x)=y^jf_{Y|X}(y|x)\Big|_{-\infty}^{\infty}-j\int y^{j-1}f_{Y|X}(y|x)dy. \tag{6.20}$$

If $f_{Y|X}(y|x)\to 0$ for $y\to\pm\infty$, then:

$$\int y^j \partial f_{Y|X}(y|x) = -j \int y^{j-1} f_{Y|X}(y|x)dy, \tag{6.21}$$

and Eq. 6.19 becomes:

$$\begin{aligned}\langle X^i Y^j g(X,Y)\rangle &= -j \iint x^i y^{j-1} f_X(x) f_{Y|X}(y|x)dydx \\ &= -j \iint x^i y^{j-1} f_{X,Y}(x,y)dxdy \\ &= -j\langle X^i Y^{j-1}\rangle.\end{aligned} \tag{6.22}$$

For $a = a_0 + a_1 Y$ and $b = b_0 + b_1 Y$, we have:

$$g(X,Y) = \frac{2}{\sigma^2}\left[4X^3 - 2(b_0 + b_1 Y)X + a_0 + a_1 Y\right]. \tag{6.23}$$

Applying this result into Eq. 6.22, we get:

$$\begin{aligned}\langle X^i Y^j\rangle a_0 + \langle X^i Y^{j+1}\rangle a_1 - 2\langle X^{i+1} Y^j\rangle b_0 - 2\langle X^{i+1} Y^{j+1}\rangle b_1 + \\ + \frac{j}{2}\langle X^i Y^{j-1}\rangle \sigma^2 = -4\langle X^{i+3} Y^j\rangle\end{aligned} \tag{6.24}$$

Applying Eq. 6.24 to different (i,j) pairs, it is possible to obtain five different equations. The following snippet shows how to implement this in Python. For this code, we need some extra libraries such as pandas for data handling and datetime:

```
import numpy as np
import matplotlib.pyplot as pl
import pandas_datareader as pdr
import pandas as pd

from datetime import datetime
from scipy.stats import norm
```

We start by obtaining the datasets and interpolating the missing values:

```
s = pdr.get_data_yahoo(symbols="^GSPC",start=datetime(2019,10,4),
                                        end=datetime(2020,5,27))
idx = pd.date_range('10-04-2019','05-27-2020')
s.index = pd.DatetimeIndex(s.index)
s = s.reindex(idx)
s = s.interpolate()

dt = pd.read_csv("coro.csv",parse_dates=['Dia'])
```

Next, we normalize the data for better numerical accuracy. Also, we create some variables that will be used to compute Eq. 6.24:

```
ts = s['Adj Close']
tc = dt['coronavirus']

m = np.mean(ts)
s = np.std(ts)
ts = [(e-m)/s for e in ts]

m = np.mean(tc)
s = np.std(tc)
tc = [(e-m)/s for e in tc]

m = [[0 for a in range(5)] for b in range(5)]
p = [[0 for a in range(5)] for b in range(5)]
n = [0 for a in range(5)]
c = [0,0,1,1,0,3]    # coefficients on X^(i+c)
d = [0,1,0,1,-1,0]   # coefficients on Y^(j+d)

i = [0,0,1,1,0]
j = [0,1,0,1,2]
```

The averages $\langle X^a Y^b \rangle$ are then computed:

```
ac = 0
for X,Y in zip(ts,tc):
        for a in range(5):
                n[a] = n[a] + (X**(i[a]+3))*(Y**j[a])

                for b in range(5):
                        m[a][b] = m[a][b] + (X**(i[a]+c[b]))*(Y**(j[a]+d[b]))

        ac = ac + 1

# Compute averages
for a in range(5):
        p[a] = [1,1,-2,-2,0.5*j[a]]
        n[a] = n[a]/ac

        for b in range(5):
                m[a][b] = m[a][b]/ac
```

The coefficients a and b can now be found. We also obtain the fitting:

```
M = [[m[a][b]*p[a][b] for b in range(5)] for a in range(5)]
b = [[-4*n[a]] for a in range(5)]

coef = np.linalg.solve(M,b)

X = ts[0]
xsp = []
dt = 1E-3
for Y in tc:
        a = ao + a1*Y
        b = bo + b1*Y
        X = X + (4*X**3 - 2*b*X + a)*dt + norm.rvs(scale=np.sqrt(sg)*dt)
        xsp.append(X)

pl.plot(idx,ts)
pl.plot(idx,xsp)
pl.show()
```

The result for our example is shown in Fig. 6.10. Once the fitting is obtained, it

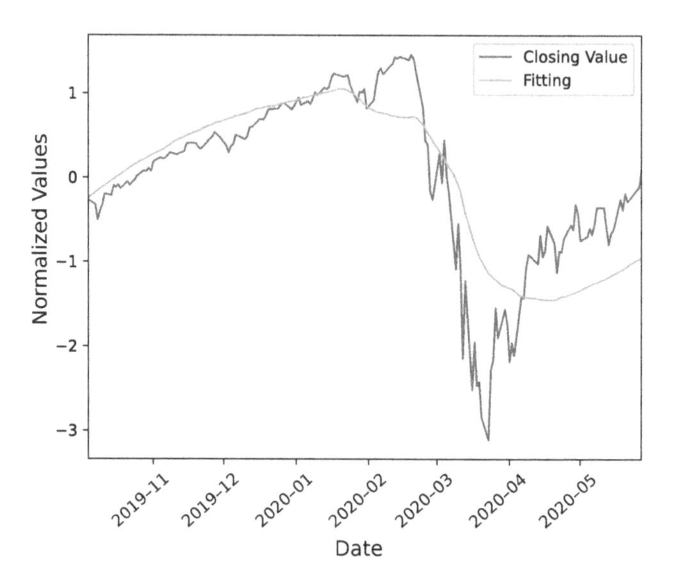

Figure 6.10: Closing values of the S&P500 index and a fitting using catastrophic dynamics

is possible to study different dynamics by comparing their coefficients. The good visual fitting observed in this particular example suggests that crises can be induced exogenously by variables other than monetary policies.

6.3 SELF-ORGANIZED CRITICALITY

We have just seen that crises may be induced exogenously. However, can a market walk by itself towards a crisis? In order to try to answer this question, we will analyze a sketch model that exaggerates this effect producing a *self-organized criticality*. But first, we must take a look at cellular automata, which is the tool that allows us to study this model.

6.3.1 CELLULAR AUTOMATA

An automaton is a tuple $A = (Z, S, N, f)$, where Z is a *lattice* of dimension d composed of discrete *cells*, $S = \{s_1, \ldots, s_r\}$ is a finite set of cell *states* (or values), $f : S^m \to S^p$ is a *transition function* (or update rule), and $N = (\mathbf{n}_1, \ldots, \mathbf{n}_m)$ is a finite neighborhood. The two most common neighborhoods are the von Neumann[11] given the set $N^r_{VN}(\mathbf{n}_0) = \left\{\mathbf{n}_k : \sum_{i=1}^{d} |(\mathbf{n}_k)_i - (\mathbf{n}_0)_i| \leq r\right\}$, and the Moore[12] given by the set $N^r_M(\mathbf{n}_0) = \{\mathbf{n}_k : |(\mathbf{n}_k)_i - (\mathbf{n}_0)_i| \leq r, \ \forall i = 1, \ldots, d\}$. These two neighborhoods are shown in Fig. 6.11.

The configuration of the automaton is described by a function $c : Z \to S$ that assigns a state to each cell. The future state of a cell is given by $c(\mathbf{n}_0)^{t+1} = f(\{\mathbf{n}_m | \mathbf{n}_m \in N(\mathbf{n}_0)\})$,

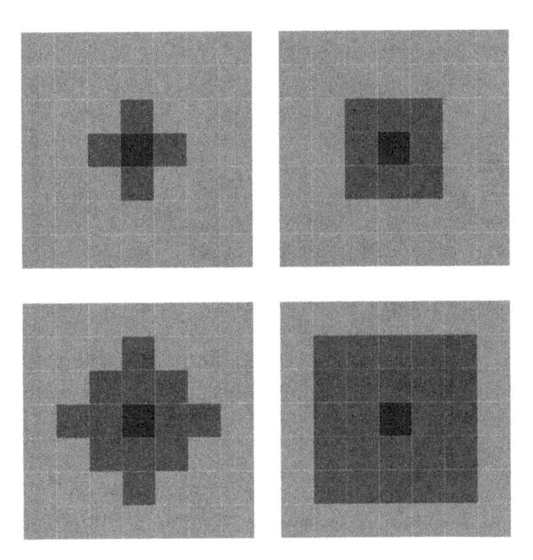

Figure 6.11: Left column: Von Neumann neighborhood, right column: Moore neighborhood, top row: $r = 1$, bottom row: $r = 2$; in all cases $\mathbf{n}_0$ is the block at the center

One of the most celebrated automaton is Conway's[13] *game of life* [149]. Given a short set of simple rules, this automaton is capable of producing elaborated emergent self-organized patterns [150, 151]. Moreover, it allows us to start thinking physics in terms of computable information [152].

The Bak[14]-Tang[15]-Wiesenfeld[16] (BTW) model [153] for self-organized criticality consists of an automaton equipped with a $Z \subset \mathbb{Z}^2$ lattice, states $S = \{1, 2, 3, 4\}$ and a von Neumann neighborhood. The transition rule consists of randomly picking a cell $\mathbf{n}_0$ and increasing its state by one:

$$c(\mathbf{n}_0)^{t+1} = c(\mathbf{n}_0)^t + 1. \tag{6.25}$$

Alternatively, one can start with a single pile of all elements that will be processed by the automaton and let it relax to its stable configuration.

If a cell has a state bigger than a threshold T corresponding to the coordination number of the lattice, the cell is said to become unstable and its state is reduced by the coordination number. Hence,

$$\begin{aligned} c(\mathbf{n}_0)^{t+1} &= c(\mathbf{n}_0)^t - T \\ c(\mathbf{n})^{t+1} &= c(\mathbf{n})^t + 1, \text{ where } \mathbf{n} \in N_{VN}^r(\mathbf{n}_0). \end{aligned} \tag{6.26}$$

When this situation occurs, the states of each of its neighbors are increased by one in a process known as *toppling*. If these neighbors also become unstable then the same procedure is executed to their neighbors recursively in a process referred to as *avalanche*. If a cell topples to the boundary of the lattice, it is lost. As illustrated in Fig. 6.12, the order in which the cells are chosen to topple does not change the final stable configuration of the grid. Hence, this model is known as *Abelian sandpile*

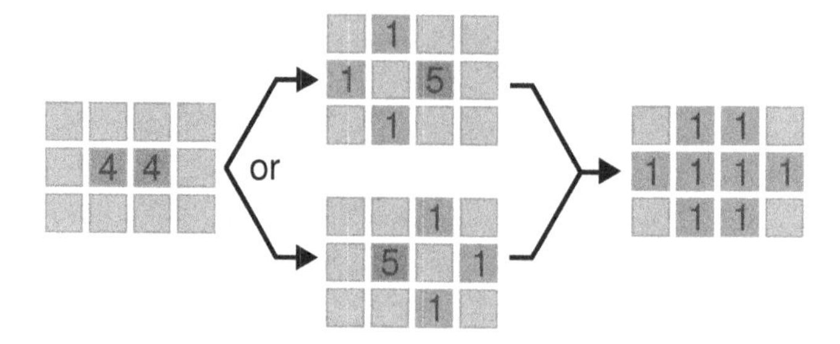

Figure 6.12: Starting with two unstable cells, the final stable grid configuration does not depend upon the order in which the cells are toppled in the Abelian sandpile model

model.

Abelian Group

A group can be defined as a set A equipped with a binary operation $\diamond$ such that for any two elements $a,b \in A \rightarrow a \diamond b \in A$ (closure). Also, this operation has to satisfy three axioms:

Associativity: $\forall a,b,c \in A \rightarrow (a \diamond b) \diamond c = a \diamond (b \diamond c)$
Identity: $\exists e \in A | \forall a \in A,\ e \diamond A = A \diamond e$
Invertibility: $\exists b \in A | \forall a \in A,\ a \diamond b = b \diamond a = e$

if this operation also satisfies:

Commutativity: $\forall a,b \in A \rightarrow a \diamond b = b \diamond a,$

then we say that this is an *Abelian group*[a].

[a]Niels Henrik Abel (1802–1829) Norwegian mathematician.

6.3.2 SIMULATIONS

The Abelian sandpile model can be simulated with the following snippet:

```
L = 500
grid = np.zeros((L,L))

xo = L » 1
yo = L » 1
grid[yo,xo] = 3E5

while(np.max(grid) >= 4):
        # Find unstable cells
        to_topple = (grid >= 4)

        # Toppling
        grid[to_topple] -= 4
        grid[1:,:][to_topple[:-1,:]] += 1
        grid[:-1,:][to_topple[1:,:]] += 1
        grid[:,1:][to_topple[:,:-1]] += 1
        grid[:,:-1][to_topple[:,1:]] += 1

        # Spillover
        grid[0,:] = 0
        grid[L-1,:] = 0
        grid[:,0] = 0
        grid[:,L-1] = 0
```

The result of this simulation is shown in Fig. 6.13. It is interesting to observe the recurring patterns in the figure. The system spontaneously evolves (self-organizes) to this critical state where the correlation length of the system diverges producing a fractal geometry. The correlation as a function of the interaction step is shown in the data collapse of Fig. 6.14. The finite size scaling was performed dividing the correlation length by L^C, where C was estimated to be around 2. Exactly the same was done with the x-axis. Until a certain number of interaction steps, the grid is dominated by the initial column. After that point, the correlation length depends on the interaction step as a power law until it eventually reaches a stable value (ξ_o) because of the finite size of the grid. The stable correlation length as a function of the grid size is shown in the inset and is also a power law, which indicates that it diverges for infinite grid sizes.

The correlation length can be calculated with the following snippet using the Wiener-Khinchin theorem (see Sec. 2.4.1.1):

Figure 6.13: Simulation of the Abelian sandpile model on a 500×500 grid and an initial pile of 3×10^5 elements

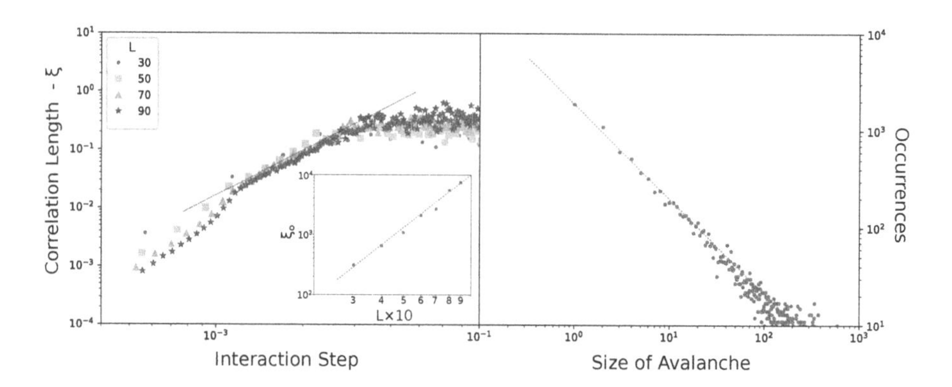

Figure 6.14: Left: Data collapse for the correlation length (ξ) as a function of the interaction step for different lattice sizes. The inset shows the stable correlation (ξ_o) as a function of the lattice size. Right: Number of occurrences as a function of the size of avalanches using a grid size of 80 units

```
def cor(M):
        L = np.shape(M)[0]»1

        F = np.fft.fft2(M)
        S = F*np.conj(F)/np.size(M)
        T = np.fft.ifft2(S).real

y = np.log(T[L,1:L])
x = range(len(y))

return 1.0/abs(np.cov(x,y)[0][1]/np.var(x))
```

Another elegant way to simulate this automaton is using recurrence as shown in the following snippet:

```
def place(array,x,y,L):
        ac = 0
        if (x >= 0 and x < L) and (y >= 0 and y < L):
                array[x][y] = array[x][y] + 1

                # Toppling
                if array[x][y] >= 4:
                        ac = ac + 1
                        array[x][y] = array[x][y] - 4
                        array,t1 = place(array, x+1, y, L)
                        array,t2 = place(array, x-1, y, L)
                        array,t3 = place(array, x, y+1, L)
                        array,t4 = place(array, x, y-1, L)
                        ac = ac + t1 + t2 + t3 + t4
                return array, ac
```

Using this approach, it is possible to easily find the number of avalanches and the mass they move. Also, the snippet considers a constant pouring of grains distributed according to a binomial distribution:

```
def dist(L):
        sand = [[0 for n in range(L)] for m in range(L)]
        massa = []
        tops = [0 for n in range(50000)]

        N = 500*L+100
        for it in range(N):
                x = np.random.binomial(L,0.5)
                y = np.random.binomial(L,0.5)
                sand,t = place(sand, x, y, L)

                massa = np.append(massa, np.mean(sand))
                tops[t] = tops[t] + 1

return massa, tops
```

The result of this simulation is shown in Fig. 6.14 for a grid size of 80 units. The dotted line indicates that the distribution of avalanches also follows a power law. This resembles experimental evidence found in earthquakes. For instance, it has been observed that the rate of secondary quakes (or replicas) after a main seismic event follows Omori's[17] law [154]:

$$n(t) = \frac{k}{c+t}, \tag{6.27}$$

where k and c are adjustable constant. The quake counting is then given by:

$$\begin{aligned} N(t) &= \int_0^t n(t)dt = \int_0^t \frac{k}{c+t}dt \\ &= k\log(c+t)|_0^t = k\left[\log(c+t) - \log(c)\right] \\ &= k\log\left(1+\frac{t}{c}\right). \end{aligned} \tag{6.28}$$

Omori's law is generalized by Utsu's[18] law [155] as:

$$n(t) = \frac{k}{(c+t)^p}, \tag{6.29}$$

where p ranges between 0.7 and 1.5. The accumulated number of events is given by:

$$N(t) = k\frac{c^{1-p} - (t+c)^{1-p}}{p-1}, \tag{6.30}$$

for $p \neq 1$. For $p > 1$, $N(t \to \infty) = k/\left[(p-1)c^{p-1}\right]$, and for $p \leq 1$, $N(t \to \infty) \to \infty$.

Another power behavior that appears in seismology is the Gutenberg[19]-Richter's[20] [156] law. This law relates the number N of events with magnitude equal or higher than a value M:

$$N = 10^{a-bM}, \tag{6.31}$$

where a and b are constants.

You might be asking yourself what seismology has to do with economics, but, surprisingly, many economic crises seem to mimic the behavior of earthquakes and follow these laws! (see for instance: [63], [157], and [158]) In other models, the returns are assumed to follow the size s of avalanches caused by agents responding to some stimulus:

$$\log(P_t) - \log(P_{t-1}) \propto \varepsilon_t s(t), \tag{6.32}$$

where ε_t is a Wiener process. Hence, the distribution of returns follows the distribution of avalanches. The latter has a power-law behavior and, consequently, shows an absence of autocorrelation.

Notes

[1]John Law (1671–1729) Scottish economist.

[2]Term developed by Friedrich Freiherr von Wieser (1851–1926) Austrian economist, advisee of Eugen Böhm Bawerk and adviser of Luwig von Mises, Friedrich Hayek and Joseph Schumpeter.

[3]Jean-Baptiste Say (1767–1832) French economist.

[4]Ian Nicholas Stewart (1945–) British mathematician.

[5]René Frédéric Thom (1923–2002) French mathematician.

[6]Erik Christopher Zeeman (1925–2016) British mathematician.

[7]Lev Davidovich Landau (1908–1968) Russian physicist, advisee of Niels Bohr and adviser of Alexei Alexeyevich Abrikosov and Evgeny Lifshitz. Landau won many prizes including the Nobel prize in physics in 1962.

[8]Aleksandr Mikhailovich Lyapunov (1857–1918) Russian mathematician, advisee of Pafnuty Chebyshev.

[9]Geronimo Cardano (1501–1576) Italian polymath.

[10]Josiah Willard Gibbs (1939–1903) American polymath, adviser of Irvin Fisher and Lee de Forst, among others.

[11]John von Neumann/Margittai Neumann János Lajos (1903–1957) Hungarian mathematician.

[12]Edward Forrest Moore (1925–2003) American mathematician.

[13]John Horton Conway (1937–2020) English mathematician winner of many prizes including the Pólya Prize in 1987.

[14]Per Bak (1948–2002) Danish physicist who coined the term *self-organized criticality*.

[15]Tang Chao (1958–) Chinese physicist, advisee of Leo Kadanoff.

[16]Kurt Wiesenfeld, American physicist.

[17]Fusakichi Omori (1868–1923) Japanese seismologist.

[18]Tokuji Utsu (1928–) Japanese geophysicist.

[19]Beno Gutenberg (1889–1960) German seismologist.

[20]Charles Francis Richter (1900–1985) American physicist.

7 Games and Competitions

Games are mathematical models that describe the interactions among rational players that try to maximize their *utility functions*. Game theory, the study about those models, finds applications in many fields such as economics, political science, sociology, and even philosophy. This is an old field that dates back to the XVIII century, but a more formal description was only given in 1838 by Cournot in a treatise about duopolies [10]. On the other hand, the unification of many concepts in game theory appears only recently in 1928 in books by von Neumann [159] and Morgenstern[1] [160]. In this chapter we will study some aspects of game theory that can be applied to economic problems through the study of some examples.

Utility

Consider the game of flipping a coin. If it falls tails you double the prize (you start with \$1) and the game continues until you obtain heads. How much would you be willing to pay to play this game? Considering that the agents are perfectly rational, they would choose to pay any amount because the expected prize of this game is:

$$P = 2\times\frac{1}{2} + 4\times\frac{1}{4} + \ldots = \sum_{i=1}^{\infty}\left(\frac{1}{2}\right)^{n} 2^{n} \to \infty. \tag{7.1}$$

Nonetheless, only a few individuals would pay more than a few dollars to play it. This is known as the *Saint Petersburg paradox*[2].

The explanation for this behavior lies, according to Daniel Bernoulli, in the *utility* that the good or service renders, instead of its rational price. Certainly, people value money, hence the *utility function* has to increase as a function of quantity. Also, it would be strange if a small extra amount of money would suddenly lead to a jump in utility. Therefore, it also has to be continuous. Finally, after you have a significant quantifiable wealth, a few extra dollars would not cause a significant increase in utility and the curve must flatten for sufficient big amounts. This is known as *diminishing marginal utility*. Functions like the logarithm show those properties.

Based on this description, von Neumann and Morgenstern stated that, in the presence of risk, rational agents act to maximize their expected value of utility [160].

DOI: 10.1201/9781003127956-7

7.1 GAME THEORY THROUGH EXAMPLES

7.1.1 THE EL FAROL BAR PROBLEM

This is a problem proposed in 1994 by William Arthur[3] [161]. Suppose that N people go out to a bar[a] regularly every Thursday night. If many people decide to go to the same bar at a specific night, the experience of these agents will not be so great (let's consider that their experiences are solely dictated by the number of people at the bar).

Given a significant big number of agents, it is not possible for them to communicate with each other to check who is going to the bar. Furthermore, the agents cannot know all direct and indirect consequences of their actions. One player deciding to go to the bar, may affect another player's decision. As Hayek pointed out, knowledge is dispersed [5]. Therefore, they do not have complete access to the information about the optimization task (logical omniscience) [162], hence a *perfect rational* strategy cannot be composed. Conversely, the agents can make some *inductive reasoning* and use their experience to judge whether it is a good strategy to go to the bar or not. Friedman[4] argued that the agents behave *as if* [163] they were rational trying to maximize their utility but under a set of constraints that force the adoption of imperfect rational choices.

This breaks the concept of *homo economicus* developed by many neoclassical authors such as Mill[5] [164], Jevons[6] [165] and Knight [113]. They considered self-interested agents trying to maximize utility functions that are perfectly known by all players. Rather, the agents have a *bounded rationality*[b] [16]. Their actions still have some level of rationality, but are mostly based on heuristics. Even worse, some of these heuristics may include biases such as imitation, preferential attachment and satisficing. The latter, for instance, consists of trying different options until finding one that reaches a specific threshold.

Kahneman and Tversky developed a *cumulative prospect theory* [166] which tries to clarify how humans make choices under this complex scenario. According to this theory, humans may adopt references and compare possible results of their actions against this reference. The further the outcomes of a strategy deviates from these references (for good or worse) the more the sensitivity of these agents reduces. Also, the agents tend to be more sensitive to losses than gains of the same magnitude, and, often, underweight high-probability events and overweight low-probability events.

We will proceed with a mathematization of the El Farol problem, known as minority game. From there, we will formalize game theory and apply it to some economic problems. The chapter will then close with another approach for competitions known as the prey-predator model.

---

[a]The bar chosen by Arthur was the *El Farol* located in Santa Fe, NM. Hence, this problem was popularized as 'the El Farol Bar problem'.

[b]Proposed by Herbert Simon as described in Chapter 1.

7.1.1.1 Minority Game

The minority game[7] [167] is a mathematization of the classic El Farol Bar problem. Here, however, we will map the same game structure to the spot market [168], where the agents can either buy (-1) or sell $(+1)$ a specific stock.

As in earlier models [169], the game begins with an odd number N of agents that have a memory about the past m events related to that stock. Therefore, there is a combination of N^m arrangements for their past actions. Furthermore, since they can only either buy or sell the stock, there are 2^{N^m} possible strategies. Each agent have a repertoire of s strategies. Thus, we will create a class for these agents with a constructor that initializes the board under this scenario:

```
class agent:
    def __init__(self,m,s):
        self.m = m
        self.s = s
        self.r = [rd.choices([-1,1],k=2**m) for a in range(s)]
        self.p = [rd.random() for a in range(s)]

    def update_points(self,pos,W,A,N):
        for strategy in range(self.s):
            r_a = self.r[strategy]
            self.p[strategy] = self.p[strategy] - (r_a[pos]*A)/(2**self.m)

    def a(self,pos):
        winning_strategy = np.argmax(self.p)
        r_beta = self.r[winning_strategy]
        return r_beta[pos]
```

We will use NumPy for numerical calculations, PyPlot for plotting, and Random for pseudo-random numbers. Therefore, we must include:

```
import numpy as np
import random as rd
import matplotlib.pyplot as pl
```

At each round t, each agent i takes an action $a_i[t] = +1 \vee -1$ based on the memory about the past winning groups and his or her strategy book. For example, if the strategy book of an agent is given by Tab. 7.1 and the past winning groups

Table 7.1
A Possible Strategy Book for an Agent with Two Strategies

History	Strategy #1	Strategy #2
−1,−1,−1	−1	−1
−1,−1,+1	−1	+1
−1,+1,−1	+1	−1
−1,+1,+1	−1	−1
+1,−1,−1	+1	+1
+1,−1,+1	−1	+1
+1,+1,−1	−1	+1
+1,+1,+1	+1	−1

were $-1,+1,-1$, then the agent will either sell the stock ($a_i[t] = +1$) according to his strategy #1 or buy the stock ($a_i[t] = -1$) according to his strategy #2. Let's say that his strategy #2 has been giving better predictions. Therefore, the ith player buys the stock ($a_i[t] = -1$). This is executed by the function a in the agent class. In our code, self.p is a vector whose elements are the points accumulated by each strategy and r_β is the strategy with the best performance. The variable *pos* gives the line in the strategy book corresponding to the memory. The overall scheme of the minority game is shown in Fig. 7.1.

After all agents have placed their bids, the minority group is elected as the winning group. For instance, if most agents decide to sell the stock but a minority decides to buy it, the latter will benefit from low prices resulting from the competition among sellers. This is computed in the function *game* below as $W[t] = -sign\left(\sum_i a_i[t]\right)$. This function is initialized with the agents and some random memory *bits*. Then 500 rounds are executed wherein μ is the line position of the memory, A_{space} is a space with the action of all players, and A is their sum. After the winning group is computed, it is pushed to the memory, and the scores of all strategies of all players are updated. This is done by the function *update_points* in the agent class. It uses the *linear payoff scheme*: $p_{i,s}[t+1] = p_{i,s}[t] - a_{i,s}[t] \cdot A[t]/2^m$, where the 's' index indicates the strategy while 'i' indicates the agent. This constitutes a *reinforced learning* structure where the agents adjust their behaviors to recent events.

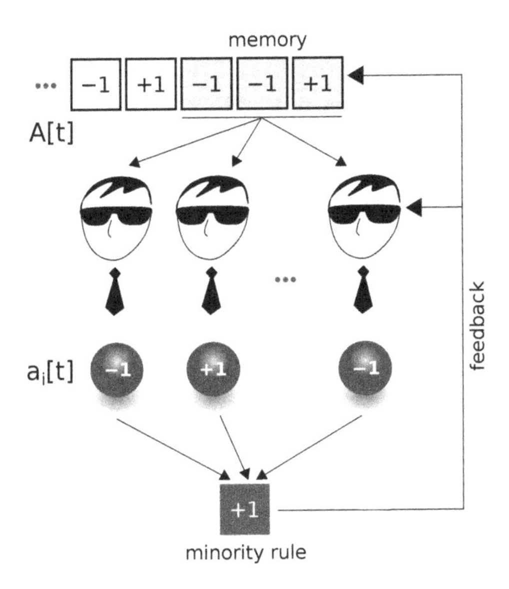

Figure 7.1: Overall scheme of a minority game: Agents take actions based on the memory of the game; the minority group wins and this information is fed back to the history of the game

```
def game(N,numStrategies,memory):
    player = [agent(memory,numStrategies) for i in range(N)]
    bits = rd.choices([-1,1],k=memory)

    # Loop over interactions
    At = []
    S = [0 for i in range(2**memory)]
    P = [0 for i in range(2**memory)]
    for it in range(500):
        # Calculate attendance
        mu = val(bits)
        A_space = [player[i].a(mu) for i in range(N)]
        A = np.sum(A_space)
        At.append(A)

        # Winning Group
        W = -np.sign(A)
        bits = push(bits,W)

        # Update virtual score
        for i in range(N):
            player[i].update_points(mu,W,A,N)
```

Finally, we calculate the information still inside this function and return the results:

```
        # Calculate information
        for nu in range(2**memory):
            if (mu == nu):
                S[nu] = S[nu] + A
                P[nu] = P[nu] + 1

    # Complete calculation of information
    H = 0
    for n in range(2**memory):
        if (P[n] > 0):
            H = H + (S[n]/P[n])**2

    return np.var(At), H/2**memory
```

In this snippet, we used a function *val* that returns an integer for a binary sequence:

```
def val(x):
    r = 0
    for i in range(len(x)):
        if (x[i] == 1):
            r = r + (1 << i)
    return r
```

Also, we need to keep a memory of the winning group. Data is entered into this list via a *push* function:

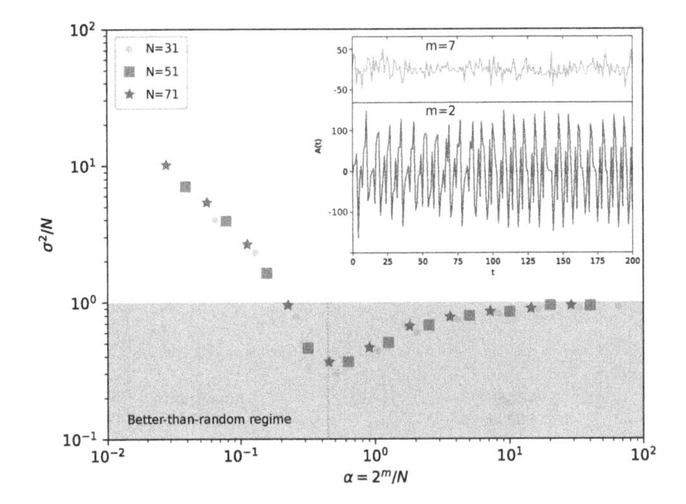

Figure 7.2: Normalized variance of A as a function of the control parameter α for different number of agents; the inset shows typical progressions of A for different memory sizes

```
def push(v,x):
    for i in range(len(v)-2,-1,-1):
        v[i+1]=v[i]
    v[0] = x

    return v
```

The variance of A (fluctuations of A around its mean) normalized by the number of agents calculated at the end of a game with 500 rounds is shown in Fig. 7.2 for agents with only two strategies. The independent variable is the control parameter $\alpha = 2^m/N$, which produces the data collapse shown in the figure for any number of agents.

If the agents took completely random choices, we would expect a volatility of A always around $1/2\left[(-1-0)^2+(1-0)^2\right] = 1$. For low values of the control parameter, though, the volatility is very large, implying that the losing group is much larger than $N/2$. However, as α increases, the volatility goes below the perfect random regime, suggesting that the agents are capable of achieving coordination and reach a point where the least resources are wasted (around $\alpha_c \approx 0.5$). Large values of α leads asymptotically to the perfect random regime.

In the region where $\alpha < \alpha_c$, the minority group is unable to predict the outcome of the game based on its history. Therefore, this corresponds to an efficient market

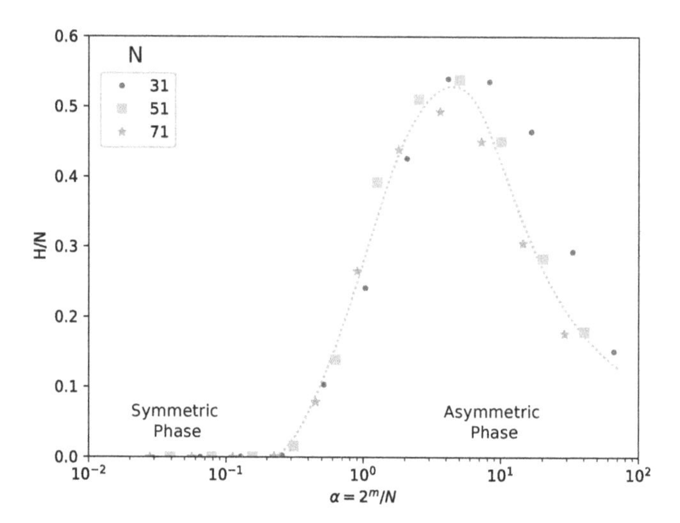

Figure 7.3: Normalized information for minority games with different number of agents as a function of the control parameter. Data was averaged with 20 samples. A dotted line is an interpolation of the data added to guide the eye

regime. The opposite happens for $\alpha > \alpha_c$. In this region, the agents can use effectively the information available to them to coordinate and predict the outcome of the game. This corresponds to an inefficient market regime. It is possible to further explore this phase transition by studying its *predictability* [170], defined as the normalized sum of conditional expectations:

$$H = \frac{1}{2^m} \sum_{\nu=1}^{2^m} \langle A | \mu = \nu \rangle^2, \tag{7.2}$$

where μ is the memory of the past m results.

In the efficient market regime, the conditional expectation of A must be zero independent on μ since both minority groups have the same likelihood. Therefore, this regime is also known as the *symmetric phase*. In the inefficient market regime, one minority group is more likely than another. Therefore, $H \neq 0$ and this implies that there is information available for the agents in A. Accordingly, this is known as the *asymmetric phase*. In case any new agent enters the game, he can exploit, and consequently reduce the information contained in A. As the control parameter increases, though, the system is carried towards the perfect random regime and we should see the information moving towards zero. This behavior is shown in Fig. 7.3.

7.1.2 COOPERATIVE GAMES

A game can be cooperative (or coalitional) if binding agreements are possible and the agents can benefit by cooperating. A game G is described by the ordered set

Table 7.2
A Cooperative Game Where Agents Want to Buy the Biggest Field in the Market, but They Have a Fixed Money Supply

Field size [acres]	Price	Player	Money
5	\$70,000	1	\$40,000
7.5	\$90,000	2	\$30,000
10	\$110,000	3	\$30,000

(N, v) where $N = \{1, \ldots, \text{number of players}\}$ is the set of agents and $v : 2^N \to \mathbb{R}$ is a characteristic function that maps every coalition of agents to a payoff.

Let's consider, for instance, a game with three agents. The power set of N is given by $\{\emptyset, \{1\}, \{2\}, \{3\}, \{1,2\}, \{1,3\}, \{2,3\}, \{1,2,3\}\}$. The characteristic function maps each subset to a specific value. Let's say these three players decide to buy a common good, the biggest field available in the market to cultivate lemons, for example. There are also three small fields being advertised on the market with sizes and prices described in Tab. 7.2. On the other hand, the players have a limited money supply also shown in the table. Alone, no player can purchase any of these fields. Players #1 and #2 together can afford field #1 and so does player #1 together with player #3. Player #2 together with player #3 cannot purchase any field. All players together can purchase the second field. Therefore, we have the results indicated in Eq. 7.3.

$$\begin{aligned} v(\emptyset) = v(\{1\}) = v(\{2\}) = v(\{3\}) &= 0 \\ v(\{1,2\}) = v(\{1,3\}) &= 5 \\ v(\{2,3\}) &= 0 \\ v(\{1,2,3\}) &= 7.5. \end{aligned} \tag{7.3}$$

We say that the game is monotone if $v(C_2) \geq v(C_1),\ \forall\, C_1 \subseteq C_2$, as in our example. Also, a game is said to be superadditive if $v(C_1 \cup C_2) \geq v(C_1) + v(C_2),\ \forall\, C_1 \cap C_2 = \emptyset$, again as in our example. In this case, two coalitions can merge without any loss and form a *grand coalition*. For example, $\{2\}$ can merge with $\{1,3\}$ to form $\{1,2,3\}$. The payoffs before merging were 0 and 7.5. After merging, however, the grand coalition has a payoff of 7.5.

The outcome of the game is given by the ordered pair (S, x), where $S = (C_1, \ldots, C_k)$ is the coalition structure, the partition of N into coalitions such that $\cup_{i=1}^{k} C_i = N$ and $C_i \cap C_j = \emptyset,\ \forall\ i \neq j$, and $x = (x_1, \ldots, x_n)$ is the payoff vector, which distributes the value of each partition such that $x_i \geq 0,\ \forall\ i \in N$, and $\sum_{i \in C} x_i = v(C),\ \forall\, C \in S$.

In our example, an outcome could be $((\{1,2,3\}, \{1,3\}), (2,3,2.5,2,3))$. Note that $2 + 3 + 2.5 = 7.5 = v(\{1,2,3\})$ and $2 + 3 = 5 = v(\{1,3\})$. If $x_i \geq v(\{i\}),\ \forall\ i \in N$

Table 7.3
Possible Permutations for the Cooperative Game of Table 7.2

π	$\mathbf{S}_\pi(\mathbf{1})$	$\mathbf{S}_\pi(\mathbf{2})$	$\mathbf{S}_\pi(\mathbf{3})$
$\pi_1 = (1,2,3)$	$\emptyset$	$\{1\}$	$\{1,2\}$
$\pi_2 = (1,3,2)$	$\emptyset$	$\{1,3\}$	$\{1\}$
$\pi_3 = (2,1,3)$	$\{2\}$	$\emptyset$	$\{2,1\}$
$\pi_4 = (2,3,1)$	$\{2,3\}$	$\emptyset$	$\{2\}$
$\pi_5 = (3,1,2)$	$\{3\}$	$\{3,1\}$	$\emptyset$
$\pi_6 = (3,2,1)$	$\{3,2\}$	$\{3\}$	$\emptyset$

then the agents are being rational in being part of this coalition and we say that this outcome is an *imputation*.

How should the players divide the field after purchasing it? What if the field is split as $(2,2,3.5)$? Clearly, players #1 and #2 could profit more from forming a coalition, purchasing the field with 5 acres and dividing it equally. Therefore, $(2,2,3.5)$ is not a stable solution. We can search those stable outcomes that no coalition would like to deviation from. These are known as the *core* of the game. Mathematically, it is given by the $core(G) = \{(S,x)|x(C) \geq v(C),\ \forall\, C \subseteq N\}$. The distribution $(2.5,2.5,2.5)$ is in the core since the players cannot get more on their own.

Although a stable distribution may appear in the core, it may not be fair. A fair payment would reward each player according to his or her contribution. Let's take, for instance, the distribution $(7.5,0,0)$. It is in the core, however it is as unfair as it can get.

In order to find a fair distribution, we must look for a solution that ensures that all profit is distributed among the players. The marginal contribution of player i to the coalition is given by $v(\{1,\ldots,i-1,i\}) - v(1,\ldots,i-1)$. This, however, is sensitive to the order at which the player appears in the coalition. The average marginal contribution fixes this problem by considering all possible orderings. The latter can be found with the set $S_\pi(i) = \{j \in N|\pi(j) < \pi(i)\}$ of the predecessors of i in the permutation π of N. For instance, for $\pi = (1,4,2,5)$, then $S_\pi(2) = \{1,4\}$. Therefore, the marginal contribution can be written as $\Delta_\pi(i) = v(S_\pi(i) \cup i) - v(S_\pi(i))$. The average marginal contribution is known as Shapley[8] value given by Eq. 7.4 [171].

$$\phi_i = \frac{1}{N!} \sum_{\pi \in \Pi_n} \Delta_\pi(i). \tag{7.4}$$

In our example, we have the permutations listed in Tab. 7.3 giving the marginal contributions shown in Eq. 7.5. Hence, the average marginal contributions are $\phi = (4.17,1.67,1.67)$, which means that in a fair division of the field, player #1 receives 4.17 acres and the other players receive 1.67 acres each.

$$\begin{aligned}
&\Delta_{\pi_1}(1) = v(\{1\}) - v(\{\emptyset\}) = 0, && \Delta_{\pi_2}(1) = v(\{1\}) - v(\{\emptyset\}) = 0,\\
&\Delta_{\pi_1}(2) = v(\{1,2\}) - v(\{1\}) = 5, && \Delta_{\pi_2}(2) = v(\{1,2,3\}) - v(\{1,3\}) = 2.5,\\
&\Delta_{\pi_1}(3) = v(\{1,2,3\}) - v(\{1,2\}) = 2.5, && \Delta_{\pi_2}(3) = v(\{1,3\}) - v(\{1\}) = 5,\\
\\
&\Delta_{\pi_3}(1) = v(\{1,2\}) - v(2) = 5, && \Delta_{\pi_4}(1) = v(\{1,2,3\}) - v(\{2,3\}) = 7.5,\\
&\Delta_{\pi_3}(2) = v(\{2\}) - v(\{\emptyset\}) = 0, && \Delta_{\pi_4}(2) = v(\{2\}) - v(\{\emptyset\}) = 0,\\
&\Delta_{\pi_3}(3) = v(\{1,2,3\}) - v(\{1,2\}) = 2.5, && \Delta_{\pi_4}(3) = v(\{2,3\}) - v(\{2\}) = 0,\\
\\
&\Delta_{\pi_5}(1) = v(\{1,3\}) - v(\{3\}) = 5, && \Delta_{\pi_6}(1) = v(\{1,2,3\}) - v(\{2,3\}) = 7.5,\\
&\Delta_{\pi_5}(2) = v(\{1,2,3\}) - v(\{1,3\}) = 2.5, && \Delta_{\pi_6}(2) = v(\{2,3\}) - v(\{3\}) = 0,\\
&\Delta_{\pi_5}(3) = v(\{3\}) - v(\{\emptyset\}) = 0, && \Delta_{\pi_6}(3) = v(\{3\}) - v(\{\emptyset\}) = 0.
\end{aligned} \tag{7.5}$$

It is known that the Shapley value is the only payoff scheme that is efficient ($\sum_i \phi_i = v(N)$), accommodates null players ($v(C \cup i) = v(C)\ \forall\, C \in 2^{\Omega}$), is symmetric ($v(C \cup i) = v(C \cup j)\ \forall\, C \in 2^{\Omega}$), and is additive ($\phi_i(G_1 + G_2) = \phi(G_1) + \phi(G_2)$) [171].

7.1.3 ULTIMATUM GAME

This is a game proposed in 1982 [172][9]. In this one-shot game, a player (proposer) proposes to split an amount of money with another player (receiver) upon the acceptance of some consequence of either accepting or rejecting the offer. The money is split in case the receiver accepts the proposal, but both players receive nothing if the receiver does not agree with the deal. This game tries to capture the human behavior under specific circumstances and hence is an example of *behavioral game theory*, other examples include the dictator game and the public goods game. Let's explore more of this game using the *extensive form* shown in Fig. 7.4.

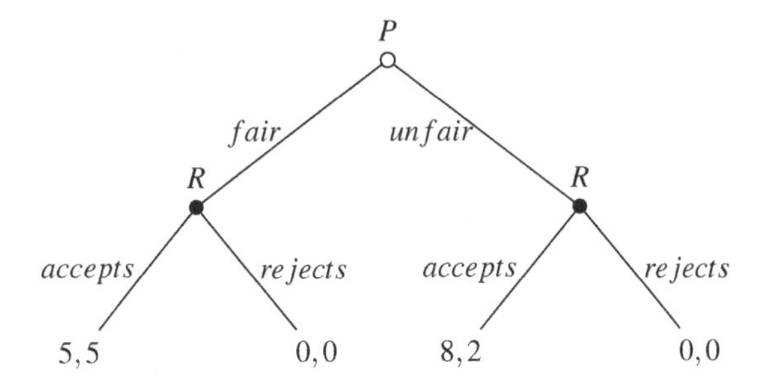

Figure 7.4: Example of an extensive form for the ultimatum game

The extensive form shown in the figure is commonly used to model sequential (or dynamical) games where one player takes action before the others. It is basically a decision tree that begins with the proposer offering a fair or unfair deal. The receiver can either accept or reject the proposal. For each of these possibilities there are utilities associated with the outcome of the game. For instance, if the receiver rejects the deal, both players receive no money and this is displayed as 0,0. If the proposer proposes a fair deal and the receiver accepts it, then both profit equally 5,5. On the other hand, if the deal is unfair and the receiver accepts it, the proposer receives a greater amount 8,2.

We say that the extensive-form presented is *complete* since it shows all players, their every possible actions, what they know for every move, and the payoffs of each player as a result of their actions. Also, the receiver only takes action after knowing exactly the proposer's action. When this happens, we say that the game has *perfect information.* The opposite case would be a game where the receiver cannot differentiate between the proposer's actions. This would happen, for instance, if they would take decisions simultaneously or if the proposer would hide his actions and demand a move from the receiver. This *imperfect information* could be denoted by a dashed line between both R nodes and this would constitute an *information set.*

7.1.3.1 Nash Equilibrium

Let's analyze some outcomes of this game. First, let's assume that the proposer made a fair proposal and the receiver only accepts a fair proposal. In this case each one would have a payoff of 5. Now that all players know each other's strategies, if the receiver changes his mind, his payoff would be zero. The same happens if the proposer changes his mind. Therefore, the proposer proposing a fair deal and the receiver only accepting a fair deal is an equilibrium situation where no player profit by changing strategies. The same happens if the proposer proposes an unfair deal and the receiver only accepts an unfair offer, and if the proposer proposes an unfair deal and the receiver accepts any offer.

Nash[10] developed this equilibrium concept in 1950 for non-cooperative games, hence it is known today as *Nash equilibria.* These situations are different from one where a player chooses the optimal move no matter how the other players act. The latter is known as *dominant strategy.* Nash equilibria correspond to optimal states where all players make optimal moves considering the moves of their opponents.

We can formalize Nash equilibria by defining a game as a triple $G = (N, \{S_i\}, \{u_i\})$, $i \in N$, where $N = \{1, 2, \ldots, n\}$ (with $n \geq 2$) is a set of players, S_i is the strategy set for the ith player, and $u_i : \prod_i S_i \to \mathbb{R}$ is the payoff (or utility function, a non-empty and finite set) for each strategy profile. Under this formalization, the *pure Nash equilibria* corresponds to strategy profiles $s^* \in S$ such that:

$$u_i(s_i^*, s_{-i}^*) \geq u_i(s_i, s_{-i}^*), \; \forall s_i \in S_i, i \in N, \tag{7.6}$$

where s_{-i} corresponds to a strategy set where the ith component is removed: $s_{-i} \equiv (s_1, s_2, \ldots, s_{i-1}, s_{i+1}, \ldots, s_n)$, or $S_{-i} \equiv \times_{j \in N \setminus \{i\}} S_j$. Also, we used the substitution $(y_i, x_{-i}) = (x_1, x_2, \ldots, x_{i-1}, y_i, x_{i+1}, \ldots, x_n)$.

In economics, we can say that the Nash equilibria is equivalent to a *self-enforcing contract* where no agent has rational incentives to break the agreement. On the other hand, Nash equilibria correspond to zero-temperature solutions where fluctuations are frozen [173]. This is a deterministic result where stochasticity is neglected and it is assumed that all agents have complete and perfect knowledge about the game.

Notwithstanding, experimental results in the ultimatum game deviate from these equilibrium points [174, 175]. Similar deviations are observed in the public goods game [176] and in the dictator game [177].

7.1.4 PRISONER'S DILEMMA

Imagine a situation where two prisoners that committed some crime together are placed in different cells and are unable to communicate with one another. As a prosecution strategy, the law enforcement agency offers some reward for each of them if they give some evidence that helps to convict the fellow prisoner (defect). If both defect, both stay two years in prison. If one defects but the other cooperate (remain silent), then the informer is released and the other is sentenced to three years in jail. But if both cooperate, then each stays one year in prison since there is no sufficient evidences to convict either. This game, invented in 1950 by Flood[11] illustrates the loss that might happen in a cooperation between rational players [178][c].

In order to explain how this game operates, let's translate it to an economic setting. Let's consider two countries that decide to undertake international trade. They have a similar dilemma: if they levy taxes on imported goods (defect) they protect their domestic companies, but the whole population loses with higher prices. Therefore, the best situation for both is a free market situation where they cooperate and do not impose taxes on each other. But what if one of them levy import taxes but expects the other to cut import taxes? This would be a dominant strategy for the former, but a terrible scenario for the other. Finally, both can impose taxes on each other trying to protect their local businesses, but this implies higher prices for both populations and possibly worse products. The payoff matrix (or *normal form*) for this game is shown in Tab. 7.4. Note that it is exactly the same payoff described by the case with two prisoners with a bias of 4. The first element inside each parenthesis is the payoff for country A whereas the second is the payoff for country B.

Table 7.4
Payoff Matrix for the Prisoner's Dilemma Adapted to the Case of International Trade

		Country B	
		coop.	*def.*
Country A	*coop.*	(3,3)	(1,4)
	def.	(4,1)	(2,2)

Nash equilibrium in this case is a situation where both defect and levy taxes. Knowing that country B is imposing taxes, it makes no sense for country A to lower

[c]The term *Prisoner's Dilemma* was coined and popularized by Tucker[12] in 1992 [179].

its import taxes, since this would produce a payoff of 1. The reciprocal is true. Knowing that country A is imposing taxes, it makes no sense for country B to lower taxes since it also produces a lower payoff. Here is the dilemma: the Nash equilibrium is both defecting and there is no rational reason why each country should change its strategy. However, it would be much better for both countries if they cooperated! This illustrates the concept of *coordination failure* where players could achieve a better result but fail because they are unable to cooperate [180].

This game also illustrates an interesting concept: This is a *non-zero-sum game*. It is actually a positive sum game! In zero-sum games, each participant's gain or loss is matched by those of other participants so that the net result is zero. In a fair trade, however, one player values more a good than money whereas the other player values more money than a good. After the interaction, both players increased their utility and the game has a positive outcome for all players.

The prisoner's dilemma can also be used to illustrate another popular concept in economics. Imagine that a company has high expenditures with marketing. If this company stops advertising, the competitor can continue its marketing campaign and win a bigger share of the market. If both, however, stop advertising, both can reduce their costs with marketing. The Nash equilibrium corresponds to both keeping their advertisements despite the fact that both would save money if they cut it. This may lead to what was coined by Stigler[13] as *regulatory capture* [181]. This is a phenomenon where companies push for state regulations that actually benefit them instead of the population in general. In our example, the companies could lobby for a regulation prohibiting advertisement for the whole industry segment under the pretense of being better for the people (see for instance [182]).

7.1.4.1 Pareto Optimum

If the game reaches a configuration where no player can change its strategy without reducing another player's payoff, we then say that this is a *Pareto optimum*. In the international trade example, the Pareto optima are all options except the Nash equilibrium. If the game is the Nash equilibrium, then both players can cooperate and both economies are improved.

Mathematically, for a set of feasible allocations Ω, an allocation $\omega \in \Omega$ is a strong Pareto optimum if $\nexists\, \omega^* \in \Omega \mid u_i(\omega^*) \geq u_i(\omega)$ for at least one player. A weak Pareto optimum is reached applying the restriction to all players.

For instance, if the best allocation of resources is the free and dispersed arrangement created by the market, then why do small central planning institutions (firms) exist and why do they succeed? Firms marginally profit from central planning at the cost of incremental knowledge problems that originate from dispersed information [183]. There are *transaction costs* in making economic trades such as agent's commissions or the cost of transporting goods across long distances. The central planning of firms helps to reduce those costs, at least internally. On the other hand, as transaction costs are reduced, the economy becomes more efficient and more capital can be allocated to the development of new goods and services. *Coase*[14] *theorem* states that if property rights are well defined and transaction costs are sufficiently low, then direct bargaining is a Pareto optimum regardless the initial allocation of property [184].

Transaction costs are also associated with the *lock-in* phenomenon [76, 185]. It is often possible to move to more efficient economic allocations, but transaction costs present a barrier through which, many actors are unwilling to pay. Therefore, the economic system remains in a state of lower efficiency. For instance, regulatory capture tends to persist because changing laws have non-pecuniary costs. Also, quite often customers are locked to a company because there are costs to switch the service to another provider.

In political science, this is known as *political entrenchment* [186]. For example, the incumbent is usually protected from the process of change because changes can be costly.

7.1.4.2 Walrasian Equilibrium

Consider two agents with limited amount of resources that want to consume two goods A and B. If they want to consume more of a good A, necessarily they need to reduce their consumption of good B. The ratio between variations in the amount of goods is known as *marginal ratio of substitution* (see Sec. 6.1), and the locus of all points with combinations of goods for which the agent has the same utility is known as *indifference curve*. The set of all indifference curves is known as *indifference map*.

Let's now take the indifference map of the second agent, rotate it π radians and overlap it with the indifference map of the first agent. This is known as the *Edgeworth*[15] *box* and helps us visualize concepts of the *general equilibrium theory* of economics[16] such as the price formation.

Walras's Law

Walras's law, based on his *general equilibrium theory* [187], states that any excess demand in one market must be matched by some excess supply in another market. This leads to the concept of Walrasian *tâtonment*, a process of trial and error through which agents coordinate towards equilibrium.

Walras' general equilibrium theory contrasts with Lachman's[17] idea of an evolutive economy guided by non-stable processes, dispersed knowledge, and subjective expectations.

The shaded region shown in Fig. 7.5, known as *exchange lens* due to its shape, shows a region where both agents would be better off if they engage in trade. Each individual would be able to obtain more goods than those limited by their indifference curves. A Pareto optimal (see Sec. 7.1.4.1) situation, though, happens when the curves are tangent to each other representing the situation where no Pareto

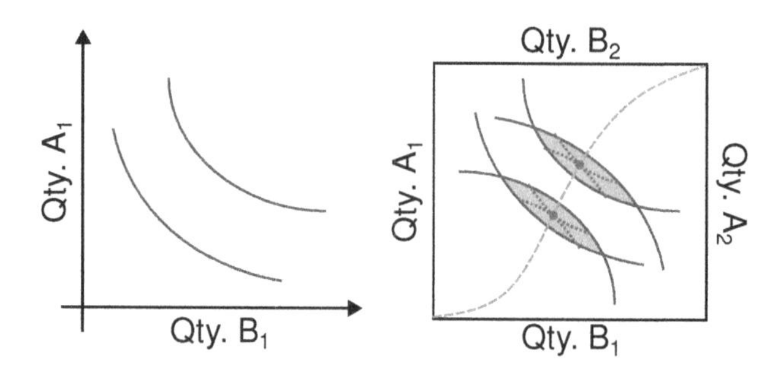

Figure 7.5: The indifference map for agent #1 (left) and the Edgeworth box (right) illustrating the exchange lenses (shaded regions), the Pareto optima (discs), and the contract curve (dashed curve)

improvement can be made. The locus of all Pareto optima constitutes a *contract curve* and shows a *Walrasian equilibrium* where the marginal rate of substitution is the same for both agents.

This also illustrates the *first fundamental theorem of welfare economics*[d] (see, for instance: [188]) that states that in a complete (see Sec. 3.1.1) and efficient market (see Sec. 1.3), market equilibria are Pareto efficient. The second theorem asserts that if the agents have convex indifference curves, then Pareto efficiency can be achieved. This decouples efficiency from distribution, and states that the market can achieve Pareto efficiency by itself. For instance, we see in Fig. 7.5 two Pareto optima in the contract curve resulting from different initial endowments. Thus, efficient allocations can be achieved if the agents face the social consequences of their actions and make choices accordingly. On the other hand, incoordination may occur if there are external attempts to correct the economy, since no agent may have perfect information about the society [5].

7.1.5 (ANTI-)COORDINATION GAMES

Let's take a scarce economic resource, ice cream for example, and let's assume that there is a group of people who want to consume it during summer. The offer of ice cream is limited, hence there is competition among these people for ice cream. Therefore, we say that this good is *rival*. Also, the ice cream shop will only give you an ice cream in exchange for money. Those who do not pay are excluded. Thus, we say that this good is *excludable*. Items that are rival and excludable are known as *private goods*.

Depending on the levels of rivalry and excludability, we may have other types of goods. For instance, a movie theater has low rivalry since it can accommodate a large number of people. Nonetheless, those who do not pay are excluded. Goods that are non-rivalrous (or have low rivalry) but are excludable are known as *club goods*. It is

[d]It dates back to Adam Smith but had contributions from Walras, Edgeworth, and Pareto, among others.

even possible that the common use of a club good creates *positive externalities*. For instance, many people prefer to go to a movie theater with friends rather than going alone. *Coordination games* capture this concept where players coordinate into the same strategy.

Consider the payoff matrix given in Tab. 7.5. Clearly, we have that $A > B$ and $D > C$ for player 1, and $a > c$ and $d > b$ for player 2. Therefore, the main diagonal corresponds to two Nash equilibria: both players choosing together the same strategy. This example is actually known as the *conflicting interest coordination game*, or more popularly, the *battle of sexes game*. This can be applied, for instance, for bank runs [189], currency crises [190], and debt restructuring [191]. For bank runs, for instance, the best option for all players is to keep their money at the bank. However, if a bank run begins, then it is also the best option for all players to withdraw their money.

Table 7.5
Payoff Matrix for the *Conflicting Interest Coordination* Game

		Player 2	
		option 1	option 2
Player 1	option 1	$(A=3, a=2)$	$(C=0, c=0)$
	option 2	$(B=0, b=0)$	$(D=2, d=3)$

One interesting point of this game is that, in a pure Nash equilibria, when one player chooses his or her favorite option, the other player does not, which is an inefficient outcome. In addition to these pure Nash equilibria, though, this game also has *mixed* Nash equilibria. In fact, using a fixed-point theorem [192][18], Nash showed in 1950 that every finite game has at least one mixed strategy equilibrium [193].

Mixed Nash equilibria are probability distributions on the strategy set S_i. This means that the player adopts some strategies with specific probabilities. Mathematically, we can express these solutions with a vector $\mathbf{p}_i = \begin{bmatrix} p_i^1 & \dots & p_i^{k_i} \end{bmatrix} \in \mathbb{R}^{k_i}$ such that $p_i^n \geq 0$, $\forall n \in \begin{bmatrix} 1 & \dots & k_i \end{bmatrix}$ and $\mathbf{p}_i \cdot \mathbf{e} = 1$. The *support* of a mixed strategy is given by $\{n | p_i^n > 0\}$.

The mixed Nash equilibria is then given by:

$$u_i(\sigma^*, \sigma_{-i}^*) \geq u_i(\sigma_i, \sigma_{-i}^*) \; \forall \sigma_i \in \Sigma_i, \tag{7.7}$$

where Σ_i is the space of mixed strategies given by $\Sigma_i = \left\{ \mathbf{p}_i \in \mathbb{R}_+^{k_i} | \mathbf{p}_i \cdot \mathbf{e} = 1 \right\}$.

Therefore, in order to compute the mixed Nash equilibria for this game, first we must find the individual payoff matrices for each player:

$$\mathbf{P}_1 = \begin{bmatrix} 3 & 0 \\ 0 & 2 \end{bmatrix}, \mathbf{P}_2 = \begin{bmatrix} 2 & 0 \\ 0 & 3 \end{bmatrix}. \tag{7.8}$$

The expected payment for player 1 is given by:

$$\mathbf{P}_1\mathbf{p}_2 = \begin{bmatrix} 3 & 0 \\ 0 & 2 \end{bmatrix} \begin{bmatrix} q \\ 1-q \end{bmatrix} = \begin{bmatrix} 3q \\ 2-2q \end{bmatrix}. \tag{7.9}$$

Hence, Nash equilibrium is achieved when:

$$\begin{aligned} 3q &= 2-2q \\ q &= 2/5. \end{aligned} \tag{7.10}$$

The payment for player 2 is given by:

$$\mathbf{P}_2^T\mathbf{p}_1 = \begin{bmatrix} 2 & 0 \\ 0 & 3 \end{bmatrix} \begin{bmatrix} p \\ 1-p \end{bmatrix} = \begin{bmatrix} 2p \\ 3-3p \end{bmatrix} \tag{7.11}$$

Nash equilibrium is reached when:

$$\begin{aligned} 2p &= 3-3p \\ p &= 3/5. \end{aligned} \tag{7.12}$$

Thus, we get:

$$\mathbf{p}_1 = \begin{bmatrix} 3/5 \\ 2/5 \end{bmatrix}, \mathbf{p}_2 = \begin{bmatrix} 2/5 \\ 3/5 \end{bmatrix}. \tag{7.13}$$

Consequently, mixed Nash equilibria is given when the players choose their preferred options with a probability of 60 %.

Returning to our discussion about types of economic goods, let's consider a good that is rivalrous but non-excludable. Such goods are known as *common-pool resources* (or 'CPR'). Take, for instance a small pond shared by two fishermen. Given that it is a natural resource, it may be difficult to exclude a player from using it. Nonetheless, one player can abuse the pond creating a negative externality for the other. In fact, given that the costs associated in maintaining the pond are distributed to all players but the benefits of overfishing are concentrated in one, it is actually expected that this situation may occur. This is known as *tragedy of the commons* [194, 195][19]. The best way to avoid the tragedy of the commons may be privatizing the good. This way, the pond is well maintained so that the owner can profit from it. Nonetheless, Ostrom [196][20] and Axelrod [197][21] showed that, under some circumstances, small groups can spontaneously circumvent the tragedy of the commons without any regulation. This happens because maintaining the good functional is of common interest.

Anti-coordination games capture this scenario where the best strategy for both players is to adopt different strategies. In a general anti-coordination game we have $B > A$ and $C > D$ for player 1, and $b > d$ and $c > a$ as happens in the *game of 'chicken'* shown in Tab. 7.6. If both players adopt option 2 (overexploit the common resource), then both lose. If one of them overexploits the resource but the other does not, then the former wins some amount, whereas the latter loses. If both decide to cooperate, however, then none has any advantage.

Table 7.6
Payoff Matrix for the Game of Chicken

		Player 2	
		option 1	option 2
Player 1	option 1	$(A=0, a=0)$	$(C=-5, c=5)$
	option 2	$(B=5, b=-5)$	$(D=-100, d=-100)$

There are two pure Nash equilibria where players adopt opposing strategies. According to Nash's theorem, there must be at least one mixed Nash equilibrium. The individual payoff matrices are given by:

$$\mathbf{P}_1 = \begin{bmatrix} 0 & -5 \\ 5 & -100 \end{bmatrix}, \mathbf{P}_2 = \begin{bmatrix} 0 & 5 \\ -5 & -100 \end{bmatrix}. \tag{7.14}$$

The expected payment for player 1 is given by:

$$\mathbf{P}_1\mathbf{p}_2 = \begin{bmatrix} 0 & -5 \\ 5 & -100 \end{bmatrix} \begin{bmatrix} q \\ 1-q \end{bmatrix} = \begin{bmatrix} -5(1-q) \\ 5q - 100(1-q) \end{bmatrix}. \tag{7.15}$$

Nash equilibrium is reached when:

$$\begin{aligned} -5(1-q) &= 5q - 100(1-q) \\ -5+5q &= 5q - 100 + 100q \\ 95 &= 100q \\ q &= 95/100. \end{aligned} \tag{7.16}$$

The expected payment of player 2 is exactly the same since $\mathbf{P}_2^T = \mathbf{P}_1$. Therefore, mixed Nash equilibrium happens when both players choose option 1 (cooperate) with a probability of 95%. In real life situations, though, choosing to dare may result in one being excluded from any future game forever. For instance, if the game of chicken consists of two cars heading straight towards one another, it may result in both players dying. In fact, this game has been used to study, for instance, brinkmanship [198][22].

There is a third possibility when $A > B$ and $C < D$ for player 1, and $a < b$ and $c > d$ for player 2 as shown in Tab. 7.7 for the *matching pennies game*. In this game each player takes a penny and chooses head or tails. After revealing the choice simultaneously, if both pennies match, player one takes both pennies. If they do not match, then player 2 takes both pennies. In this situation there is no pure Nash strategy, since either player is better off switching options no matter what side of the coin is chosen. This is a case of *discoordination game*.

The mixed Nash equilibrium can be computed by first identifying the individual payoff matrices:

Table 7.7
Payoff Matrix for the Matching Pennies Game

		Player 2	
		option 1	option 2
Player 1	option 1	$(A=+1, a=-1)$	$(C=-1, c=+1)$
	option 2	$(B=-1, b=+1)$	$(D=+1, d=-1)$

$$\mathbf{P}_1 = \begin{bmatrix} 1 & -1 \\ -1 & 1 \end{bmatrix}, \mathbf{P}_2 = \begin{bmatrix} -1 & 1 \\ 1 & -1 \end{bmatrix}. \tag{7.17}$$

The expected payment for player 1 is given by:

$$\mathbf{P}_1\mathbf{p}_2 = \begin{bmatrix} 1 & -1 \\ -1 & 1 \end{bmatrix} \begin{bmatrix} p \\ 1-p \end{bmatrix} = \begin{bmatrix} 2p-1 \\ -2p+1 \end{bmatrix}. \tag{7.18}$$

Nash equilibrium is given when:

$$\begin{aligned} 2p-1 &= -2p+1 \\ p &= 1/2. \end{aligned} \tag{7.19}$$

Thus, the mixed Nash equilibrium happens when each player chooses heads or tails with a 50% probability.

7.1.5.1 Ratchet Effect

Another illustration of the tragedy of the commons can be found in the *ratchet effect*. This effect is found in the social domain where the reversal of human processes are restricted after some limiting point is reached [199]. For instance, controllers tend to base their policies based on the results of the previous year (theory of adaptive expectations in Ch. 1.1). Another example is the difficulty in dismantling bureaucratic structures once they are provisionally created. In the tragedy of the commons, the rational incentive is for the agents to progressively deplete the common without the possibility of returning to a better state.

This ratchet effect is also used to illustrate the *Parrondo*[23] *paradox*[e]. Consider the problem illustrated in Fig. 7.6. Game A consists of applying a small clockwise rotation to a line that supports a set of disks, whereas game B consists of transferring these disks to the sawtooth structure maintaining their horizontal position. We play game A until the disks get close to valley number 1 and then switch to game B. Some disks will then fall to valley number 2 and the remaining disks will fall to valley

[e] The original idea, called "How to cheat a bad mathematician" was not published, but an early reference to this game is given in [200].

number 0. If we continue playing this sequential pattern repeatedly, eventually the disks will move to the left of the figure.

Thus, even two losing games, such as the tragedy of the commons, can produce positive outcomes when they are coupled and played together. A good description of the Parrondo paradox using Brownian ratchets is given in [201].

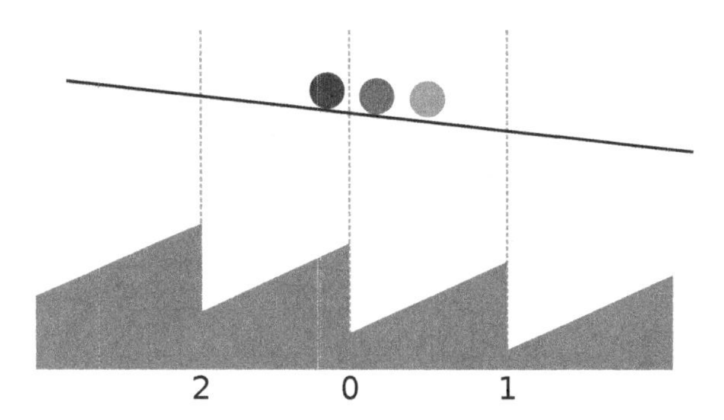

Figure 7.6: Illustration of the Parrondo paradox

7.2 EVOLUTIONARY GAME THEORY

Let's now abandon the premise that the agents are rational. In fact, let's consider that once an agent is born with some behavior, it will last forever and cannot be changed. Although this can be more evident, for instance, to the genetic component of biological species, it has being adapted to some economic problems [202–206]. Indeed, the adapted game theory for this situation was first proposed[24] to explain evolutionary biological behaviors [207]. Hence, it is known as *evolutionary game theory*.

Given an efficient market, the price of a security already impounds all relevant information about it. But what would happen to this market if all agents stopped trying to forecast prices? This is exactly what gives information to prices, so the market would be ceased.

In order to find an answer to this problem, we may consider a liquid market with two types of players: agents that either analyze or not (at some fixed cost) the price of a security being traded. In this scenario, two agents randomly meet and start a trade. Also, given the size of the market, it is unlikely that the same pair is picked more than once. If one agent is of the category that analyzes the prices and the other is not, then the former receives a return discounting transaction costs and expenditures to acquire the information. If the analysis of a player belongs to this category but the other does not, then the former has an advantage. Thus, it receives a return multiplied by a premium discounting only transaction costs.

If the situation is the opposite, then the trader receives the return divided by this multiplicative premium. If, however, neither player belongs to the category that analyzes prices, then the players receive a return discounting only transaction costs. This is exactly the situation where two agents dispute a resource (the premium of finding a good deal) at some cost (of acquiring information) that appears in the biological literature as the *Hawk-Dove* game [207]. In this analogy, a 'hawk' behavior would be assigned to those traders that are willing to pay for information, whereas the 'dove' behavior would be assigned to those that are not. The payoff matrix for this game is shown in Tab. 7.8.

Table 7.8
Fitness Function for the Hawk-dove Game

		Player 2	
		Hawk	*Dove*
Player 1	*Hawk*	$\left(\frac{V-C}{2}, \frac{V-C}{2}\right)$	$(V, 0)$
	Dove	$(0, V)$	$\left(\frac{V}{2}, \frac{V}{2}\right)$

Instead of a payoff, we need a *fitness function* that describes the aptitude of each strategy. Let's consider that the market is composed mostly of individuals with hawk behavior (a fraction of $1-x$ individuals) and some individuals start to show a dove behavior (a fraction of x individuals).

According to the payoff matrix in Tab. 7.8, when a player with a hawk behavior encounters another player with a hawk behavior, the hawk behavior receives an aptitude score of $(V-C)/2$. If the player with a hawk behavior encounters a player with a dove behavior, the former receives an aptitude score of V. Therefore, the fitness function for the hawk behavior is:

$$\begin{aligned} \mathscr{F}_h(hawk) &= (1-x)\frac{V-C}{2} + xV \\ &= \frac{V-C}{2} + \frac{x}{2}(V+C). \end{aligned} \tag{7.20}$$

A player with a dove behavior can meet a player with a hawk behavior and scores 0 aptitude points for the dove behavior. Or, a player with a dove behavior can meet another player with the same behavior and scores $V/2$ aptitude points for the dove behavior. Therefore, the fitness function for the dove behavior is:

$$\mathscr{F}_h(dove) = x\frac{V}{2}. \tag{7.21}$$

The hawk strategy has a higher fitness than the dove strategy when:

$$\begin{aligned} \mathscr{F}_h(hawk) &> \mathscr{F}_h(dove) \\ \frac{V-C}{2} + \frac{x}{2}(V+C) &> x\frac{V}{2} \\ x &> 1 - \frac{V}{C}. \end{aligned} \tag{7.22}$$

Since x is a fraction, $C > V$. The fraction x is a small fraction of individuals that begin to behave differently. Therefore, when $x \to 0$ we get $V \geq C$, which is false. Therefore, the hawk market cannot withstand a dove emergence and we say that the hawk behavior is not an *evolutionary stable strategy* (ESS). Let's analyze the opposite behavior when a fraction x of individuals begin to adopt a hawk behavior in a dove market. The fitness function for a hawk behavior is:

$$\begin{aligned}\mathscr{F}_d(hawk) &= x\frac{V-C}{2} + (1-x)V \\ &= V - \frac{x}{2}(V+C).\end{aligned} \tag{7.23}$$

Similarly, the fitness function for a dove behavior is:

$$\begin{aligned}\mathscr{F}_d(dove) &= (1-x)\frac{V}{2} \\ &= \frac{V}{2} - x\frac{V}{2}.\end{aligned} \tag{7.24}$$

The dove behavior has a higher fitness than the hawk behavior when:

$$\begin{aligned}\mathscr{F}_d(dove) &> \mathscr{F}_d(hawk) \\ \frac{V}{2} - x\frac{V}{2} &> V - \frac{x}{2}(V+C) \\ x &> V/C.\end{aligned} \tag{7.25}$$

Again, taking the limit when $x \to 0$, we get $V < 0$, which is again false. Therefore, the dove market cannot withstand a hawk emergence either, and consequently is not an ESS.

7.2.1 MIXED STRATEGY

Although this game does not have a pure evolutionary stable strategy, it may have a mixed strategy, similar to what we have seen with Nash equilibria. In order to look for a mixed strategy let's consider the case of a market where individuals have a hawk behavior with probability p and, consequently, also a dove behavior with probability $1-p$ (a *p-market*). Another group of traders begin to display a hawk behavior with probability q and a dove behavior with probability $1-q$ (a *q-market*). The payoff matrix is shown in Tab. 7.9.

In this case, the probability of pairing two individuals with hawk behaviors is pq and this gives a aptitude points to player 1. The probability of pairing hawk and dove behaviors is $p(1-q)$ and this gives b aptitude points to player 1. Symmetrically, the probability of pairing dove and hawk behaviors is $(1-p)q$ giving c aptitude points to player 1. Finally, the probability of pairing two dove behaviors is $(1-p)(1-q)$ and it gives d aptitude points to player 1. The expected payoff for player 1 is then:

$$W(p,q) = pq \cdot a + p(1-q) \cdot b + (1-p)q \cdot c + (1-p)(1-q) \cdot d, \tag{7.26}$$

Table 7.9
Payoff Matrix for a Generic Hawk-dove Game Considering Mixed Strategies

		Player 2	
		Hawk, q	*Dove*, $1-q$
Player 1	*Hawk*, p	(a,a)	(b,c)
	Dove, $1-p$	(c,b)	(d,d)

which can be written in matrix notation as:

$$W(p,q) = \mathbf{p}^T \mathbf{A}\mathbf{q}, \tag{7.27}$$

where $\mathbf{p} = \begin{bmatrix} p & 1-p \end{bmatrix}^T$ and $\mathbf{A}$ is the individual payoff matrix given in Tab. 7.9. For arbitrary p and q, the expected payoff is:

$$\begin{aligned} W(p,q) &= pq\cdot a + p(1-q)\cdot b + (1-p)q\cdot c + (1-p)(1-q)\cdot d \\ &= pq\frac{V-C}{2} + p(1-q)V + (1-p)(1-q)\frac{V}{2} \\ &= \frac{V}{2}\left(1+p-q-pq\frac{C}{V}\right). \end{aligned} \tag{7.28}$$

Let's now consider that a fraction x of q-individuals begin to appear in a p-market $(1-x)$. The fitness function for the p and q-individuals are:

$$\begin{aligned} \mathscr{F}_p(p) &= (1-x)W(p,p) + xW(p,q) \\ \mathscr{F}_p(q) &= (1-x)W(q,p) + xW(q,q). \end{aligned} \tag{7.29}$$

For p to be an ESS, we must have:

$$\mathscr{F}_p(p) > \mathscr{F}_p(q). \tag{7.30}$$

For small fractions x, this happens when $W(p,p) > W(q,p)$ or when $W(p,p) = W(q,p)$ and $W(p,q) > W(q,q)$.

The mixed Nash equilibrium for this game can be calculated by first finding the individual payoff matrices:

$$\mathbf{P}_1 = \begin{bmatrix} \frac{V-C}{2} & V \\ 0 & \frac{V}{2} \end{bmatrix}, \mathbf{P}_2 = \begin{bmatrix} \frac{V-C}{2} & 0 \\ V & \frac{V}{2} \end{bmatrix}. \tag{7.31}$$

Since the payoff matrix is symmetric, the expected payment for either player is given by:

$$\mathbf{P}_1\mathbf{p}_2 = \begin{bmatrix} \frac{V-C}{2} & V \\ 0 & \frac{V}{2} \end{bmatrix} \begin{bmatrix} p \\ 1-p \end{bmatrix} = \begin{bmatrix} p\frac{V-C}{2} + V(1-p) \\ \frac{V}{2}(1-p) \end{bmatrix}. \tag{7.32}$$

Nash equilibrium is reached when:

$$\begin{aligned} -\frac{V+C}{2}p+V &= \frac{V}{2}-\frac{V}{2}p \\ p=q &= \frac{V}{C}. \end{aligned} \tag{7.33}$$

For $p = V/C$ we get:

$$\begin{aligned} W(p,p) &= \frac{V}{2}(1-p^2/p) = \frac{V}{2}(1-p), \\ W(q,p) &= \frac{V}{2}(1+q-p-pq/p) = \frac{V}{2}(1-p). \end{aligned} \tag{7.34}$$

Therefore, in order for the Nash equilibrium be an ESS, we must have:

$$\begin{aligned} W(p,q) &> W(q,q) \\ 1+p-q-pq/p &> 1-q^2/p \\ q^2-2pq+p^2 &> 0. \end{aligned} \tag{7.35}$$

This is a parabola whose point of minimum touches p. Therefore, it is always positive (as long as $q \neq p$), and the Nash equilibrium point $p = V/C$ is an ESS, which implies that, for this probability, individuals that begin to use another strategy will not succeed in flipping this market.

7.2.2 REPLICATOR DYNAMICS

Consider the population of a particular species N_i. It takes, on average, a period τ for each individual to generate an offspring. Therefore, after a short period Δ, the population of this species will increase by $(\Delta/\tau)N_i(t)$ and we get:

$$\begin{aligned} N_i(t+\Delta) &= N_i(t) + \frac{\Delta}{\tau}N_i(t) \\ N_i(t+\Delta) - N_i(t) &= f_i N_i(t)\Delta, \end{aligned} \tag{7.36}$$

where $f_i = 1/\tau$ is the frequency at which the population of that species is increased. Taking the limit of a very short interval:

$$dN_i(t) = f_i N_i(t)dt. \tag{7.37}$$

We identify the Markov propagator $G[N_i(t),dt] = f_i N_i(t)dt$. We can always add a stochastic term to account for fluctuations on this frequency, which leads to $G[N_i(t),dt] = N_i(t)(f_i dt + \sigma_i dW_i)$, where dW_i is an uncorrelated Wiener process. Under this new scenario, the evolution of the population is given by the geometric Brownian motion:

$$dN_i(t) = f_i N_i(t)dt + \sigma_i N_i(t)dW_i(t). \tag{7.38}$$

Considering a total population of $N = \sum_i N_i$ individuals of all species, we can make questions about the dynamics of the fractions $x_i(t) = N_i/N$. To find this dynamics we must use the multidimensional Ito's lemma (Eq. 7.42) with:

$$\begin{aligned}\frac{\partial x_i}{\partial t} &= 0\\ \frac{\partial x_i}{\partial N_j} &= \frac{\delta_{ij}}{N} - \frac{x_i}{N}\\ \frac{\partial^2 x_i}{\partial N_j^2} &= -\frac{\partial \delta_{ij}}{N^2} - \frac{1}{N^2}\left(\frac{\partial x_i}{\partial N_j}N - x_i\right)\\ &= -2\frac{\delta_{ij}}{N^2} + 2\frac{x_i}{N^2}.\end{aligned} \tag{7.39}$$

Multidimensional Ito's Lemma

For a doubly differentiable function $f(t, S_1, S_2, \ldots, S_N)$, its Taylor's expansion is:

$$df = \frac{\partial f}{\partial t}dt + \sum_{i=1}^{N}\frac{\partial f}{\partial S_i}dS_i + \frac{1}{2}\sum_{i=1}^{N}\sum_{j=1}^{N}\frac{\partial^2 f}{\partial S_i \partial S_j}dS_i dS_j + \ldots \tag{7.40}$$

Substituting the differential dS_i by a stochastic differential equation of the form $dS_i = a_i(S,t)dt + b_i(S,t)dW_i$, we get:

$$\begin{aligned}df = &\frac{\partial f}{\partial t}dt + \sum_{i=1}^{N}\frac{\partial f}{\partial S_i}(a_i dt + b_i dW) +\\ &+\frac{1}{2}\sum_{i=1}^{N}\sum_{j=1}^{N}\frac{\partial^2 f}{\partial S_i \partial S_j}(a_i dt + b_i dW_i)(a_j dt + b_i dW_j) + \ldots\end{aligned} \tag{7.41}$$

Eliminating high order terms, we obtain the multidimensional obtain Ito's lemma:

$$\begin{aligned}df &\approx \frac{\partial f}{\partial t}dt + \sum_{i=1}^{N}\frac{\partial f}{\partial S_i}(a_i dt + b_i dW_i) + \frac{1}{2}\sum_{i=1}^{N}\sum_{j=1}^{N} b_i b_j \frac{\partial^2 f}{\partial S_i \partial S_j}\rho_{ij}dt\\ &= \left(\frac{\partial f}{\partial t} + \sum_{i=1}^{N} a_i \frac{\partial f}{\partial S_i} + \frac{1}{2}\sum_{i=1}^{N}\sum_{j=1}^{N} b_i b_j \rho_{ij}(S,t)\frac{\partial^2 f}{\partial S_i \partial S_j}\right)dt\\ &+ \sum_{i=1}^{N} b_i \frac{\partial f}{\partial S_i}dW_i,\end{aligned} \tag{7.42}$$

where ρ_{ij} is the correlation between both Wiener processes.

Using these results in the multidimensional Ito's lemma, we get:

$$
\begin{aligned}
dx_i = & \left[\sum_{j=1}^{N} f_j N_j \left(\frac{\delta_{ij}}{N} - \frac{x_i}{N}\right) + \frac{1}{2}\sum_{j=1}^{N} \sigma_j^2 N_j^2 \left(-2\frac{\delta_{ij}}{N^2} + 2\frac{x_i}{N^2}\right)\right] dt + \\
& + \sum_{j=1}^{N} \sigma_j N_j \left(\frac{\delta_{ij}}{N} - \frac{x_i}{N}\right) dW_j \qquad (7.43) \\
= & \, x_i \left(f_i - \sum_{j=1}^{N} f_j x_j - \sigma_i^2 x_i + \sum_{j=1}^{N} \sigma_j^2 x_j^2\right) dt + x_i \left(\sigma_i dW_i - \sum_{j=1}^{N} \sigma_j x_j dW_j\right),
\end{aligned}
$$

which is the *stochastic replicator dynamics equation*. The *deterministic replicator dynamics equation* is found by setting $\sigma_k = 0$:

$$\dot{x}_i = x_i \left[f_i - \phi(x)\right], \tag{7.44}$$

where $\phi(x) = \sum_{j=1}^{N} f_j x_j$.

In evolutionary game theory, the frequency f_i is given by the expected payoff $(\mathbf{Ax})_i$. Therefore, we get:

$$\dot{\mathbf{x}} = \mathbf{x} \odot \left[\mathbf{Ax} - \left(\mathbf{x}^T \mathbf{Ax}\right) \mathbf{e}\right]. \tag{7.45}$$

We can consider the population of individuals with a hawk behavior to be p. Therefore, the population of individuals with a dove behavior is $1-p$. For the payoff matrix given in Tab. 7.8 we have:

$$\begin{bmatrix} \dot{p} \\ -\dot{p} \end{bmatrix} = \begin{bmatrix} p \\ 1-p \end{bmatrix} \odot \left(\begin{bmatrix} p\frac{V-C}{2} + (1-p)V \\ (1-p)\frac{V}{2} \end{bmatrix} - \frac{1}{2}\left(V - p^2 C\right) \begin{bmatrix} 1 \\ 1 \end{bmatrix} \right). \tag{7.46}$$

From the second line of this matrix equation we get:

$$\dot{p} = 1/2 p(p-1)(pC - V). \tag{7.47}$$

In steady state, setting $\dot{p} = 0$ we get the trivial solutions $p = 0, 1$ and $p = V/C$, which is also an ESS. These are known as *evolutionary stable states*. This behavior is illustrated in Fig. 7.7. For values of $V > C$ the final population consists only of individuals with hawk-behavior. For other values, the final population consists of a population of V/C individuals with a hawk behavior and $1 - V/C$ individuals with a dove behavior.

This dynamics can be simulated with the following Runge-Kutta[25] code. The function to be simulated is given by:

```
def fun(p,V,C):
    return 0.5*p*(p-1)*(p*C-V)
```

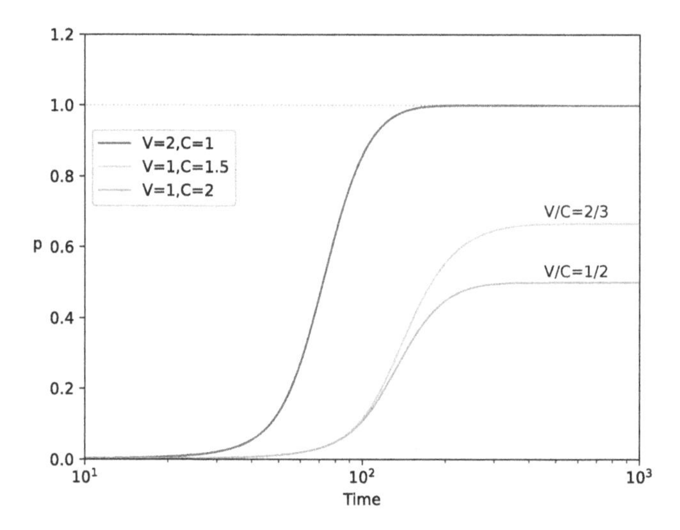

Figure 7.7: Fraction p of individuals with a hawk behavior as a function of time for different values of V/C

The Runge-Kutta itself is calculated with:

```
def RK(po,V,C):
    h = 0.1
    p = po

    x = []
    for i in range(1000):
        k1 = fun(p,V,C)
        k2 = fun(p + 0.5*h*k1,V,C)
        k3 = fun(p + 0.5*h*k2,V,C)
        k4 = fun(p + h*k3,V,C)
        p = p + (h/6)*(k1+2*k2+2*k3+k4)
        x.append(p)
    return x
```

7.3 LOTKA-VOLTERRA EQUATIONS

A concept similar to the replicator dynamics is given by the famous Lotka[26]-Volterra[27] equations. This is a model initially conceived to study autocatalytic reactions [208], organic systems [209] and prey-predator systems [210].

Lotka-Volterra equations have also been used to model many socioeconomic models such as the competition between oppressive governments and opposing rebels [211]. One situation where the Lotka-Volterra equations are used in economics is given by the Goodwin[28] model [212]. This is a model that tries to capture endogenous fluctuations in an economic system.

Aggregate output (q) is defined in Keynesian[29] economics as the total amount of goods and services produced and supplied in an economy during a specific period. The economic activity coordinates these goods in order to achieve good allocations. Therefore, it is not trivial to sum the contribution of heterogeneous good [213]. Nonetheless, the Goodwin model assumes that the aggregate output is given by the ratio between a homogeneous capital stock (k) and a constant capital-output ratio (σ). In the model, this capital is fully used by the labor to produce these goods. Therefore, $q = k/\sigma = al$, where a is the productivity of the labor. For wages wl, profits $q - wl$ are generated, but since $l = q/a$, we get a profit $q(1 - w/a)$ that leads to a profit rate $1 - w/a$. Hence, the growth rate of the aggregate output is given by:

$$\frac{\dot{q}}{q} = \frac{1}{\sigma}\left(1 - \frac{w}{a}\right). \tag{7.48}$$

Philips observed that in the short term, as the economy grows, companies expand and need more workers. In order to attract a good labor force, companies increase wages[f] [214]. Therefore, the growth of wages $\dot{w}/w$ must be a function of the unemployment rate. Note, however, that this relationship only holds for short periods. For instance, the Phillips curve fails to explain stagflation[g]. In the long-run, the unemployment rate seems to converge to a *natural rate of unemployment*. As pointed by Friedman [215] and Phelp[30] [216, 217], that is unaffected by changes in wages and rises of price levels[h]. This is known as *Non-accelerating inflation rate of unemployment* (NAIRU) [215]. In the Goodwin model, a linear approximation of the Philips for the employment curve is used:

$$\frac{\dot{w}}{w} = \rho\frac{l}{n} - \eta, \tag{7.49}$$

where l is the employment of labor, n is the total labor force such that $v = l/n$ is the employment rate. Also ρ and η are positive constants.

The growth rate of the employment level ($l = q/a$) is given by:

$$\begin{aligned} \dot{l} &= \frac{\dot{q}}{a} - \frac{q}{a^2}\dot{a} \\ \frac{\dot{l}}{l} &= \frac{\dot{q}}{q} - \frac{\dot{a}}{a}. \end{aligned} \tag{7.50}$$

[f]Economists call it a *tight labor market* because it is difficult to find new labor.

[g]Increase of prices and unemployment at the same time.

[h]This does not imply that the natural rate should be pursued, as it could lead to other phenomena such as a high increase of price levels [218].

Assuming that the productivity grows at a constant rate λ and using the result of Eq. 7.48 we get:

$$\frac{\dot{l}}{l} = \frac{1}{\sigma}\left(1 - \frac{w}{a}\right) - \lambda \tag{7.51}$$

The growth rate of the employment rate ($v = l/n$) is given by:

$$\begin{aligned}\dot{v} &= \frac{\dot{l}}{n} - \frac{l}{n^2}\dot{n} \\ \frac{\dot{v}}{v} &= \frac{\dot{l}}{l} - \frac{\dot{n}}{n}.\end{aligned} \tag{7.52}$$

Assuming that the labor force grows at a constant rate θ and using the result of Eq. 7.51, we get:

$$\frac{\dot{v}}{v} = \frac{1}{\sigma}\left(1 - \frac{w}{a}\right) - \lambda - \theta. \tag{7.53}$$

The wage share u is defined in Keynesian economics as the ratio between wages and the output $(w/q)l = w/a$. Its growth rate is given by:

$$\begin{aligned}\dot{u} &= \frac{\dot{w}}{a} - \frac{w}{a^2}\dot{a} \\ \frac{\dot{u}}{u} &= \frac{\dot{w}}{w} - \frac{\dot{a}}{a}.\end{aligned} \tag{7.54}$$

Using the result of Eq. 7.49 we get:

$$\frac{\dot{u}}{u} = \rho v - \eta - \lambda. \tag{7.55}$$

Equations 7.53 and 7.55 are exactly the Lotka-Volterra equations:

$$\begin{aligned}\frac{\dot{v}}{v} &= \frac{1}{\sigma}(1 - u) - \lambda - \theta \\ \frac{\dot{u}}{u} &= \rho v - \eta - \lambda,\end{aligned} \tag{7.56}$$

that can be generalized as:

$$\begin{aligned}\frac{\dot{x}}{x} &= \alpha - \beta y \\ \frac{\dot{y}}{y} &= \delta x - \gamma,\end{aligned} \tag{7.57}$$

where:

$$\begin{aligned} x &= v, \quad \alpha = \tfrac{1}{\sigma} - \lambda - \theta, \quad \delta = \rho \\ y &= u, \quad \beta = \tfrac{1}{\sigma}, \qquad\qquad \gamma = \eta + \lambda.\end{aligned} \tag{7.58}$$

It is possible to find an analytical solution dividing one equation by the other and then integrating:

$$\frac{dx}{dy} = \frac{x(\alpha - \beta y)}{y(\delta x - \gamma)}$$
$$dy\frac{\alpha - \beta y}{y} = dx\frac{\delta x - \gamma}{x}$$
$$\int dy\frac{\alpha - \beta y}{y} = \int dx\frac{\delta x - \gamma}{x}$$
$$\alpha \log y - \beta y = \delta x - \gamma \log x + V$$
$$V = \alpha \log y + \gamma \log x - \beta y - \delta x. \tag{7.59}$$

The fixed points correspond to those values of x and y that produce constant solutions $\dot{x} = \dot{y} = 0$:

$$\begin{aligned} \alpha x - \beta xy = 0 \quad &\rightarrow \quad x(\alpha - \beta y) = 0 \\ \delta xy - \gamma y = 0 \quad &\rightarrow \quad y(\delta x - \gamma) = 0. \end{aligned} \tag{7.60}$$

Therefore:

$$(x, y) = (0, 0) \text{ and } \left(\frac{\gamma}{\delta}, \frac{\alpha}{\beta}\right). \tag{7.61}$$

In order to check whether these solutions are stable, we can use Hartman[31]-Grobman[32] theorem.

Hartman-Grobman Theorem

The Hartman-Grobman linearization theorem [219, 220] states that a diffeomorphism[a] is topologically equivalent to its linearization around a hyperbolic fixed point[b] (x^*, y^*) with the form:

$$\begin{bmatrix} \dot{u} \\ \dot{v} \end{bmatrix} = \mathbf{A} \begin{bmatrix} u \\ v \end{bmatrix}, \tag{7.62}$$

where $u = x - x^*$, $v = y - y^*$ and $\mathbf{A}$ is the Jacobian matrix[c] evaluated at the equilibrium point.

The set of eigenvectors for which the corresponding eigenvalues $Re\{\lambda\} < 0$ define a *stable manifold*, whereas those eigenvectors whose corresponding eigenvalues $Re\{\lambda\} > 0$ define an *unstable manifold*. Finally, those eigenvectors whose corresponding eigenvalues $Re\{\lambda\} = 0$ define a *center manifold*.

---

[a] A differentiable map between manifolds for which the inverse map is also differentiable.

[b] The fixed point of a map is a hyperbolic fixed point (HFP) if the stability matrix (the Jacobian matrix evaluated at these points) has no eigenvalues in the unit circle. HFPs receive this name because the orbits of the system around these points resemble hyperbolas. HFPs are also known as *saddle points* [77].

[c] Also known as *community matrix* by the mathematical biology community.

The Jacobian matrix for the Lotka-Volterra equations is given by:

$$\mathbf{J} = \begin{bmatrix} \frac{\partial f_1}{\partial x} & \frac{\partial f_1}{\partial y} \\ \frac{\partial f_2}{\partial x} & \frac{\partial f_2}{\partial y} \end{bmatrix} = \begin{bmatrix} \alpha - \beta y & -\beta x \\ \delta y & \delta x - \gamma \end{bmatrix}. \tag{7.63}$$

Its eigenvalues are:

$$\begin{aligned} \begin{vmatrix} \alpha - \beta y - \lambda & -\beta x \\ \delta y & \delta x - \gamma - \lambda \end{vmatrix} &= 0 \\ [(\alpha - \beta y) - \lambda]\,[(\delta x - \gamma) - \lambda] + \beta x \delta y &= 0 \\ \lambda^2 - \lambda\,[\delta x - \gamma + \alpha - \beta y] + \alpha \delta x - \alpha \gamma + \beta y \gamma &= 0. \end{aligned} \tag{7.64}$$

For $(x^*, y^*) = (0,0)$ we have $\lambda^2 - \lambda(\alpha - \gamma) - \alpha\gamma = 0$. Therefore:

$$\begin{aligned} \lambda &= \frac{(\alpha - \gamma) \pm \sqrt{(\alpha - \gamma)^2 + 4\alpha\gamma}}{2} \\ &= \frac{\alpha - \gamma \pm (\alpha + \gamma)}{2} \\ &= \alpha,\ -\gamma. \end{aligned} \tag{7.65}$$

Since both parameters α and γ are positive, this is an unstable saddle point. On the other hand, when $(x^*, y^*) = (\gamma/\delta, \alpha/\beta)$, we have $\lambda^2 - \lambda\,[\gamma - \gamma + \alpha - \alpha] + \alpha\gamma - \alpha\gamma + \alpha\gamma = 0$. Therefore $\lambda = \pm i\sqrt{\alpha\gamma}$. Since the eigenvalues are purely complex, the center subspace is composed of closed orbits. This corresponds to periodic oscillations with frequency $\omega = \sqrt{\lambda_1\lambda_2} = \sqrt{\alpha\gamma}$ in the time domain.

The Lotka-Volterra equations can be solved numerically using the symplectic Euler method:

$$\begin{aligned} x_{n+1} &= x_n + h x_{n+1}(\alpha - \beta y_n) \\ x_{n+1} &= \frac{x_n}{1 - h(\alpha - \beta y_n)} \end{aligned} \tag{7.66}$$

and

$$\begin{aligned} y_{n+1} &= y_n + h y_n(\delta x_{n+1} - \gamma) \\ y_{y+1} &= y_n\,[1 + h(\delta x_{n+1} - \gamma)], \end{aligned} \tag{7.67}$$

where h is the time step for the simulation.

In Python we can define a function:

```
def LV(xo,yo,p):
    alpha = p[0]
    beta = p[1]
    delta = p[2]
    gamma = p[3]
    h = 0.04

    x = xo
    y = yo
    xsp = []
    ysp = []
    for i in range(750):
        x = x/(1-h*(alpha-beta*y))
        y = y*(1+h*(delta*x-gamma))
        xsp.append(x)
        ysp.append(y)

    return xsp,ysp
```

This function can be invoked in the main function with:

```
sigma = 0.5
lbd = 0.8
theta = 0.3
rho = 0.5
eta = 0.6

alpha = 1.0/sigma - lbd - theta
beta = 1.0/sigma
delta = rho
gamma = eta + lbd

p = [alpha,beta,delta,gamma]

xi,yi = 0.25,0.37
x,y = LV(xi,yi,p)
pl.plot(x,y)
```

A simulation for the Goodwin model showing endogenous oscillatory behavior is shown in Figs. 7.8 and 7.9.

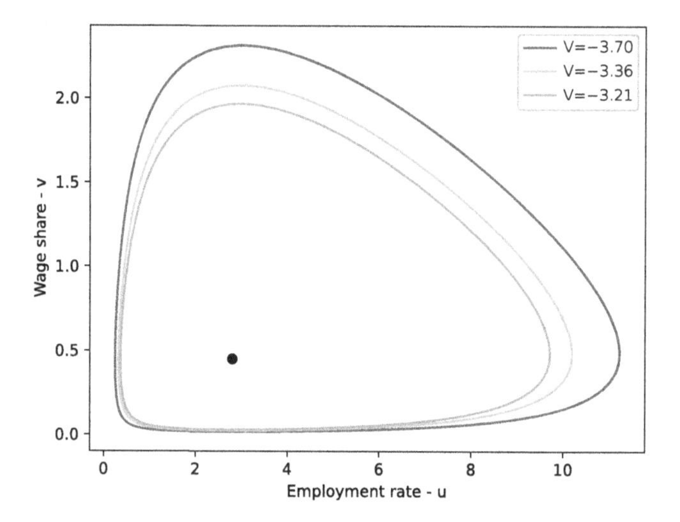

Figure 7.8: The phase diagram for the Goodwin model with different initial values leading to the values of V shown in the legend. The black dot indicates the attractor of the system

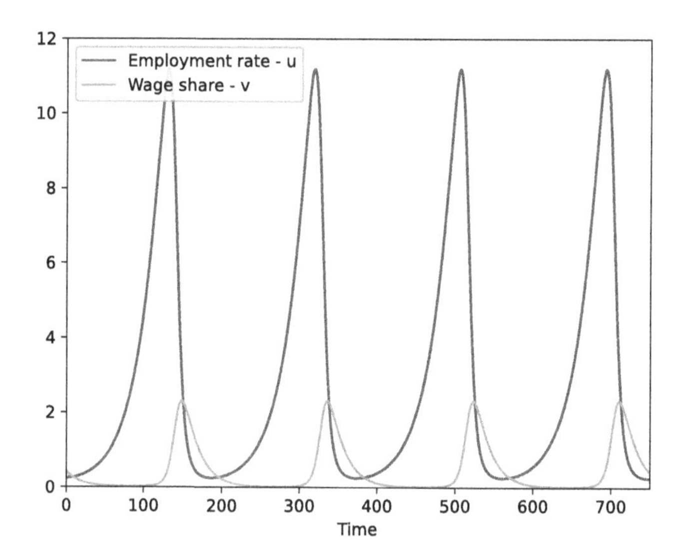

Figure 7.9: The time evolution of employment rate and wage share exhibiting oscillatory behavior in the Goodwin model

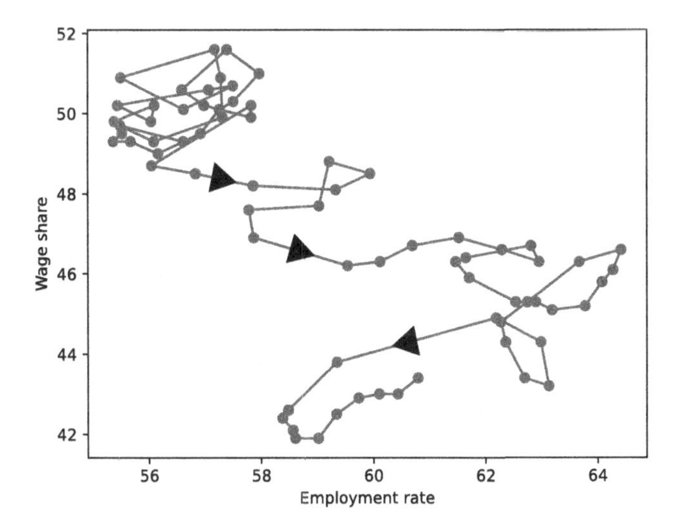

Figure 7.10: Phase space of wage shares as a function of employment rate with real data obtained from the Federal Reserve Bank of St. Louis between 1948 and 2019 for the United States (the triangles indicate the direction of the curve)

It is interesting to contrast the model with real data as shown in Fig. 7.10. The figure shows the share of gross domestic income as a percentage of the gross domestic product as a function of the employment-population ratio for the US between 1948 and 2019. For short periods, the phase space produces Goodwin-like oscillations corresponding to the employment rate lagging behind the wage share, but this tendency is completely lost for long periods.

The increase in outsourcing, the stagnation of productivity and the shift to more specialized jobs has increased mono- and oligopsony[i] power leaning the bargaining balance towards the companies, which forced a reduction of wage shares over the past four decades [221]. On the other hand, this does not capture the emergence of better and more efficient products, the increase of consumer choice and other benefits directly paid by companies to retain workers [222]. All this factors have actually increased economic well being.

Notes

[1]Oskar Morgenstern (1902–1977) Austrian economist, advisee of Ludwig von Mises.

[2]Term coined by Daniel Bernoulli (1700–1782) Swiss mathematician and physicist. Bernoulli, the son of Johann Bernoulli, used to live in the eponymous city. The problem, though, was firstly proposed by his nephew Nicolaus Bernoulli (1687–1759) Swiss mathematician.

[3]William Brian Arthur (1945–) Irish economist.

[4]Milton Friedman (1912–2006) American economist, adviser of Harry Markowitz among others. Friedman was awarded many prizes, including the Nobel memorial prize in economic sciences in 1976.

---

[i]Mono/oligo-psony: A market structure where one/a few buyer(s) control(s) most of the market.

[5]John Stuart Mill (1806–1873) British philosopher and economist.

[6]William Stanley Jevons (1835–1882) English economist and logician, advisee of Augustus de Morgan.

[7]Proposed by Damien Challet (1974–) Swiss physicist and his adviser of Yi-Cheng Zhang from the Université de Fribourg.

[8]Lloyd Shapley (1923–2016) American mathematician, advisee of Albert Tucker. Shapley was awarded many prizes including the Nobel prize in 2012 for his contribution to cooperative games.

[9]Developed by Werner Güth (1944–) German economist.

[10]John Forbes Nash Jr. (1928–2015) American mathematician, advisee of Albert Tucker. Nash won many prizes including the Nobel Memory Prize in Economic Sciences in 1994.

[11]Merrill Meeks Flood (1908–1911) American mathematician.

[12]Albert William Tucker (1905–1995) Canadian mathematician, adviser of John Nash among others.

[13]George Joseph Stigler (1911–1991) American economist, advisee of Frank Knight and adviser of Thomas Sowell, among others. Stigler received the Nobel memorial prize in economic sciences in 1982.

[14]Ronald Harry Coase (1910–2013) British economist, awarded the Nobel prize in economics in 1991.

[15]Francis Ysidro Edgeworth (1845–1926) Irish political economist, awarded the gold Guy Medal by the Royal Statistical Societyl in 1907.

[16]Conceptualized by Marie-Esprit-Léon Walras (1834–1910) French economist.

[17]Ludwig Lachmann (1906–1990) German economist.

[18]Nash originally used Brouwer's fixed-point theorem stated by Luitzen Egbertus Jan Brouwer (1881–1966) Dutch mathematician and philosopher.

[19]Conceptualized by William Forster Lloyd (1794–1852) British economist, and popularized by Garrett Hardin (1915–2003) American ecologist.

[20]Elinor Claire Ostrom (1993–2012) American economist. Ostrom won several prizes including the Nobel Memorial Prize in economics in 2009.

[21]Robert Marshall Axelrod (1943–) American political scientist.

[22]Thomas Crombie Schelling (1921–2016) American economist, awarded the Nobel memorial prize in economics in 2005 has been called it "the threat that leaves something to chance".

[23]Juan Manuel Rodríguez Parrondo (1964–) Spanish physicist.

[24]Proposed by John Maynard Smith (1920–2004) British mathematician and biologist and George Robert Price (1922–1975) Americn geneticist. Smith was awarded many prizes including the Royal Medal in 1995.

[25]Carl David Tolmé Runge (1856–1927) German mathematician, advisee of Karl Weierstrass and adviser of Max Born among others. Martin Wilhelm Kutta (1867–1944) German mathematician.

[26]Alfred James Lotka (1880–1949) American mathematician.

[27]Vito Volterra (1860–1940) Italian mathematician and physicist. Advisee of Enrico Betti and adviser of Paul Lévy among others.

[28]Richard Murphey Goodwin (1913–1996) American mathematician and economist.

[29]John Maynard Keynes (1883–1946) English economist.

[30]Edmund Strother Phelps (1933–) American economist, advisee of Arthur Okun. Phelps won many awards including the Nobel Memorial Prize in Economic Sciences in 2006.

[31]Philip Hartman (1915–2015) American mathematician.

[32]Vadim Matveevich Grobman, Russian mathematician.

8 Agent-Based Simulations

"There's no love in a carbon atom, no hurricane in a water molecule, and no financial collapse in a dollar bill." P. Dodds[1]

The word *complex* comes from the Latin *com*, which means "together" and *plectere*, which means "to weave". The real meaning of the word *complex*, though, is more *complex* than that. Complex systems can partially be understood as collections of agents that interact non-trivially among themselves and their environments producing novel and rich phenomena that, typically, cannot be anticipated from the study of their individual units [223, 224]. Some topics related to complexity were already studied in the previous chapters without an explicit reference to it. For instance, cellular automata was already studied in Sec. 6.3.1 and game theory was seen in Chapter 7.

As stated in the first chapter, economics is a discipline that deals with interacting people that are subject to emotions, conformity, collective motion, and many other complex phenomena. As Wolfram[2] once put it, in order to develop models that capture these complex behaviors, one must look for novel tools beyond the standard mathematical descriptions that we are used to [225].

This chapter deals exactly with the interaction of agents and the emergence that appears in these processes. It begins exploring the intricate connections that agents make and then moves towards some socioeconomic models of opinion dynamics and segregation. The book ends with a study of kinetic models, linking trade and wealth to the Boltzmann[3] equation.

8.1 COMPLEX NETWORKS

If you paid attention to the endnotes so far, you probably noticed that many important authors make networks of collaboration. The same is valid for banks interacting through credit or firms interacting through trade. Depending on the topology of the network, some behaviors may emerge and affect properties such as performance and fragility. In this section we will study these networks, some of their metrics and applications in econophysics.

In order to study networks, we must study *graphs*. These are ordered pairs $G = (V, E)$ where V is a set of points called *vertices* (or *nodes*), and $E \subseteq \{\{x,y\} | x,y \in V,\ x \neq y\}$ is a set of lines called *edges* (or *links*) that connect those nodes. The graph may be *weighted* if there is a function $w : E \rightarrow \mathbb{R}$ associated with all edges of the graph. On the other hand, if the graph is unweighted, then $w : E \rightarrow \mathbb{B}$, where $\mathbb{B}$ is the Boolean[4] domain $\mathbb{B} = \{0, 1\}$. We say that the graph is *undirected* if the edges are comprised of unordered vertices, and *directed* otherwise. All graphs presented here will be unweighted and undirected, unless explicitly stated otherwise.

DOI: 10.1201/9781003127956-8

A graph can be *simple* if it allows only one edge between a pair of nodes or *multigraph* otherwise. A multigraph with loops is known as *pseudograph*. Finally, a graph can be *complete* if all pair of nodes are connected by edges.

A graph can be represented by an *adjacency matrix* whose elements are the number of edges that directly connect two nodes. In the case of a simple graph, the adjacency matrix is a Boolean matrix[a]. We can also define the *neighborhood* $N_G(v)$ of a node v as a subgraph of G induced by the neighbors (adjacent nodes, or nodes connected by an edge to v) of v.

A *walk* of length n is defined as an alternating sequence of nodes and edges $v_0, e_1, v_1, e_2 \ldots, e_k, v_k$ such that edge $e_i = \{v_{i-1}, v_i\}$, for $1 \leq i \leq n$. In Fig. 8.1, a walk could pass sequentially through nodes *abacd*. This walk can be *closed* if the first and last nodes are the same. An example would be a walk through the sequence *abca*. It is considered *open* otherwise. An example would be *abc*. Also, a *trail* is defined as a walk with no repeated edges such as *abcd*. A *path* is defined as an open trail with no repeated nodes such as *abcd*. Furthermore, a *cycle* is defined as a closed trail with no repeated vertices except the first and last nodes such as *acdfa*. Finally, a *circuit* is a closed trail with no repeated edges that may have repeated nodes. An example would be *abcfdca*.

Observe that the number of 1-walks between nodes is given by the adjacency matrix. The number of 2-walks between two nodes is given by $\sum_n a_{in} a_{nj}$ but this leads to the product AA. Therefore, we find by induction that the elements of the nth power of the adjacency matrix gives the number of n-walks between these nodes. For instance, for the graph in Fig. 8.1 the adjacency matrix is given by Eq. 8.1. The 2-walks between a and a are *aba*, *aca*, and *afa* and this is captured by the element $A^2_{11} = 3$. Also note that the number of triangles in a graph can be found from the diagonal of A^3. Since a triangle has three nodes and every node is counted, we must divide the trace of A^3 by 3. Also, both clockwise and counterclockwise walks are computed. Therefore the number of triangles is given by $Tr\left(A^3\right)/6$.

$$A = \begin{bmatrix} 0 & 1 & 1 & 0 & 0 & 1 \\ 1 & 0 & 1 & 0 & 1 & 0 \\ 1 & 1 & 0 & 1 & 0 & 1 \\ 0 & 0 & 1 & 0 & 0 & 1 \\ 0 & 1 & 0 & 0 & 0 & 1 \\ 1 & 0 & 1 & 1 & 1 & 0 \end{bmatrix}, \; A^2 = \begin{bmatrix} 3 & 1 & 2 & 2 & 2 & 1 \\ 1 & 3 & 1 & 1 & 0 & 3 \\ 2 & 1 & 4 & 1 & 2 & 2 \\ 2 & 1 & 1 & 2 & 1 & 1 \\ 2 & 0 & 2 & 1 & 2 & 0 \\ 1 & 3 & 2 & 1 & 0 & 4 \end{bmatrix} \tag{8.1}$$

The *degree matrix*, a related concept, is a diagonal matrix whose elements are the node degrees. A *Laplacian matrix* can be constructed as $L = D - A$.

A graph can be *regular* if all nodes have the same number of connections (have the same number of neighbors, or the same *degree*, or even the same *coordination number*, depending the audience) as shown in Fig. 8.2. If the regular graph (discounting its external nodes, or *leaves*, that have degree 1) contains no cycles and is connected, then it is a *Cayley*[5] *tree*. If this tree has an infinite number of nodes (no

[a]A matrix whose elements are in the Boolean domain $\mathbb{B}$.

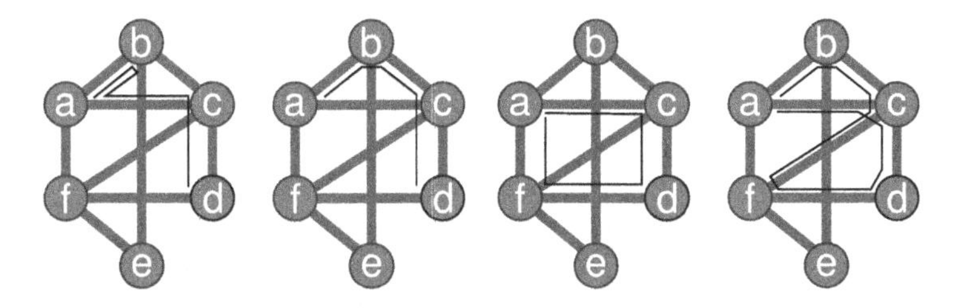

Figure 8.1: A random graph used to illustrate the concepts of (from left to right) walk, trail, cycle, and circuit

leaves), then it is a *Bethe*[6] *lattice*. A *spanning tree* of a graph G is a subgraph of G that is a tree and contains all nodes of G.

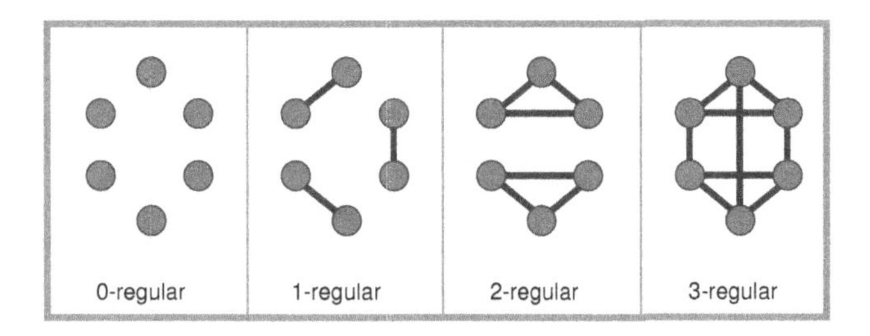

Figure 8.2: Graphs with the same number of nodes but with different regularities

If a graph has non-trivial topological properties, then we call it a *complex network*. A non-trivial topological property could be, for instance, a degree[b] distribution with a long tail. In order to explore these properties we must study some graph metrics.

8.1.1 METRICS

A *metric space* is a tuple (X,d) where X is a set, and $d : X \times X \to \mathbb{R}^+$ is a function (called metric or distance) such that for any $x,y,z \in X$, d satisfies:

1. Leibniz's[7] law of indiscernibility of identicals: $d(x,y) = 0 \iff x = y$,
2. The symmetry axiom: $d(x,y) = d(y,x)$, and
3. The triangle inequality: $d(x,z) \leq d(x,y) + d(x,z)$.

Considering graphs, the *shortest path* (or *geodesic*) between nodes is a metric that can be found in different algorithms such as Dijkstra's[8] [226], Bellman[9]-Ford[10] [227] and Floyd[11]-Warshall[12] [228, 229]. The number of edges in a geodesic is known as *distance* between two nodes. The *eccentricity* of a node is the longest distance between this node and any other in the graph. The *radius* of a graph is the minimum eccentricity of any node. On the other hand, the diameter of a graph is the maximum eccentricity of any node.

[b]Number of connections of a node makes with other nodes.

8.1.1.1 Clustering Coefficient

The node neighborhood of a node v_i is defined as:

$$N_i = \{v_j : e_{ij} \in E \vee e_{ji} \in E\}. \tag{8.2}$$

The *local clustering coefficient* for a node measures the propensity of the neighborhood of a node to form a clique[c], or cluster together. It is defined as the ratio between the existing number of edges among its neighbors (or similarly, the number of triangles that they form) and the total number of edges that there could be among them:

$$C_i = 2\frac{\left|\{e_{jk} : v_j, v_k \in N_i, e_{jk} \in E\}\right|}{|N_i|(|N_i|-1)}. \tag{8.3}$$

In terms of the adjacency matrix, it can be computed as:

$$C_i = \frac{\sum_{j,k} A_{ij}A_{jk}A_{ki}}{\sum_j A_{ij}\left(\sum_j A_{ij} - 1\right)}. \tag{8.4}$$

The *global clustering coefficient*, on the other hand, is the ratio between the number of closed triplets[d] (or three times the number of triangles) and the number of all triples in the graph. In terms of the adjacency matrix, it is given by:

$$C = \frac{\sum_{i,j,k} A_{ij}A_{jk}A_{ki}}{\sum_i \left[\sum_j A_{ij}\left(\sum_j A_{ij} - 1\right)\right]} = \frac{Tr(A^3)}{\sum_i k_i(k_i - 1)}, \tag{8.5}$$

where

$$k_i = \sum_j A_{ij}. \tag{8.6}$$

Figure 8.3 shows three graphs with the same number of nodes but with different topologies. The local clustering coefficients for node 1 are $C_1 = 1$, 1/3, and 0 for the graphs a, b, and c, respectively. On the other hand, the global clustering coefficients are $C = 1$, 0.44, and 0, respectively.

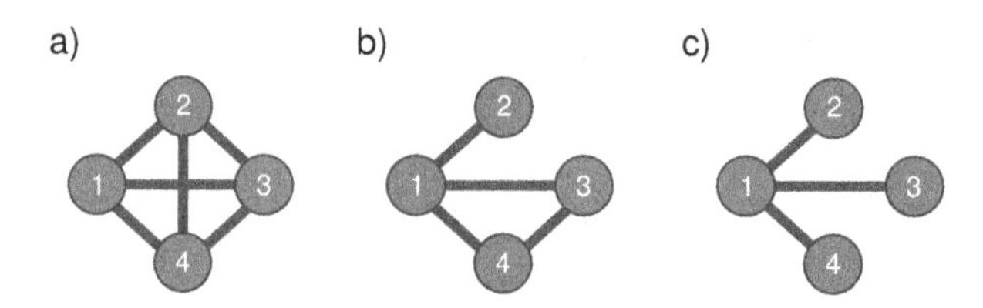

Figure 8.3: Three graphs with the same number of nodes but with distinct topologies

[c] A subgraph where every two distinct nodes are adjacent.

[d] An undirected subgraph consisting of three nodes connected by either two edges (open) or three edges (closed).

A snippet to compute the global clustering coefficient is given bellow.

```
import numpy as np:

def gclust(A):
    N = np.shape(A)[0]

    num = np.trace(np.linalg.matrix_power(A,3))

    k = [0 for j in range(N)]
    for i in range(N):
    for j in range(N):
        k[i] = k[i] + A[i,j]

    den = 0
    for i in range(N):
        den = den + k[i]*(k[i]-1)

    return num/den
```

Some models link the clustering of agents with respect to their demands to many stylized facts such as the fat tails observed in the distribution of returns in the stock market [230].

8.1.1.2 Centrality

How important is a node compared to all others in the network? This is quantified by the *centrality* [231] of a node. The *degree centrality* is simply defined as the ratio between its degree and the total number of nodes subtracted by one. Figure 8.4 shows a *centralized network* where a central node works as a *hub*, a *decentralized network* with multiple hubs, and a *distributed network* with a few or no hubs.

It is also possible to define a *closeness centrality* [232] as the reciprocal of the *farness*, or the sum of the distances d between a specific node and all other nodes of a network:

$$C_c(p) = \frac{|V|-1}{\sum_{(p \neq q) \in V} d(p,q)}. \tag{8.7}$$

There are many other centrality metrics such as *betweenness centrality* [233] and *eigenvector centrality* [234].

The concept of centrality helps us understand some interesting phenomena such as the *friendship paradox*. Consider a network made of symmetrical friendships. The average number of friends a person has is the average degree of the network:

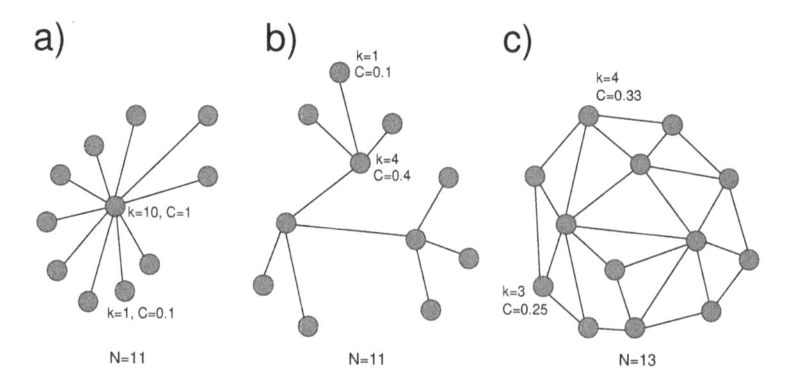

Figure 8.4: a) Centralized network, b) decentralized network, and c) distributed network. k indicates the degree of a node, and C is its degree centrality

$$\mu = \frac{\sum_{v \in V} d(v)}{|V|}. \tag{8.8}$$

The average number of friends that a friend of a person has, though, can be found by randomly choosing an edge and one of its endpoints. One endpoint is the original person, while the other is a friend. The average degree of the latter node is the value we seek. The probability of selecting a node with a specific degree is:

$$p(v) = \frac{d(v)}{2|E|} = \frac{d(v)}{\sum_{v \in V} d(v)}. \tag{8.9}$$

Hence, the average number of friends of friends is:

$$\begin{aligned} \mu_{ff} &= \sum_{v \in V} p(v) d(v) = \sum_{v \in V} \frac{d^2(v)}{\sum_{v \in V} d(v)} \\ &= \frac{|V|}{|V| \sum_{v \in V} d(v)} \sum_{v \in V} d^2(v). \end{aligned} \tag{8.10}$$

Cauchy-Schwarz Inequality

Consider the polynomial function $f : \mathbb{R} \to \mathbb{R}^+$:

$$\begin{aligned} f(x) &= \sum_i (a_i x - b_i)^2 \\ &= \left(\sum_i a_i^2\right) x^2 - 2\left(\sum_i a_i b_i\right) x + \sum_i b_i^2. \end{aligned} \tag{8.11}$$

Since it is a nonnegative function, its determinant has to be less than or equal to zero:

$$\begin{aligned} \Delta_x = 4\left(\sum_i a_i b_i\right)^2 - 4\left(\sum_i a_i^2\right)\left(\sum_i b_i^2\right) \leq 0 \\ \left(\sum_i a_i^2\right)\left(\sum_i b_i^2\right) \geq \left(\sum_i a_i b_i\right)^2. \end{aligned} \tag{8.12}$$

This is known as the *Cauchy*[a]*-Schwarz*[b]*inequality.*

[a] Augustin-Louis Cauchy (1789–1857) French polymath.

[b] Karl Hermann Amandus Schwarz (1853–1921) Prussian mathematician, advisee of Karl Weierstrass.

According to the Cauchy-Schwarz inequality, we get:

$$\begin{aligned} \mu_{ff} &= \frac{1}{|V|\sum_{v\in V} d(v)}\left(\sum_{v\in V} 1^2\right)\left(\sum_{v\in V} d^2(v)\right) \geq \frac{1}{|V|\left(\sum_{v\in V} d(v)\right)}\left(\sum_{v\in V} d(v)\right)^2 \\ \mu_{ff} &\geq \frac{\sum_{v\in V} d(v)}{|V|} = \mu \\ \mu_{ff} &\geq \mu. \end{aligned} \tag{8.13}$$

This implies that, on average, people tend to make friends with people who already have a number of friends higher than the average centrality. This appears to be a paradox since the original person that we picked could now be a friend's friend. The solution, however, is in the fact that we are talking about averages. This problem, related to structure of the social network, also illustrates the *class size paradox* where a person can experience a much more crowded environment than it really is.

8.1.1.3 Assortativity

In many networks, there is often a tendency of similar nodes to preferentially attach to each other. The *assortativity* (or *homophily*) quantifies this tendency through the correlation of nodes.

Let's define e_{ij} as the probability that an edge connects a node of type i to another node of type j, such that $\sum_{ij} e_{ij} = 1$. Also, let $a_i = \sum_j e_{ij}$ be the probability that an edge comes from a node of type i and $b_j = \sum_i e_{ij}$ be the probability that an edge connects to a node of type j. The assortativity coefficient is then given as:

$$r = \frac{\sum_i e_{ii} - \sum_i a_i b_i}{1 - \sum_i a_i b_i} = \frac{Tr(\mathbf{e}) - \|\mathbf{e}^2\|}{1 - \|\mathbf{e}^2\|}, \tag{8.14}$$

where $\|\mathbf{a}\|$ is the sum of all elements of $\mathbf{a}$. If there is no assortative mixing, then $e_{ij} = a_i b_j$ and, according to Eq. 8.14, $r = 0$. On the other hand, if there is perfect assorativity, then $\sum_i e_{ii} = 1$ and $r = 1$. If the network is perfectly *disassortative* (every node connects to a node of a different type), then $e_{ii} = 0$ and:

$$r = \frac{\sum_i a_i b_i}{\sum_i a_i b_i - 1} = \frac{\|\mathbf{e}^2\|}{\|\mathbf{e}^2\| - 1}. \tag{8.15}$$

The graph in Fig. 8.4a is not assortative, whereas the graphs in Fig. 8.4b and c have assortativity coefficients of -0.6 and -0.2, respectively, implying some level of disassortativeness. A regular graph, on the other hand, is perfectly assortative. The following snippet computes the assortativity.

```
import numpy as np

def assort(A):
        # Degree matrix
        D = np.sum(A,1)

        # Total number of edges
        T = int(np.sum(D)/2)

        # Number of nodes
        N = np.shape(A)[0]

        # Prob. edge from
        e = np.array([[0.0 for i in range(max(D)+1)] for j in range(max(D)+1)])
        for i in range(N-1):
                for j in range(i+1,N):
                        e[D[i],D[j]] += A[i,j]/T

        p = np.sum(np.dot(e,e))
        num = np.trace(e)-p
        den = 1-p

        if (np.trace(e) >= 0.9999):
                r = 1
        else:
                r = num/den

return(r)
```

8.1.2 RANDOM NETWORKS

One of the first attempts to understand social networks was given by the random network model proposed by Erdös-Rényi [235][13]. In this model, we start with N unconnected nodes and pick pairs randomly, connecting them by an edge with constant probability p.

The probability of finding a node with degree k is given by the product of three terms: i) the possibility of selecting k links among the total number $N-1$, ii) the probability that k nodes are present, and iii) the probability that the remaining nodes are not chosen. Mathematically, this gives us a binomial distribution:

$$P(k) = \binom{N-1}{k} p^k (1-p)^{N-k}. \tag{8.16}$$

The expected degree of the network can be found expanding the binomial expression:

$$\begin{aligned}
(p+q)^N &= \sum_k \binom{N}{k} p^k q^{N-k-1} \\
\frac{\partial}{\partial p}(p+q)^N &= \sum_k \binom{N}{k} k p^{k-1} q^{N-k-1} \\
N(p+q)^{N-1} &= \frac{1}{p}\sum_k k \binom{N}{k} p^k q^{N-k-1} \\
Np &= \sum_k k \binom{N}{k} p^k (1-p)^{N-k-1}, \text{ where } q = 1-p \\
\langle k \rangle &= Np.
\end{aligned} \tag{8.17}$$

Similarly, we can find the variance of this degree:

$$\begin{aligned}
\frac{\partial^2}{\partial p^2}(p+q)^N &= \sum_k \binom{N}{k} k(k-1) p^{k-2} q^{N-k-1} \\
&= \sum_k \binom{N}{k} k^2 p^{k-2} q^{N-k-1} - \sum_k \binom{N}{k} k p^{k-2} q^{N-k-1} \\
p^2 N(N-1)(p+q)^{N-2} &= \langle k^2 \rangle - \langle k \rangle \\
\langle k^2 \rangle &= p^2 N(N-1) + Np \\
\langle k^2 \rangle - \langle k \rangle^2 &= p^2 N(N-1) + Np - N^2 p^2 \\
\sigma_k^2 &= Np - Np^2 \\
&= Np(1-p).
\end{aligned} \tag{8.18}$$

Therefore, the bigger the network is, the more the distribution shifts and spreads towards larger values. Social networks, though, do not empirically show this behavior [236].

8.1.2.1 Average Path Length

It is fair to say that between two near nodes there can be, on average, $\langle k \rangle$ paths. For second neighbors, there may be $\langle k \rangle^2$ paths, and so on. Therefore, for a distance d, the number of paths is:

$$N(d) = 1 + \langle k \rangle + \langle k^2 \rangle + \ldots + \langle k^d \rangle = \sum_{i=0}^{d} \langle k^i \rangle = \frac{\langle k \rangle^{d+1} - 1}{\langle k \rangle - 1}. \tag{8.19}$$

However, the maximum path length (diameter of the network) cannot be longer than the number of nodes. Therefore, $N(d_{\max}) \approx N$ and we deduce that:

$$\begin{aligned}\langle k\rangle^{d_{\max}} &\approx N \\ d_{\max}\log(\langle k\rangle) &\approx \log(N) \\ d_{\max} &\approx \frac{\log(N)}{\log(\langle N\rangle)}.\end{aligned} \tag{8.20}$$

This sublinear relationship between the diameter and the number of nodes is known as *small world phenomenon*. This was initially measured in 1967 by Milgram[14] using letters [237,238]. In short, a recipient would receive a letter addressed to some person and then forward this letter to a friend who was likely to know him or her. After the letter was received by the final contact person, the researcher would count how many steps were necessary for completing this path. The result average path length was about six, which gave rise to the popular expression *six degrees of separation*.

8.1.2.2 Clustering Coefficient

The k_i neighbors of a node i can make a maximum number of connections among themselves:

$$\binom{k_i}{2} = \frac{k_i!}{2(k_i-2)!} = \frac{k_i(k_i-1)}{2}. \tag{8.21}$$

Its neighbors, however, are connected with probability p, creating $p\binom{k_i}{2}$ connections. Hence, the expected number of links among the neighbors of node i is:

$$\langle L_i\rangle = p\frac{k_i(k_i-1)}{2}. \tag{8.22}$$

The clustering coefficient is exactly this probability:

$$C_i = p = \frac{2\langle L_i\rangle}{k_i(k_i-1)} = \frac{\langle k\rangle}{N}, \tag{8.23}$$

as previously derived in Eq. 8.17.

This implies that the clustering coefficient should be inversely proportional to the size of the network. However, this is not observed experimentally either. Rather, social networks show a high clustering coefficient nearly independent on the size of the network.

8.1.3 SCALE-FREE NETWORKS

A network is considered *scale invariant* if its degree distribution follows a power law $P(k) \propto k^{-\gamma}$ with no characteristic scale. Thus, when rescaling its distribution, we get the same distribution (except for a multiplicative factor):

$$P(ak) \propto a^{-\gamma}k^{-\gamma} \propto P(k). \tag{8.24}$$

This distribution can be normalized as:

$$P(k) = Ck^{-\gamma}. \tag{8.25}$$

Since, it is a distribution:

$$\sum_k P(k) = C\sum_k k^{-\gamma} = 1 \rightarrow C = \left(\sum_k k^{-\gamma}\right)^{-1} = \zeta^{-1}(\gamma), \tag{8.26}$$

where $\zeta(\gamma)$ is the Riemann[15] zeta function.

One way of producing scale-free networks is using a *preferential attachment* procedure such as the Yule-Simon's urn process [239, 240] in which the probability of adding a ball to a growing number of urns is linearly proportional to the number of balls already in an urn. In the preferential attachment process in networks proposed by Barabasi[16] and Albert[17] [241, 242] we start with m_0 nodes and a node with $m < m_0$ links is progressively added to the network. Every new node is connected to an existing node with probability proportional to the number of connections it already has:

$$p_i = \frac{k_i}{\sum_n k_n}. \tag{8.27}$$

8.1.3.1 Degree Distribution

The sum in the denominator of the last equation can be computed with the *degree sum formula*[e].

Degree sum formula

Consider a pair (v, e), where v is a node and e is an edge. The number of edges that connect to a node v is simply its degree: $deg(v)$. Therefore, the sum of all degrees is the sum of all incident pairs (v, e). Each edge, though, is connected to two nodes. Therefore, the total number of pairs is twice the number of edges. Since both sums are the same, we conclude that the sum of all degrees of a graph is twice the number of edges:

$$\sum_v deg(v) = 2|E|. \tag{8.28}$$

This implies that the sum of degrees of all nodes (even if they are odd) is always even. Consequently, if we imagine a group of people, the number of those who have shaken hands with people from a subgroup with an odd number of individuals is always even. Hence, this is also known as the *handshaking lemma*.

[e] A nice derivation of the properties of scale-free networks can be found in [243].

If a new node makes m connections at every instant, then the number of edges is mt. If we discount this new node and use the degree sum formula, we find that the denominator gives:

$$\sum_n k_n = 2mt - m. \tag{8.29}$$

The temporal change of the degree of a node has to be propotional to the number of connections that are added and the probability that we find a node with this degree:

$$\begin{aligned}
\frac{dk_i}{dt} &= mp_i \approx \frac{mk_i}{2mt} \text{ for large } t \\
\int_{t_i}^{t} \frac{dk_i}{k_i} &= \int_{t_i}^{t} \frac{dt}{2t} \\
\ln\left(\frac{k_i}{m}\right) &= \frac{1}{2}\ln\left(\frac{t}{t_i}\right), \text{ since } k_i(t) = m \\
k_i(t) &= m\left(\frac{t}{t_i}\right)^{1/2}.
\end{aligned} \tag{8.30}$$

The probability of finding a node with degree smaller than k is:

$$\begin{aligned}
P(k_i(t) < k) &= P\left(m\left(\frac{t}{t_i}\right)^{1/2} < k\right), \text{ from the previous equation.} \\
&= P\left(t_i > \frac{m^2 t}{k^2}\right) \\
&= 1 - P\left(t_i \leq \frac{m^2 t}{k^2}\right).
\end{aligned} \tag{8.31}$$

Since we are adding a node at fixed time steps, the number of nodes with degree smaller than k is just $N_< = t\frac{m^2}{k^2}$. On the other hand, the total number of nodes grows linearly as $N_T = m_0 + t \approx t$ for $t \to \infty$. Therefore, the probability of finding a node with a degree smaller than k is:

$$P(k_i(t) < k) = 1 - \frac{m^2}{k^2}. \tag{8.32}$$

Therefore, the probability of finding a node with degree k is:

$$p(k) = \frac{\partial P(k_i(t) < k)}{\partial k} = 2m^2 k^{-3}. \tag{8.33}$$

Most of real social networks show a power-law behavior similar to this.

8.1.3.2 Clustering Coefficient

Let's define the *preferential attachment* of a new node j to an existing node i with degree k_i as:

$$\Pi\left(k_i(j)\right) = \frac{k_i(j)}{\sum_n k_n(j)}. \tag{8.34}$$

If the new node makes m connections, then:

$$p_{ij} = m\Pi\left(k_i(j)\right) = m\frac{k_i(j)}{\sum_l k_l(j)} = \frac{k_i(j)}{2j}. \tag{8.35}$$

Considering that the arrival time of the ith node is i and using the result from Eq. 8.30:

$$p_{ij} = \frac{m\left(\frac{j}{i}\right)^{1/2}}{2j} = \frac{m}{2}(ij)^{-1/2}. \tag{8.36}$$

Assuming now a continuum, the number of connections among neighbors is given by:

$$\begin{aligned}
N_\triangle &= \int_{i=1}^{N}\int_{j=1}^{N} P(i,j)P(i,l)P(j,l)\,di\,dj \\
&= \frac{m^3}{8}\int_{i=1}^{N}\int_{j=1}^{N} (ij)^{-1/2}(il)^{-1/2}(jl)^{-1/2}\,di\,dj \\
&= \frac{m^3}{8l}\int_{i=1}^{N}\frac{di}{i}\int_{j=1}^{N}\frac{dj}{j} \\
&= \frac{m^3}{8l}\left(\ln(N)\right)^2.
\end{aligned} \tag{8.37}$$

Therefore, the clustering coefficient is given by:

$$\begin{aligned}
C_l &= \frac{2N_\triangle}{k_l(k_l-1)} \\
&= \frac{\frac{m^3}{4l}\left(\ln(N)\right)^2}{k_l(k_l-1)}.
\end{aligned} \tag{8.38}$$

Using once again the result from Eq. 8.30:

$$\begin{aligned}
k_l &= m\left(\frac{N}{l}\right)^{1/2} \\
k_l(k_l-1) &\approx k_l^2 = m^2\frac{N}{l},
\end{aligned} \tag{8.39}$$

we get:

$$C_l \approx \frac{m}{4N}\left(\ln(N)\right)^2. \tag{8.40}$$

Therefore, the Barabasi-Albert network has a higher clustering coefficient when compared to the random network. The clustering behavior found in real social networks, on the other hand, tends to be higher.

8.1.4 SMALL WORLD NETWORKS

The small world property of social networks is well captured by the Watts[18]-Strogatz[19] model [244].

This network can be constructed from a regular ring network composed of N nodes connected to K neighbors symmetrically to each side. A node is picked randomly and $K/2$ neighboring nodes are reconnected with any other node of the ring with probability β[f]. Hence, for $\beta = 0$, we end up with a regular network, whereas for $\beta = 1$ the resulting network is random, as shown in Fig. 8.5. Intermediate values of β generate networks with high clustering coefficients, small diameter, and small world property. The degree distribution for this model, however, does not reflect that of real social networks.

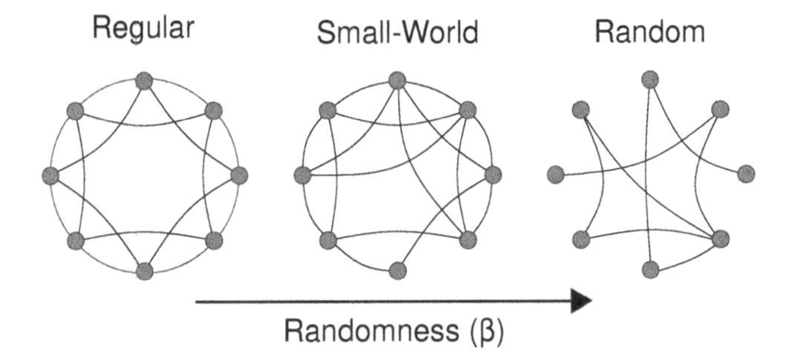

Figure 8.5: Varying parameter β in the Watts-Strogatz model, it is possible to generate regular, small-world, and random networks

8.2 SOCIOECONOMIC MODELS

The Ising[20] model [245, 246] is a popular tool in statistical mechanics used to study ferromagnetic systems. In this model, we have a Hamiltonian[21] given by:

$$\mathscr{H} = -h\sum_{i} s_i - \sum_{i,j} J_{ij} s_i s_j, \tag{8.41}$$

where h and J are coupling constants, and s is a spin state. Typically J_{ij}, which represents a spin-spin interaction, is such that it is a constant for nearest neighbors and 0 otherwise. The constant h can represents the presence of an external field, and the first sum is related to this field trying to align the spins in a specific direction.

[f]This model is also known as the beta model because of this parameter.

In a typical algorithm to find the equilibrium state of a Ising model, single spin states are randomly created and tested in a Monte Carlo approach (see Appendix B). This is generally a slow process, but alternatives such as the Swendsen[22] [247] and Wolff[23] algorithms [248] are available. In the latter, for example, we create clusters of spins as test states.

In this section we will study how similar ideas using agents instead of spins can be used to model some socioeconomic models. We will start with a model for social segregation and then move to opinion dynamics and will finish the chapter with a simple, yet elegant, model for the formation of prices in a market.

8.2.1 SCHELLING'S MODEL OF SEGREGATION

The Schelling[24] model [249] is an Ising-like agent-based model [250–252] proposed to study social segregation. It is based on an automaton with a $Z \subset \mathbb{Z}^2$ lattice of size N and a neighborhood (originally Moore). The states of the automaton are $S = \{A, B, 0\}$, where A and B are two types of agents that may occupy a grid cell, and 0 indicates an empty one. Only one agent may occupy a grid cell at a time.

The simulation starts with a fraction $\rho = N_0/N^2$ of unoccupied cells. The remaining fraction $1 - \rho$ is randomly occupied by agents of either group with equal probability. This can be created with the following snippet:

```
def createGrid(rho,N):
        return np.array([[np.random.choice([0,-1,1],p=[rho,(1-rho)/2,(1-rho)/2]) \
                for i in range(N)] for j in range(N)])
```

At each round, the agents study their neighborhoods and check the fraction of neighbors that are of the same type:

```
def neighborhood(t):
        p = np.zeros(np.shape(t))

        # Von Neumann neighborhood
        p[:,:-1] = t[:,1:]
        p[:,1:] = p[:,1:] + t[:,:-1]
        p[:-1,:] = p[:-1,:] + t[1:,:]
        p[1:,:] = p[1:,:] + t[:-1,:]

        p = np.where(t != 0, p*t,0)

return p
```

If this fraction is below a certain threshold f, then the agent is unsatisfied and relocates to an empty grid cell[g]. If only unsatisfied agents are allowed to migrate, then we say that is a *constrained* (or *solid*) simulation, whereas if all agents are allowed to migrate (as long as they do not worsen their situations), then we say that it is an *unrestricted* (or *liquid*) simulation. Note that the agents can improve their satisfaction even if their migration may reduce the satisfaction of their neighbors (see Pareto efficiency—Sec. 7.1.4.1). Also, even though the global polarization of the lattice is preserved, the local polarization is not.

We can define two other functions, one for moving an agent to a new destination and another one that finds a new destination:

```
def move(t,frm,to):
        k = t[frm]
        t[frm] = 0
        t[to] = k

def destination(t):
        unoccupied = np.where(t == 0)
        l = len(unoccupied[0])
        p = np.random.randint(l)

        m = unoccupied[0][p]
        n = unoccupied[1][p]

        return(m,n)
```

Given these functions, the simulation itself is performed by:

```
grid = createGrid(0.3,80)

for it in range(50000):
        nbr = neighborhood(grid)
        frm = np.unravel_index(nbr.argmin(),nbr.shape)
        to = destination(grid)
        move(grid,frm,to)
```

[g]See homophily in Sec. 8.1.1.3.

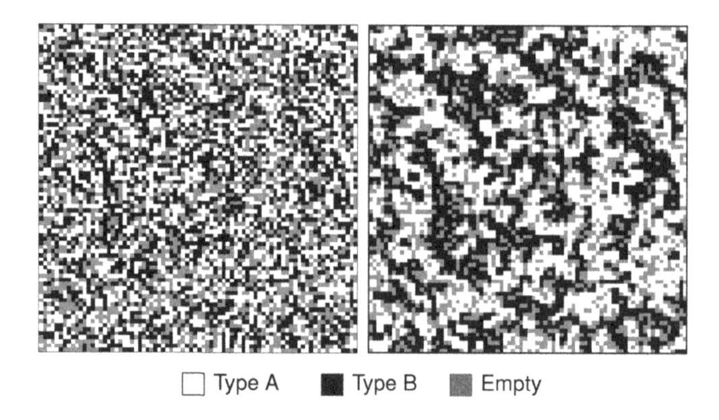

Figure 8.6: Initial grid configuration (left) and steady state grid configuration (right) for the Schelling model with $\rho = 0.3$ on a 80×80 lattice

The result of this simulation is shown in Fig. 8.6. It is clear that the steady state solution exhibits the formation of clusters with agents of different types.

In order to quantify the formation of clusters, we can use the *segregation coefficient* S [253]. This is an order parameter corresponding to the weighted average cluster size:

$$S = \sum_{\{i\}} n_i p_i, \tag{8.42}$$

where $p_i = n_i/M$ is the probability of finding a cluster of mass n_i, and $M = N^2(1-\rho)$ is the total number of agents. A normalized segregation coefficient is often calculated as:

$$s = \frac{S}{M/2} = \frac{2}{M}\sum_{\{i\}} n_i \frac{n_i}{M} = \frac{2}{[N^2(1-\rho)]^2}\sum_{\{i\}} n_i^2, \tag{8.43}$$

where we have used the fact that the biggest cluster can only be $M/2$. In the extreme situation where there are only two clusters, the normalized segregation coefficient is 1, whereas it is $1/M$ if there is no cluster formation.

Clusters can be identified with the following flood fill algorithm:

```
def cluster(t,y,x,c):
    L = np.shape(t)[0]
    mass = 0

    candidates = [(y,x)]
    while(len(candidates)>0):
        y,x = candidates.pop()
        if (t[y,x] == c):
            if (y > 0):
                candidates.append((y-1,x))
            if (y < L-1):
                candidates.append((y+1,x))
            if (x > 0):
                candidates.append((y,x-1))
            if (x < L-1):
                candidates.append((y,x+1))

            mass = mass + 1
            t[y,x] = 3

    return mass
```

The segregation coefficient can be found with:

```
def segregation(t):
    L = np.shape(t)[0]
    n = []

    for j in range(L):
        for i in range(L):
            mass = cluster(t,j,i,1)
            if (mass > 0):
                n = np.append(n,mass)
            mass = cluster(t,j,i,-1)
            if (mass > 0):
                n = np.append(n,mass)

    return 2*np.sum(n**2)/np.sum(t!=0)**2
```

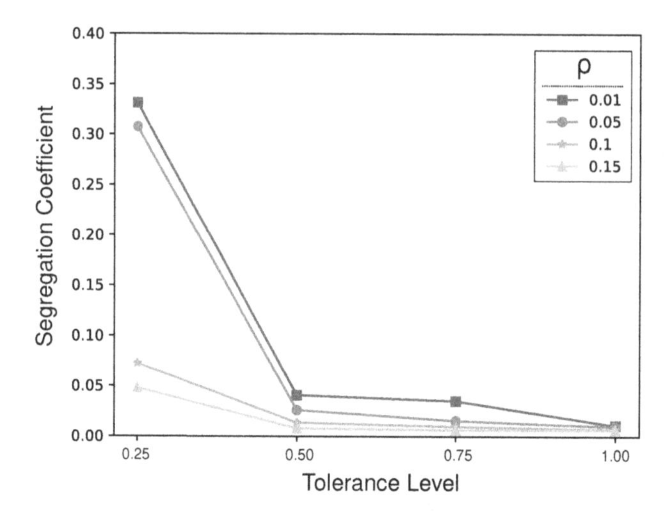

Figure 8.7: Single shot simulation of the segregation coefficient as a function of the tolerance level for the Schelling model on a 80×80 grid and different values of ρ

A single shot simulation of the segregation coefficient as a function of the *tolerance level*[h] is shown in Fig. 8.7. The tolerance level can only assume four values when using von Neumann neighborhood: 1/4, 2/4, 3/4, and 4/4. It is interesting to note that a tolerance level higher than 1/4 is enough to cause a significant lowering of the segregation coefficient.

Although Schelling model does not capture many restrictions such as financial barriers, it is supported by many empirical evidence (see [254, 255], for instance).

8.2.2 OPINION DYNAMICS

In 1951 Asch[25] proposed the following experiment [256]: given a card A with a single line drawn on it and another card B with three lines, a college student would have to answer which of the three lines in card B had the same length as the one in card A. The participant would be in a group with seven confederates that would purposefully choose a wrong answer. Asch also had a control condition where the participant would be tested alone. The experiment showed that approximately 75 % of the participants conformed at least once with the group even if the answer was completely wrong, whereas less than 1 % of the participants gave wrong answers in the control condition. This is a classic experiment that shows the tendency of humans to conform with a group.

When the coordination of behaviors occurs spontaneously without a central authority, we call it *herd behavior*. This is widely seen in nature as a form of *collective behavior*[i]. This happens, for example, during an *information cascade* when investors

[h]The fraction of neighbors of a different type one tolerates before moving.

[i]One of the most popular models for collective motion is the Vicsek[26] model.

tend to follow the investment strategies of other agents rather then their own [257]. This partially explains the dot-com bubble of 2001, for example. After seeing the commercial potential of the internet, investors put aside their personal believes and conformed with a common tendency of investing in e-commerce start-ups. Between 1995 and 2000, the Nasdaq index rose more than 400 %, but after this *irrational exuberance*[27], the index lost all its gains leading to the liquidation of a vast number of companies and a general glut in the job market for programmers.

The Hegselmann[28]-Krause[29] (HK) is an agent-based model [258] that tries to capture this behavior. The simulation starts with an opinion profile $x_i \in [0;1]$, $i = 1,\ldots,N$, where N is the number of agents:

```
import numpy as np
import matplotlib.pyplot as pl

NA = 50
NI = 10

x = np.zeros((NA,NI))
x[:,0] = [np.random.uniform() for i in range(NA)]
```

At each simulation step, a neighborhood for each individual is formed with agents that have similar opinions:

$$\mathcal{N}_t(n) = \{m : |x_t(m) - x_t(n)| \leq \varepsilon\}, \tag{8.44}$$

where ε is a *confidence level*. Therefore, the neighborhood is bounded by this level and the model is often known as a *bounded confidence model*. The following snippet finds the neighborhood and calculate the average opinion:

```
def neighborhood(S,el):
        ac = 0
        z = 0.0
        for y in S:
                if abs(y-el) <= 0.05:
                        z = z + y
                        ac = ac + 1

        return z/ac
```

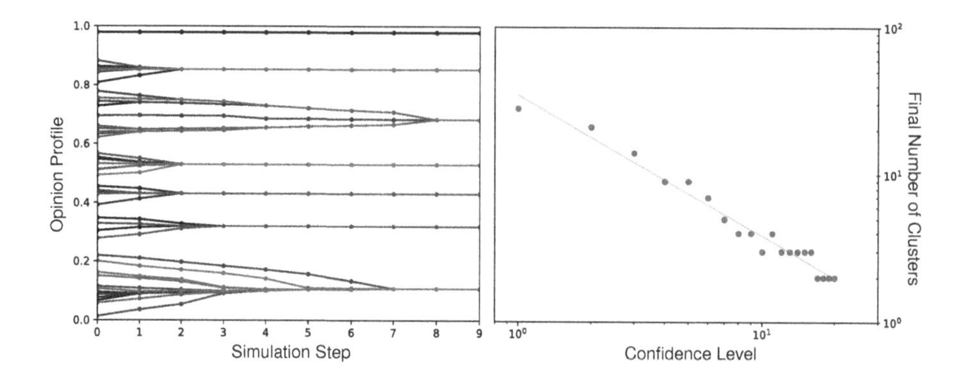

Figure 8.8: The evolution of the opinion profile in the HK model (left) and the final number of clusters as a function of the confidence level

The opinion profile is updated by the average opinion of the neighborhood of each agent:

$$x_n^{t+1} = \frac{1}{|\mathcal{N}_t(n)|} \sum_{m \in \mathcal{N}_t(n)} x_m^t. \tag{8.45}$$

```
for t in range(NI-1):
    for i in range(NA):
        x[i,t+1] = neighborhood(x[:,t],x[i,t])
```

The result for a simulation with 50 agents in shown in Fig. 8.8. Regardless of the initial distribution, clusters of agents tend to be formed and opinions tend to a small set. The final number of clusters approximately depends on the confidence level as $N_{final} \sim \varepsilon^{-1}$. In other words, more groups are formed as the agents restrict their neighborhood of individuals with similar opinions.

A variant of the HK is the Deffuant[30] model [259]. There, pairs of agents are picked randomly and adjust their opinions if they are relatively close. Otherwise, communication is believed not to be possible and they keep their old believes. The update rule is given by:

$$\begin{aligned} x_i(t+\Delta) &= x_i(t) - \mu\left[x_i(t) - x_j(t)\right] \\ x_j(t+\Delta) &= x_j(t) - \mu\left[x_j(t) - x_i(t)\right], \end{aligned} \tag{8.46}$$

where $\mu \in [0, 1/2]$ is a *convergence parameter.*

The initialization for this simulation is identical to that of the HK model, but the dynamics is now given by:

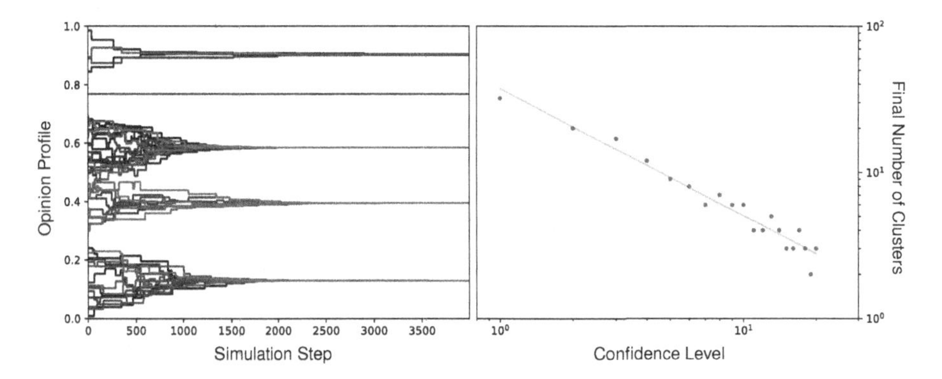

Figure 8.9: The evolution of the opinion profile in the Deffuant model (left) and the final number of clusters as a function of the confidence level

```
for t in range(NI-1):
        i = np.random.randint(NA)
        j = np.random.randint(NA)

        x[:,t+1] = x[:,t]
        if abs(x[i,t]-x[j,t]) < 0.1:
                x[i,t+1] = x[i,t] - mu*(x[i,t]-x[j,t])
                x[j,t+1] = x[j,t] - mu*(x[j,t]-x[i,t])
```

The result of a simulation with 50 agents and $\mu = 0.35$ is shown in Fig. 8.9.

As in the HK model, the final number of clusters approximately depends on the confidence level as $N_{final} \sim \varepsilon^{-0.9}$.

8.2.2.1 Kirman Model

Imagine two nearly identical securities. It is not uncommon that a large majority of investors end up choosing one rather than the other. The same happens, for instance, with ants presented with two sources of similar foods. Instead of the ants exploring both sources equally, one source is consumed first. The Kirman[31] model [260] is a Markov chain agent-based model created to answer this kind of problem.

Let's start with two distinct sources A and B and a total population of N agents. At each simulation step, two random agents meet and one is converted to the other's opinion with a chance $1 - \delta$. There is also a probability ε that an agent changes its opinion independently of meeting another agent. This could happen, for instance, as a reaction to an exogenous information.

The state of the system $k \in (0,1,\ldots,N)$ is defined as the number of agents that prefer source A. Therefore, the probability p_1 that k increases by one agent is given by:

$$p_1 = p(k,k+1) = \left(1 - \frac{k}{N}\right)\left(\varepsilon + (1-\delta)\frac{k}{N-1}\right), \tag{8.47}$$

whereas the probability that k is decreased by one agent is given by:

$$p_2 = p(k,k-1) = \frac{k}{N}\left(\varepsilon + (1-\delta)\frac{N-k}{N-1}\right). \tag{8.48}$$

There is also a probability $p_3 = 1 - p_1 - p_2$ that k remains unchanged. Note that this model resembles a Polya urn process (see Sec. 3.1.2). If $\varepsilon = 1/2$ and $\delta = 1$ then it is just an Ehrenfest[32] urn process where the agents change opinions without any interaction. Also, when $\varepsilon = \delta = 0$, the expected value of k is:

$$\langle k_{n+1}|\mathscr{F}_n\rangle = (k_n+1)p_0 + (k_n-1)p_0 + k_n(1-2p_0) = k_n, \tag{8.49}$$

where

$$p_0 = \left(1 - \frac{k}{N}\right)\frac{k}{N-1} = \frac{k}{N}\frac{N-k}{N-1}. \tag{8.50}$$

Therefore, under these parameters, the process becomes a martingale (see Sec. 3.1).

The Kirman model can easily be simulated with the following snippet:

```
N = 100
eps = 0.15
delta = 0.3
k = N/2
x = []

for it in range(10000):
    p1 = (1-float(k)/N)*(eps+(1-delta)*float(k)/(N-1))
    p2 = (float(k)/N)*(eps+(1-delta)*(N-k)/(N-1))

    r = np.random.rand()

    if (r <= p1):
        k = k + 1
    if (r > p1 and r <= p1+p2):
        k = k - 1

x.append(k)
```

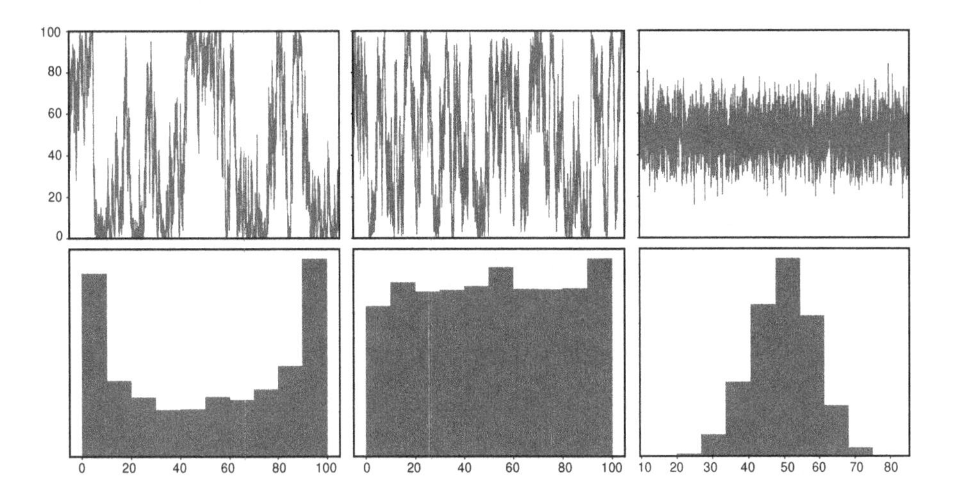

Figure 8.10: The number of agents k that prefer source A as a function of the simulation step (top) and their respective histograms for $\varepsilon = 5\times10^{-3}$, $\delta = 10^{-2}$ (left), $\varepsilon = 10^{-2}$, $\delta = 2\times10^{-2}$ (center), and $\varepsilon = 0.15$, $\delta = 0.3$ (right)

Some results of this simulation are shown in Fig. 8.10. It is interesting to note that it is possible to adjust the type of distribution of k by changing the values of ε and δ.

8.2.3 MARKET SPIN MODELS

The Bornholdt model[33] [261] is another Ising-like automaton on a $Z \subset \mathbb{Z}^2$ lattice of size N and a Von Neumann neighborhood. The states of this automaton are $S = \{+1, -1\}$, where $+1$ corresponds to a buyer, and -1 corresponds to a seller state.

The simulation starts with a random field and the dynamics is given by a stochastic transition function that assigns the state $+1$ to a cell with probability p_i, and the state -1 with a probability $1 - p_i$. The probability p_i is given by:

$$p_i = \left[1 + \exp\left(-2\beta h_i(t)\right)\right]^{-1}, \tag{8.51}$$

where β is the inverse temperature. $h_i(t)$ is a local field representing the influence to conform with the majority of nearest neighbors in accordance with earlier models [262, 263]:

$$h_i(t) = J \sum_{j \in N^1_{VN}(i)} S_j(t) - \alpha C_i(t) \langle S \rangle, \tag{8.52}$$

where J is a *disagreement* constant that indicates the tendency for the agent to conform, and $\alpha > 0$ is a *demagnetizing* constant that indicates the tendency for the agent to seek an anti-ferromagnetic order. This latter term represents the preference towards the minority group (see Sec. 7.1.1.1). C_i is a second spin available to each cell that relates to the strategy of the agent with respect to the global magnetization.

$C_i = 1$, for instance, designates a desire for the agent to join the global minority group that is interested in future returns, a *fundamentalist* behavior. The other situation $C_i = -1$ points to the desire to follow the majority group, a *chartist* behavior. Thus, the dynamics of the *strategy spin* is given by:

$$C_i(t+1) = \begin{cases} -C_i(t) & \text{if } \alpha S_i(t) C_i(t) \sum_j S_j(t) < 0, \\ C_i(t) & \text{otherwise.} \end{cases} \tag{8.53}$$

If, however, the strategy spin is allowed to change instantaneously, then Eq. 8.52 becomes:

$$h_i(t) = J \sum_{j \in N_{VN}^1(i)} S_j(t) - \alpha S_i(t) \left| \langle S(t) \rangle \right|. \tag{8.54}$$

The magnetization of the system $M(t) = \langle S(t) \rangle$ is identified as the price, from which it is possible to obtain the logarithmic returns.

To simulate the Bornholdt model we start with the following snippet:

```
import numpy as np
import random as rd
import matplotlib.pyplot as pl

N = 32
J = 1
beta = 1.0/1.5
alpha = 4

S = np.array([rd.choices([1,-1],k=N) for i in range(N)])
Sn = np.zeros((N,N))

r = []
M = 1
```

The simulation itself is an update loop:

```
for it in range(2000):
        Ml = M
        M = np.average(S)
        for i in range(N):
                for j in range(N):
                        sm = 0
                        for x in neig(i,j,N):
                                sm = sm + S[x[0],x[1]]
                        h = J*sm - alpha*S[i,j]*abs(M)
                        p = 1.0/(1+np.exp(-2*beta*h))

                        if (rd.random() < p):
                                Sn[i,j] = 1
                        else:
                                Sn[i,j] = -1

                S = Sn
                r = np.append(r,np.log(abs(M))-np.log(abs(Ml)))
```

In the snippet, *neig* is a function that returns the Von Neumann neighborhood:

```
def neig(i,j,N):
        z = []
        if (i > 0):
                z.append([i-1,j])
        if(i < N-1):
                z.append([i+1,j])
        if (j > 0):
                z.append([i,j-1])
        if (j < N-1):
                z.append([i,j+1])

        return np.array(z)
```

The result of the simulation shows metastable phases as shown in Fig. 8.11. Moreover, the log-returns in Fig. 8.12 show fat tails as indicated by the CCDF. Bornholdt also showed that his model also shows some stylized facts such as volatility clustering.

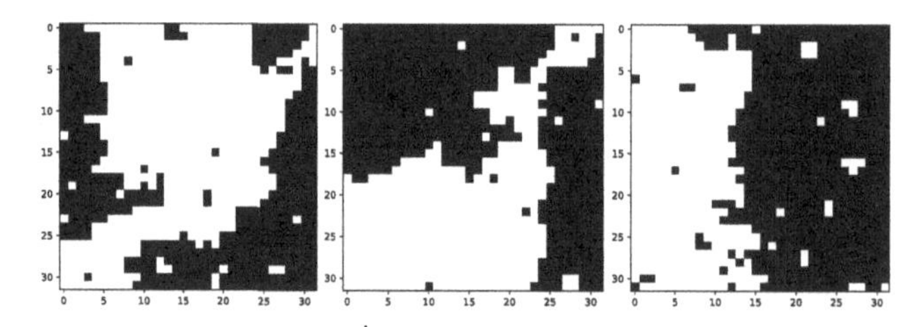

Figure 8.11: Grid configuration on a 32×32 lattice at undercritical temperature after t=100, 200, and 300 simulation steps (from left to right)

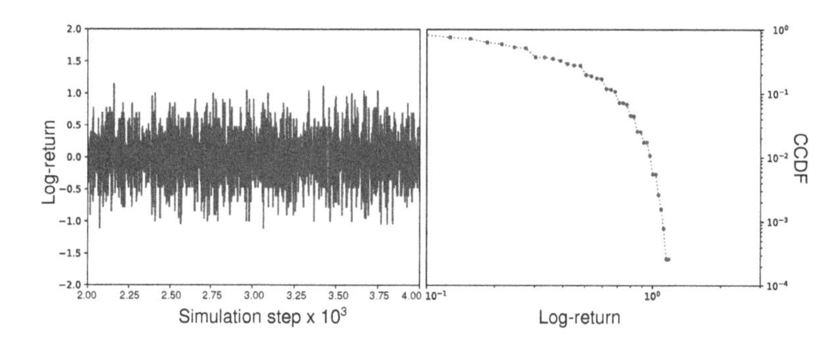

Figure 8.12: Left: Log-returns for the Bornholdt model with $\alpha = 4.0$, $J = 1.0$, and $\beta = 2/3$. Right: The corresponding complementary cumulative distribution function (CCDF)

8.3 KINETIC MODELS FOR WEALTH DISTRIBUTION

In the case of socioeconomic problems, we can use *ad hoc* models [264] where particles maps to agents, energy maps to wealth[j], and the binary collisions map to trade interactions. Under this scheme, we can invoke the famous Boltzmann equation to model the interaction among agents (see Appendix F).

In simulations of wealth distribution, for instance, two traders with initial possessions $v_1, v_2 \in \Omega \subseteq \mathbb{R}$ meet randomly and are distributed as:

$$\begin{bmatrix} v_1'(t) \\ v_2'(t) \end{bmatrix} = \begin{bmatrix} p_1 & q_1 \\ p_2 & q_2 \end{bmatrix} \begin{bmatrix} v_1(t) \\ v_2(t) \end{bmatrix}, \tag{8.55}$$

where $v_1', v_2' \in \Omega$ as well.

In order to work with the framework of random processes, we can adopt associated independent random variables X and Y that are distributed in accordance with:

$$P(X \in S) = P(Y \in S) = \int_S f(v,t)dv, \ \forall S \subseteq \Omega. \tag{8.56}$$

Their interaction rules are:

$$\begin{bmatrix} X'(t) \\ Y'(t) \end{bmatrix} = \begin{bmatrix} p_1 & q_1 \\ p_2 & q_2 \end{bmatrix} \begin{bmatrix} X(t) \\ Y(t) \end{bmatrix}, \tag{8.57}$$

After a small period Δ, the random variables are updated to $X(t+\Delta) = X'(t)$ if there was a binary interaction and $X(t+\Delta) = X(t)$ otherwise. The probability of interaction can be assigned to a Bernoulli distributed random variable T such that $P(T=1) = 1 - P(T=0) = \mu\Delta$, where μ is an *interaction kernel.*

Therefore, we can write:

$$\begin{aligned} X(t+\Delta) &= TX'(t) + (1-T)X(t) \\ Y(t+\Delta) &= TY'(t) + (1-T)Y(t). \end{aligned} \tag{8.58}$$

For any linear observable φ we can calculate expected values:

$$\begin{aligned} \langle \varphi(X(t+\Delta)) \rangle &= \langle T\varphi(X'(t)) \rangle + \langle (1-T)\varphi(X(t)) \rangle \\ \langle \varphi(Y(t+\Delta)) \rangle &= \langle T\varphi(Y'(t)) \rangle + \langle (1-T)\varphi(Y(t)) \rangle \end{aligned}. \tag{8.59}$$

This can be rewritten as:

$$\begin{aligned} \langle \varphi(X(t+\Delta)) - \varphi(X(t)) \rangle &= \mu\Delta \left[\langle \varphi(X'(t)) \rangle - \langle \varphi(X(t)) \rangle \right] \\ \langle \varphi(Y(t+\Delta)) - \varphi(Y(t)) \rangle &= \mu\Delta \left[\langle \varphi(Y'(t)) \rangle - \langle \varphi(Y(t)) \rangle \right] \end{aligned}. \tag{8.60}$$

Taking the limit when $\Delta \to 0$:

$$\begin{aligned} \frac{\partial}{\partial t} \langle \varphi(X(t)) + \varphi(Y(t)) \rangle = \mu \big[&\langle \varphi(X'(t)) \rangle + \langle \varphi(Y'(t)) \rangle \\ &- \langle \varphi(X(t)) \rangle - \langle \varphi(Y(t)) \rangle \big]. \end{aligned} \tag{8.61}$$

[j]Not to be confused with income, which is an inflow of resources.

We are assuming that the possessions are uncorrelated (*Stosszahlansatz*). Therefore, we take the joint probability distribution as the product of the individual ones (see Appendix G). Thus,

$$2\partial_t \left\langle \int_\Omega \varphi(v) f(v,t) dv \right\rangle = \\ = \mu \left\langle \iint_{\Omega\times\Omega} \left[\varphi(v_1') + \varphi(v_2') - \varphi(v_1) - \varphi(v_2)\right] f(v_1,t) f(v_2,t) dv_1 dv_2 \right\rangle_{\substack{p_1,q_1\\ p_2,q_2}}. \tag{8.62}$$

It is possible to obtain a more physics-friendly equation considering a constant interaction kernel and a Dirac[34] delta observable $\varphi(\star) = \delta(v - \star)$ in this weak form of the Boltzmann equation:

$$\partial_t f(v,t) = \frac{1}{2} \left\langle \iint_{\Omega\times\Omega} \left[\delta(v - v_1) + \delta(v - v_2)\right] f(v_1,t) f(v_2,t) dv_1 dv_2 \right\rangle - f(v,t). \tag{8.63}$$

This equation can be written as:

$$\partial_t f + f = Q_+(f,f), \tag{8.64}$$

where $Q_+(f,f)$ is the *collision operator* given by:

$$Q_+(f,f) = \frac{1}{2} \left\langle \iint_{\Omega\times\Omega} \left[\delta(v - v_1) + \delta(v - v_2)\right] f(v_1,t) f(v_2,t) dv_1 dv_2 \right\rangle. \tag{8.65}$$

8.3.1 CONSERVATIVE MARKET MODEL

Some models assume that trade is conservative [265, 266], implying that the wealth is preserved in a transaction. This considers wealth as having an objective nature quantifiable by the stock of some scarce resource such as gold. In this case we have:

$$\int_\Omega f(v,t) dv = m, \tag{8.66}$$

where m is a finite and constant wealth. Also, we have that the collisions are perfectly elastic: $\langle p_1 + p_2 \rangle = \langle q_1 + q_2 \rangle = 1$. In the Chakraborti-Chakrabarti[35] model [265], for example, the distribution of wealth is given by:

$$\begin{bmatrix} v_1' \\ v_2' \end{bmatrix} = 1/2 \begin{bmatrix} 1+\lambda & 1-\lambda \\ 1-\lambda & 1+\lambda \end{bmatrix} \begin{bmatrix} v_1 \\ v_2 \end{bmatrix}, \tag{8.67}$$

where $\lambda \in [0,1]$ is a parameter that relates to the saving tendency of the agents. According to this rule, we have conservation since $v_1' + v_2' = v_1 + v_2$.

The variance of the wealth in this model can be found using the observable $\varphi(v) = (v-m)^2$ in the weak form (Eq. 8.62):

$$\begin{aligned}
\partial_t \left\langle \int_\Omega (v-m)^2 f(v,t)dv \right\rangle &= \\
&= \frac{1}{2} \left\langle \iint_{\Omega\times\Omega} \left[(v_1'-m)^2 + (v_2'-m)^2 - (v_1-m)^2 - (v_2-m)^2\right] \right. \\
&\qquad\qquad \left. f(v_1,t)f(v_2,t)dv_1dv_2 \right\rangle \\
&= -\frac{(1-\lambda^2)}{4} \left\langle \iint_{\Omega\times\Omega} (v_1-v_2)^2 f(v_1,t)f(v_2,t)dv_1dv_2 \right\rangle \\
&= -\frac{(1-\lambda^2)}{4} \left\langle \iint_{\Omega\times\Omega} \left[(v_1-m)-(v_2-m)\right]^2 f(v_1,t)f(v_2,t)dv_1dv_2 \right\rangle \\
&= -\frac{(1-\lambda^2)}{4} \left\langle \iint_{\Omega\times\Omega} \left[(v_1-m)^2 + (v_2-m)^2 - 2(v_1-m)(v_2-m)\right] \right. \\
&\qquad\qquad \left. f(v_1,t)f(v_2,t)dv_1dv_2 \right\rangle .
\end{aligned}$$

Since the two processes are uncorrelated, we end up with:

$$\partial_t \int_\Omega (v-m)^2 f(v,t)dv = -\frac{(1-\lambda^2)}{2} \int_\Omega (v-m)^2 f(v,t)dv. \tag{8.68}$$

Hence, the variance of the distribution approaches zero at an exponential rate $-(1-\lambda^2)/2$. This implies that every agent ends up with the same wealth, which is not what is observed in real scenarios.

8.3.2 NON-CONSERVATIVE MARKET MODEL

If wealth is preserved in a transaction, why would anyone engage in trade? People trade because they give more value to the good they are receiving rather than the one that is being given. This imposes big limitations to the use of conservative models (see, for instance: [267]).

Value is subjective[k] [268]. Thus, making a cardinal measurement of wealth troublesome. Rather, one usually resorts to ordinal descriptions, or in a wider context, wealth could be defined as the ability to have one's desires fulfilled. Therefore, trade can only occur if there is an increase in wealth for both players, producing a positive sum game. The quantification of marginal utility and possession, nonetheless, has to be accepted and some non-conservative models that try to incorporate an increase of wealth in economic transactions have been devised [269, 270].

Slalina's[36] model, for instance, is a model inspired by dissipative gases where the exchange rule is given by:

---

[k]Consider this famous paradox attributed to Adam Smith: What would the values of water and diamond be for a person who is dying of dehydration in a desert?

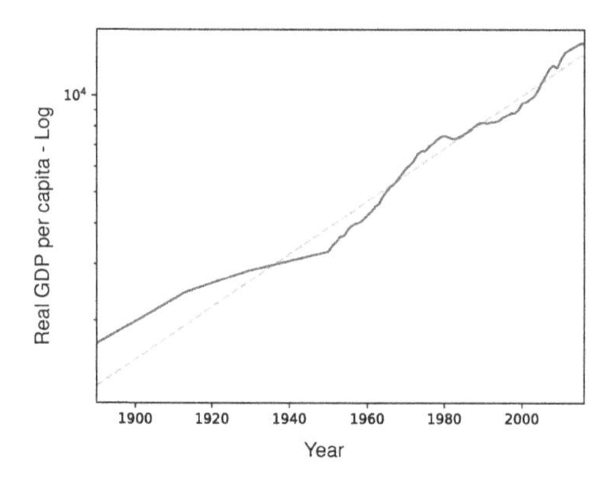

Figure 8.13: Real global GDP *per capita* adjusted for the value of money in 2011-US$ (created with data from [1] through ourworldindata.org; the dashed line is an exponential fit)

$$\begin{bmatrix} v_1'(t) \\ v_2'(t) \end{bmatrix} = \begin{bmatrix} 1-\beta+\varepsilon & \beta \\ \beta & 1-\beta+\varepsilon \end{bmatrix} \begin{bmatrix} v_1(t) \\ v_2(t) \end{bmatrix}, \tag{8.69}$$

where ε is a positive growth rate parameter, and $\beta \in [0,1]$ plays the role of the saving propensity.

The evolution of the average wealth can be found using $\varphi(v) = v$ in Eq. 8.62:

$$\begin{aligned} 2\partial_t \left\langle \int_\Omega v f(v,t) dv \right\rangle &= \\ \Big\langle \iint_{\Omega\times\Omega} [(1-\beta+\varepsilon)v_1 + \beta v_2 + \beta v_1 + (1-\beta+\varepsilon)v_2 - v_1 - v_2] \\ f(v_1,t)f(v_2,t)dv_1dv_2 \Big\rangle \\ \frac{\partial \bar{v}}{\partial t} &= \frac{1}{2}\left\langle \iint_{\Omega\times\Omega} \varepsilon(v_1+v_2) f(v_1,t)f(v_2,t)dv_1dv_2 \right\rangle \\ \frac{\partial \bar{v}}{\partial t} &= \varepsilon \left\langle \int_\Omega v f(v,t) dv \right\rangle = \varepsilon \bar{v}. \end{aligned} \tag{8.70}$$

Consequently, the average wealth grows as $\bar{v} = \bar{v}_0 e^{\varepsilon t}$. The GDP[1] *per capita* indeed shows an exponential growth as illustrated in Fig. 8.13.

Since the wealth grows exponentially, there is no steady state solution. Nonetheless, it is possible to seek self-similar solutions rescaling the wealth distribution as:

$$f(v,t) = \frac{1}{\bar{v}(t)} g\left(\frac{v}{\bar{v}(t)}, t\right). \tag{8.71}$$

[1]Gross domestic product, a measure of all goods and services produced during a specific period in an economy.

The temporal derivative of this equation is given by:

$$\begin{aligned}
\frac{df}{dt} &= -\frac{1}{\bar{v}^2}\frac{d\bar{v}}{dt}g + \frac{1}{\bar{v}}\left(\frac{\partial g}{\partial t} + \frac{\partial g}{\partial x}\frac{\partial x}{\partial \bar{v}}\frac{\partial \bar{v}}{\partial t}\right), \; x = \frac{v}{\bar{v}} \\
&= -\frac{\varepsilon}{\bar{v}}g + \frac{1}{\bar{v}}\left[\frac{\partial g}{\partial t} + \bar{v}\frac{\partial g}{\partial v}\left(-\frac{v}{\bar{v}^2}\right)\varepsilon\bar{v}\right] \\
&= \frac{\varepsilon}{\bar{v}}g + \frac{1}{\bar{v}}\left(\frac{\partial g}{\partial t} - \varepsilon v\frac{\partial g}{\partial v}\right) \\
&= \frac{1}{\bar{v}}\left[\frac{\partial g}{\partial t} - \varepsilon\left(g + v\frac{\partial g}{\partial v}\right)\right] \\
&= \frac{1}{\bar{v}}\left[\frac{\partial g}{\partial t} - \varepsilon\frac{\partial}{\partial v}(vg)\right].
\end{aligned} \tag{8.72}$$

Using this result in Eq. 8.64 we get:

$$\begin{aligned}
&\frac{1}{\bar{v}}\left[\frac{\partial g}{t} - \varepsilon\frac{\partial}{\partial v}(vg)\right] + f = Q_+(f,f) \\
&\frac{\partial g}{\partial t} = Q_+(g,g) - g + \varepsilon\frac{\partial}{\partial v}(vg),
\end{aligned} \tag{8.73}$$

which now shows a drift term related to the growth of wealth.

Following the steps showed at the beginning of this section backwards, we get the equation:

$$\begin{aligned}
&\left\langle \frac{\partial}{\partial t}\int_\Omega \varphi(v)g(v,t)dv \right\rangle - \varepsilon\left\langle \int_\Omega \varphi(v)\frac{\partial}{\partial v}(vg)dv \right\rangle = \\
&\quad = \frac{1}{2}\left\langle \int_{\Omega\times\Omega}\left[\varphi(v_1') + \varphi(v_2') - \varphi(v_1) - \varphi(v_2)\right] g(v_1,t)g(v_2,t)dv_1dv_2 \right\rangle.
\end{aligned} \tag{8.74}$$

Making $\varphi(v) = e^{-sv/\bar{v}}\Theta(v)$[m], the integrals become Laplace transforms and we get, according to the transition rule (Eq. 8.69):

$$\frac{\partial}{\partial t}G(s) + \varepsilon s\frac{\partial}{\partial s}G(s) = G([1-\beta+\varepsilon]s)\,G(\beta s) - G(s). \tag{8.75}$$

In steady state:

$$\varepsilon s\frac{\partial}{\partial s}G(s) = G([1-\beta+\varepsilon]s)\,G(\beta s) - G(s). \tag{8.76}$$

By expanding the parameters β and ε in Taylor series and taking the inverse Laplace transform, Slalina showed that the asymptotic wealth distribution for the tails displays a Pareto power-law behavior [264, 269]. This is observed in real data (see for instance [271]).

[m] Θ is the Heaviside[37] function defined as $\Theta(x) = \begin{cases} 1 & \text{if } x > 0 \\ 0 & \text{otherwise.} \end{cases}$

Notes

[1]Peter Sheridan Dodds, Australian mathematician.

[2]Stephen Wolfram (1959–) British physicist.

[3]Ludwig Eduard Boltzmann (1844–1906) advisee of Josef Stefan, Gustav Kirchhoff and Herman von Helmholtz among others. Boltzmann advised Paul Ehrenfest among others.

[4]George Boole (1815–1864) British philosopher.

[5]Arthur Cayley (1821–1895) British mathematician winner of many awards, including the De Morgan Medal in 1884.

[6]Hans Albrecht Bethe (1906–2005) German physicist winner of many awards, including the Nobel Prize in Physics in 1967. Bethe was advised by Arnold Sommerfeld and advised many notable students including Jun John Sakurai, David James Thouless, and Freeman Dyson.

[7]Gottfried Wilhem von Leibniz (1646–1716) German polymath advised by Christian Huygens (among others) and adviser of Jacob Bernoulli.

[8]Edsger Wybe Dijkstra (1930–2002) Dutch computer scientist.

[9]Richard Ernest Bellman (1920–1984) American mathematician winner of many prizes including the John vou Neumann Theory Prize in 1976.

[10]Lester Randolph Ford Jr. (1927–2017) American mathematician.

[11]Robert W. Floyd (1936–2001) American computer scientist, winner of the Turing Award in 1978.

[12]Stephen Warshall (1935–2006) American computer scientist.

[13]Paul Erdös (1913–1996) and Alfréd Rényi (1921–1970) Hungarian mathematicians.

[14]Stanley Milgram (1933–1984) American social psychologist.

[15]Georg Friedrich Bernhard Riemann (1826–1866) German mathematician, advisee of Carl Friedrich Gauss.

[16]Albert-Lászl6 Barabási (1967–) Romanian physicist advisee of Eugene Stanley and Tamás Vicsek.

[17]Réka Albert (1972–) Romanian physicist.

[18]Duncan James Watts (1971–) Canadian physicist, advisee of Steven Strogatz.

[19]Steven Henry Strogatz (1959–) American mathematician, adviser of Duncan Watts.

[20]Ernst Ising (1900–1998) German physicist.

[21]William Rowan Hamilton (1805–1865) Irish mathematician.

[22]Robert Swendsen, American physicist.

[23]Ulli Wolff, German physicist.

[24]Thomas Crombie Schelling (1921–2016) American economist winner of the Nobel Memorial Prize in Economic Sciences in 2005.

[25]Solomon Eliot Asch (1907–1996) Polish social psychologist, adviser of Stanley Milgram.

[26]Tamás Vicsek (1948–) Hungarian physicist, adviser of Albert-Lászl6 Barabási.

[27]Phrase firstly used by Alan Greenspan (1926–) American economist, chair of the Federal Reserve of the United States between 1987 and 2006.

[28]Rainer Hegselmann (1950–) German philosopher.

[29]Ulrich Krause, German mathematician and economist.

[30]Guillaume Deffuant, French complexity scientist.

[31]Alan Kirman (1939–) British economist.

[32]Paul Ehrenfest (1880–1933) Austrian physicist, advisee of Ludwig Boltzmann and adviser of George Uhlenbeck among others.

[33]Stefan Bornholdt, German physicist.

[34]Paul Adrien Maurice Dirac (1902–1984) English physicist, adivisee of Ralph Fowler. Dirac was the recipient of the Nobel Prize in Physics in 1933.

[35]Anirban Chakraborti and Bikas Kanta Chakrabarti (1952–), Indian physicists.

[36]František Slanina, Czech physicist.

[37]Oliver Heaviside (1850–1925) English polymath.

A Simulation of Stochastic Processes

Here we will see how to numerically solve stochastic differential equations of the type:

$$dX_t = a(X_t,t)dt + b(X_t,t)dW_t, \tag{A.1}$$

where W is a Wiener process.

Let's first divide an interval $\mathscr{I} = [0,T]$ in N subintervals such that each has a temporal width $\Delta t = T/N$ that begins in X_n and ends in X_{n+1}. Therefore, the solution of Eq. A.1 within an interval is given by:

$$X_{t_{n+1}} = X_{t_n} + \int_{t_n}^{t_{n+1}} a(X_s)ds + \int_{t_n}^{t_{n+1}} b(X_s)dW_s. \tag{A.2}$$

In Milstein's[a] method, we apply Ito's lemma to the functions $a(X_s)$ e $b(X_s)$:

$$df \approx \frac{\partial f}{\partial t}dt + \frac{\partial f}{\partial X}dX + \frac{1}{2}\frac{\partial^2 f}{\partial X^2}dX^2. \tag{A.3}$$

Connecting it with Eq. A.1 we get:

$$\begin{aligned} df &\approx \frac{\partial f}{\partial t}dt + \frac{\partial f}{\partial X}a(X)dt + \frac{\partial f}{\partial X}b(X)dW + \frac{1}{2}b^2(X)\frac{\partial^2 f}{\partial X^2}dW^2 \\ &\approx \left[f'(X)a(X) + \frac{1}{2}f''(X)b^2(X)\right]dt + f'b(X)dW. \end{aligned} \tag{A.4}$$

Integrating:

$$\begin{aligned} f(t) &\approx f_0 + \int_0^t \left[f'(X)a(X) + \frac{1}{2}f''(X)b^2(X)\right]dt + \int_0^t f'(X)b(X)dW \\ &\approx f_0 + \int_0^t f'(X)a(X)dt + \int_0^t \frac{1}{2}f''(X)b^2(X)dt + \int_0^t f'(X)b(X)dW \end{aligned} \tag{A.5}$$

Let's apply this result to the proper functions $a(X_s)$ e $b(X_s)$:

[a]Grigori Noichowitsch Milstein, Russian mathematician.

DOI: 10.1201/9781003127956-A

$$\begin{aligned}a(X_{t_{n+1}}) &\approx a(X_{t_n}) + \int_{t_n}^{t_{n+1}} \left[a'(X_u)a(X_u) + 1/2a''(X_u)b^2(X_u)\right] du + \\ &\qquad + \int_{t_n}^{t_{n+1}} \left[a'(X_u)b(X_u)\right] dW_u \\ b(X_{t_{n+1}}) &\approx b(X_{t_n}) + \int_{t_n}^{t_{n+1}} \left[b'(X_u)a(X_u) + 1/2b''(X_u)b^2(X_u)\right] du + \\ &\qquad + \int_{t_n}^{t_{n+1}} \left[b'(X_u)b(X_u)\right] dW_u.\end{aligned} \tag{A.6}$$

Plugging these results back in Eq. A.2 we find:

$$\begin{aligned}X_{t_{n+1}} \approx X_{t_n} &+ \int_{t_n}^{t_{n+1}} \left(a(X_{t_n}) + \int_{t_n}^{t_s} \left[a'(X_u)a(X_u) + 1/2a''(X_u)b^2(X_u)\right] du + \right. \\ &\qquad \left. + \int_{t_n}^{t_s} a'(X_u)b(X_u)dW_u\right) ds \\ &+ \int_{t_n}^{t_{n+1}} \left(b(X_{t_n}) + \int_{t_n}^{t_s} \left[b'(X_u)a(X_u) + 1/2b''(X_u)b^2(X_u)\right] du + \right. \\ &\qquad \left. + \int_{t_n}^{t_s} b'(X_u)b(X_u)dW_u\right) dW_s.\end{aligned} \tag{A.7}$$

$dtdt$ terms are of order $\mathcal{O}(dt^2)$ while $dtdW$ terms are of order $\mathcal{O}(dt^{3/2})$. Finally, $dWdW$ terms are of order $\mathcal{O}(dt)$. Neglecting high order terms, we get:

$$\begin{aligned}X_{t_{n+1}} &\approx X_{t_n} + \int_{t_n}^{t_{n+1}} a(X_{t_n})ds + \int_{t_n}^{t_{n+1}} b(X_{t_n})dW_s + \int_{t_n}^{t_{n+1}} \int_{t_n}^{t_s} b'(X_u)b(X_u)dW_u dW_s \\ &\approx X_{t_n} + a(X_{t_n})\Delta_t + b(X_{t_n})\Delta_{W_n} + b'(X_{t_n})b(X_{t_n}) \int_{t_n}^{t_{n+1}} W_s dW_s,\end{aligned} \tag{A.8}$$

where we have used the fact that in a Lévy process $W_0 = 0$.

For the resulting integral, we can write:

$$\begin{aligned}\int_0^t W_s dW_s &= \int_0^t W_s \frac{dW}{ds} ds \\ &= \frac{1}{2} \int_0^t \frac{d}{ds} W_s^2 ds.\end{aligned} \tag{A.9}$$

Since we got the integral of a derivative, we would like to write the result as $1/2W_t^2$. However, W is a non-differentiable function and this result is incorrect. Alternatively, we can use a Riemann sum to obtain the right answer. In order to do so, we apply the following algebraic trick:

$$W_{t_n} = 1/2\left(W_{t_{n+1}} + W_{t_n}\right) - 1/2\left(W_{t_{n+1}} - W_{t_n}\right). \tag{A.10}$$

Applying the Riemann sum, the integral becomes:

$$\begin{aligned}\int_{t_n}^{t_{n+1}} W_s dW_s &= \sum_{t_k<t_n} W_{t_k}\left(W_{t_{k+1}} - W_{t_k}\right) \\ &= \sum_{t_k<t_n} 1/2\left(W_{t_{n+1}} + W_{t_n}\right)\left(W_{t_{n+1}} - W_{t_n}\right) - \\ &\qquad - \sum_{t_k<t_n} 1/2\left(W_{t_{n+1}} - W_{t_n}\right)\left(W_{t_{n+1}} - W_{t_n}\right) \\ &= \sum_{t_k<t_n} 1/2\left(W_{t_{n+1}}^2 - W_{t_n}^2\right) - \sum_{t_k<t_n} 1/2\left(W_{t_{n+1}} - W_{t_n}\right)^2.\end{aligned} \tag{A.11}$$

For the first summation, we have a telescopic sum $(W_1^2 - W_0^2) + (W_2^2 - W_1^2) + \ldots + (W_n^2 - W_{n-1}^2) + (W_{n+1}^2 - W_n^2) = W_{n+1}^2 - W_0^2 = W_{n+1}^2 \to W_n^2$ for $\Delta_t \to 0$. Since this summation occurs within a subinterval, we get Δ_W^2.

An integral of the form:

$$\int_0^T f(t)dW_t = \lim_{\Delta t\to 0} \sum_{i=0}^{k-1} f(t_i)(W_{i+1} - W_i) \tag{A.12}$$

is known as *Ito's integral*. It contrasts with *Stratonovich's integral*[b] given by:

$$\int_0^T f(t)dW_t = \lim_{\Delta t\to 0} \sum_{i=0}^{k-1} f\left(\frac{1}{2}[t_{i+1} + t_i]\right)(W_{i+1} - W_i). \tag{A.13}$$

For the second term we have:

$$\begin{aligned} 1/2 \sum_{t_k<t_n} \Delta W^2 &= 1/2 \sum_{t_k<t_n} \delta_t \\ &= 1/2 \cdot \Delta_t.\end{aligned} \tag{A.14}$$

Note that the summations are within a subinterval, hence $\sum \delta_t = \Delta_t$. The Riemann sum becomes:

$$\int_{t_n}^{t_{n+1}} W_s dW_s = 1/2(W_n^2 - \Delta_t). \tag{A.15}$$

Finally, we arrive at an expression for $X_{t_{n+1}}$:

$$X_{t_{n+1}} \approx X_{t_n} + a(X_{t_n})\Delta_t + b(X_{t_n})\Delta_{W_n} + 1/2 b'(X_{t_n})b(X_{t_n})\left(\Delta_{W_n}^2 - \Delta_t\right) \tag{A.16}$$

Maruyama[c] extended Euler's method to approximate stochastic differential equations numerically [272] as:

[b]Developed by the Russian physicist Leontévich Stratonovich (1930-1997).

[c]Gisiro Maruyama (1916-1986) Japanese mathematician.

$$\begin{aligned}\int_{t_n}^{t_{n+1}} a(X_s,s)ds &\approx a(X_n,t_n)\delta_t \\ \int_{t_n}^{t_{n+1}} b(X_s,s)dW_s &\approx b(X_n,t_n)\Delta W_n,\end{aligned} \tag{A.17}$$

where $\Delta W_n = W_{t_{n+1}} - W_{t_n}$. Hence, we use:

$$X_{n+1} = X_n + a(X_n,t_n)\delta_t + b(X_n,t_n)\Delta W_n. \tag{A.18}$$

This can be easily achieved considering that $b'(X_{t_n})$ is zero in Milstein's approximation.

B Monte Carlo Simulations

A Metropolis-Hastings[a] Markov chain Monte Carlo simulation [273, 274] begins with the probability of finding the system in a state ϕ given by the Boltzmann factor:

$$p(\phi) = \frac{1}{Z} e^{-\beta H(\phi)}, \tag{B.1}$$

where Z is the partition function, β is the inverse temperature $(1/k_B T)$, and H is the Hamiltonian of the system. In equilibrium, the system must obey the detailed balance condition (see Sec. 3.2.2.1):

$$\begin{aligned}
\pi_i^* p_{ij} &= \pi_j^* p_{ji} \\
\frac{p_{ji}}{p_{ij}} &= \frac{\pi_i^*}{\pi_j^*} = \frac{e^{-\beta H(\phi_i)}}{e^{-\beta H(\phi_j)}} \\
&= e^{-\beta\left[H(\phi_i) - H(\phi_j)\right]} \\
&= e^{-\beta \Delta\varepsilon}.
\end{aligned} \tag{B.2}$$

This can be written as:

$$\frac{p_{ji}}{p_{ij}} = \frac{\pi_i^*}{\pi_j^*} = \frac{g(j \to i) A(j \to i)}{g(i \to j) A(i \to j)} = e^{-\beta \Delta\varepsilon}, \tag{B.3}$$

where g is the probability of selecting a state, and A is the acceptance rate of such state.

For the Ising model (see Sec. 8.2), this corresponds of randomly picking a site. If there are N spins in the grid, then $g = 1/N$, and:

$$A(i \to j) = \begin{cases} \exp\{-\beta(\varepsilon_i - \varepsilon_j)\}, & \text{if } \varepsilon_i > \varepsilon_j \\ 1, & \text{otherwise.} \end{cases} \tag{B.4}$$

Thus, new test states are generated. If a state decreases the energy of the system, it is accepted. Otherwise, it can still be accepted with probability $\exp\{-\beta\Delta\varepsilon\}$. This is equivalent of having an acceptance probability $p(\phi_j) = \min(1, \exp\{-\beta\Delta\varepsilon\})$.

Another possibility for the acceptance rate is the so-called *heat bath*:

$$A(i \to j) = \frac{e^{-\beta\varepsilon_j}}{e^{-\beta\varepsilon_j} + e^{-\beta\varepsilon_i}} = \frac{e^{-\beta\Delta\varepsilon}}{1 + e^{-\beta\Delta\varepsilon}}. \tag{B.5}$$

[a]Nicholas Constantine Metropolis (1915-1999) Greek physicist, and Wilfred Keith Hastings (1930-2016) Canadian statistician.

DOI: 10.1201/9781003127956-B

C Fokker-Planck Equation

The Fokker[a]-Planck[b] equation is a partial differential equation that describes the evolution of diffusive systems [275, 276].

In order to derive this equation, let's consider a smooth function $h(Y)$ with a compact support and make the following approximation:

$$\begin{aligned}\int_{-\infty}^{\infty} h(Y)\frac{\partial P(Y,t|X)}{\partial t}dY &= \lim_{\Delta\to 0}\int_{-\infty}^{\infty} h(Y)\left[\frac{P(Y,t+\Delta|X)-P(Y,t,X)}{\Delta}\right]dY \\ &= \lim_{\Delta\to 0}\frac{1}{\Delta}\left[\int_{-\infty}^{\infty} h(Y)P(Y,t+\Delta|X)dY - \int_{-\infty}^{\infty} h(Y)P(Y,t|X)dY\right].\end{aligned} \tag{C.1}$$

Let's now use the Kolmogorov-Chapman equation (Eq. 3.75) in the first integral on the right hand side of the equation. Let's also change the name of the integrating variable in the second integral from Y to Z and use the identity $\int_{-\infty}^{\infty} P(Y,\Delta|Z)dY = 1$. This way, we get:

$$\begin{aligned}\int_{-\infty}^{\infty} h(Y)\frac{\partial P(Y,t|X)}{\partial t}dY &= \lim_{\Delta\to 0}\frac{1}{\Delta}\left[\int_{-\infty}^{\infty} h(Y)P(Y,\Delta|Z)P(Z,t|X)dZdY\right. \\ &\quad \left. - \int_{-\infty}^{\infty} h(Z)P(Z,t,X)dZ\int_{-\infty}^{\infty} P(Y,\Delta|Z)dY\right] \\ &= \lim_{\Delta\to 0}\frac{1}{\Delta}\left[\int_{-\infty}^{\infty} P(Z,t|X)\int_{-\infty}^{\infty} P(Y,\Delta|Z)\left(h(Y)-h(Z)\right)dYdZ\right]\end{aligned} \tag{C.2}$$

Doing a Taylor expansion for $h(Y)$ around Z:

$$h(Y) = h(Z) + \sum_{n=1}^{\infty} h^{(n)}(Z)(Y-Z)^n/n! \tag{C.3}$$

Let's also create the following variable:

$$Q^{(n)}(Z) = \frac{1}{n!}\lim_{\Delta\to 0}\frac{1}{\Delta}\int_{-\infty}^{\infty}(Y-Z)^n P(Y,\Delta|Z)dY. \tag{C.4}$$

Therefore:

---

[a] Adriaan Daniël Fokker (1887-1972) Dutch physicist, advisee of Hendrik Lorentz.

[b] Max Karl Ernst Ludwig Planck (1858-1947) German physicist, advisee of Gustav Kirchhoff and Hermann von Helmholtz, adviser of Gustav Ludwig Hertz, Max von Laue, Walter Schottky, Moritz Schlick and Julius Edgar Lilienfeld among others. Planck won the Nobel prize of physics in 1918.

DOI: 10.1201/9781003127956-C

$$\int_{-\infty}^{\infty} h(Y)\frac{\partial P(Y,t|X)}{\partial t}dY = \int_{-\infty}^{\infty} P(Z,t|X)\sum_{n=1}^{\infty} Q^{(n)}(Z)h^{(n)}(Z)dZ. \tag{C.5}$$

The integration by parts for high order derivatives can be written as:

$$\begin{aligned}\int_{-\infty}^{\infty} u\frac{d^n v}{dx^n}dx &= \int_{-\infty}^{\infty} u\frac{d}{dx}\left(\frac{d^{n-1}v}{dx^{n-1}}\right)dx \\ &= u\frac{d^{n-1}v}{dx^{n-1}}\Big|_{-\infty}^{\infty} - \int_{-\infty}^{\infty}\left(\frac{d^{n-1}v}{dx^{n-1}}\right)\frac{du}{dx}dx.\end{aligned} \tag{C.6}$$

If the function u, though, has a compact support, both the function itself and its derivatives tend to zero at $\pm\infty$. Therefore,

$$\int_{-\infty}^{\infty} u\frac{d^n v}{dx^n}dx = -\int_{-\infty}^{\infty}\frac{du}{dx}\left(\frac{d^{n-1}v}{dx^{n-1}}\right)dx. \tag{C.7}$$

Applying this procedure n times we get:

$$\int_{-\infty}^{\infty} u\frac{d^n v}{dx^n}dx = \int_{-\infty}^{\infty} v\left(-\frac{d}{dx}\right)^n udx. \tag{C.8}$$

Consequently, we get:

$$\int_{-\infty}^{\infty} h(Y)\frac{\partial P(Y,t|X)}{\partial t}dY = \int_{-\infty}^{\infty} h(Z)\sum_{n=1}^{\infty}\left(-\frac{\partial}{\partial Z}\right)^n\left[Q^{(n)}(Z)P(Z,t|X)\right]dZ. \tag{C.9}$$

Changing the name of the integrating variable on the left hand side of the equation to Z:

$$\int_{-\infty}^{\infty} h(Z)\left(\frac{\partial P(Z,t,X)}{\partial t} - \sum_{n=1}^{\infty}\left(-\frac{\partial}{\partial Z}\right)^n\left[Q^{(n)}(Z)P(Z,y|X)\right]\right)]dZ = 0. \tag{C.10}$$

Since $h(Z)$ is a generic function, it follows that:

$$\frac{\partial P(Z,t,X)}{\partial t} = \sum_{n=1}^{\infty}\left(-\frac{\partial}{\partial Z}\right)^n\left[Q^{(n)}(Z)P(Z,y|X)\right]. \tag{C.11}$$

We can consider $p(X,t)$ as the PDF of $X(t)$. Therefore, $p(X,t)$ can be chosen as the solution for the last equation for a unitary excitation at $t=0, X=X_0$:

$$\frac{\partial p(x,t)}{\partial t} = \sum_{n=1}^{\infty}\left(-\frac{\partial}{\partial x}\right)^n\left[Q^{(n)}(x)p(x,t)\right]. \tag{C.12}$$

This procedure that transforms the integro-differential equation in a differential equation through an expansion is known as the Kramers[c]-Moyal[d] expansion [277, 278].

Considering only terms up to the second order:

$$\begin{aligned}\frac{\partial p(x,t)}{\partial t} &= -\frac{\partial}{\partial x}\left[\frac{\partial Q(x)}{\partial x}p(x,t)\right]+\frac{\partial^2}{\partial x^2}\left[\frac{\partial^2 Q(x)}{\partial x^2}p(x,t)\right] \\ \frac{\partial p(x,t)}{\partial t} &= -\frac{\partial}{\partial x}\left[\mu(x,t)p(x,t)\right]+\frac{\partial^2}{\partial x^2}\left[D(x,t)p(x,t)\right],\end{aligned} \tag{C.13}$$

which is the Fokker-Plank equation and:

$$\mu(x,t) = \frac{\partial Q(x)}{\partial x}. \tag{C.14}$$

is known as the *drift coefficient*, and:

$$D(x,t) = \frac{\partial^2 Q(x)}{\partial x^2} = \frac{1}{2}\sigma^2(X,t). \tag{C.15}$$

is known as *diffusion coefficient.*

Given a stochastic differential equation (SDE):

$$dx = f(x,t)dt + g(x,t)dW, \tag{C.16}$$

it is possible to find the corresponding Fokker-Planck equation.

In order to obtain this representation we assume a function $\varphi = \delta(x-X)$. Let's Taylor expand it:

$$d\varphi = \frac{\partial\varphi}{\partial t}dt + \frac{\partial\varphi}{\partial s}ds + \frac{1}{2}\frac{\partial^2\varphi}{\partial s^2}ds^2 + \ldots \tag{C.17}$$

and substitute it back in the SDE:

$$\begin{aligned}d\varphi &= \frac{\partial\varphi}{\partial t}dt + \frac{\partial\varphi}{\partial s}\left(f(x,t)dt+g(x,t)dW\right) + \frac{1}{2}\frac{\partial^2\varphi}{\partial s^2}\left(f(x,t)dt+g(x,t)dW\right)^2 + \ldots \\ &\approx \frac{\partial\varphi}{\partial t}dt + f(x,t)\frac{\partial\varphi}{\partial s}dt + g(x,t)\frac{\partial\varphi}{\partial s}dW + \frac{1}{2}g^2(x,t)\frac{\partial^2\varphi}{\partial s^2}dt \\ &\approx \left(\frac{\partial\varphi}{\partial t} + f(x,t)\frac{\partial\varphi}{\partial s} + \frac{1}{2}g^2(x,t)\frac{\partial^2\varphi}{\partial s^2}\right)dt + g(x,t)\frac{\partial\varphi}{\partial s}dW,\end{aligned} \tag{C.18}$$

where we kept only low order terms and used the variance of a Wiener process $dW^2 \to dt$.

Only f and g are time dependent. Therefore:

---

[c]Hans Anthony Kramers (1894-1952) Dutch physicist, advisee of Niels Bohr and Paul Ehrenfest.

[d]José Enrique Moyal (1910-1998) Australian mathematician.

$$\begin{aligned}
\frac{d\varphi}{dt} &\approx \left(f(x,t)\frac{\partial \varphi}{\partial s} + \frac{1}{2}g^2(x,t)\frac{\partial^2 \varphi}{\partial s^2} \right) + g(x,t)\frac{\partial \varphi}{\partial s}\eta \\
\left\langle \frac{d\varphi}{dt} \right\rangle &\approx \left\langle f(x,t)\frac{\partial \varphi}{\partial s} + \frac{1}{2}g^2(x,t)\frac{\partial^2 \varphi}{\partial s^2} \right\rangle \\
&\approx \left\langle f(x,t)\frac{\partial \varphi}{\partial s} \right\rangle + \frac{1}{2}\left\langle g^2(x,t)\frac{\partial^2 \varphi}{\partial s^2} \right\rangle.
\end{aligned} \tag{C.19}$$

$$\begin{aligned}
\left\langle \frac{d\varphi}{dt} \right\rangle = \frac{d}{dt}\langle \varphi \rangle &= \frac{d}{dt}\int P(z,t)\delta(z-X)dz \\
&= \frac{\partial}{\partial t}P(X,t),
\end{aligned}$$

where the Leibniz rule[e] was used

For the first expected value on the right hand side of the equation:

$$\left\langle f(x,t)\frac{\partial \varphi}{\partial z} \right\rangle = \int P(z,t)f(z,t)\frac{\partial \delta(z-X)}{\partial z}dz. \tag{C.20}$$

Integrating by parts:

$$\begin{aligned}
&= P(z,t)f(z,t)\delta(z-X)|_{-\infty}^{\infty} - \int \delta(z-X)\frac{\partial}{\partial z}\left(P(z,t)f(z,t)\right)dz \\
&= -\frac{\partial}{\partial X}\left(P(X,t)f(x,t)\right).
\end{aligned} \tag{C.21}$$

For the remaining expected value, we get:

$$\begin{aligned}
\frac{1}{2}\left\langle g^2(x,t)\frac{\partial^2 \varphi}{\partial s^2} \right\rangle &= \frac{1}{2}\int P(z,t)g(z,t)\frac{\partial^2 \delta(z-X)}{\partial z^2}dz \\
&= \frac{1}{2}\int \left(P(z,t)g^2(z,t)\right)\frac{\partial}{\partial z}\frac{\partial \delta(z-X)}{\partial z}dz \\
&= -\frac{1}{2}\int \frac{\partial \delta(z-X)}{\partial z}\frac{\partial\left(P(z,t)g^2(z,t)\right)}{\partial z}dz \\
&= \frac{1}{2}\int \delta(z-X)\frac{\partial^2 P(z,t)g^2(z,t)}{\partial z^2}dz \\
&= \frac{1}{2}\frac{\partial^2 P(X,t)g^2(X,t)}{\partial X^2}.
\end{aligned} \tag{C.22}$$

Combining Eqs. C.19, C.21 and C.22:

$$\frac{\partial}{\partial t}P(X,t) = -\frac{\partial}{\partial X}\left(P(X,t)f(x,t)\right) + \frac{1}{2}\frac{\partial^2}{\partial X^2}\left(P(X,t)g^2(X,t)\right). \tag{C.23}$$

---

[e] $\frac{d}{dx}\left(\int_a^b f(x,t)dt\right) = f(x,b)\frac{db}{dx} - f(x,a)\frac{da}{dx} + \int_a^b \frac{\partial}{\partial x}f(x,t)dt.$

D Girsanov Theorem

Let W_t be a Brownian motion on a probability space $(\Omega, \Sigma, \mathbb{P})$ and $\theta(t)$, $0 \leq t \leq T$ be an adapted process to a corresponding filtration $\mathscr{F}$. Now, for an Ito process:

$$d\tilde{W}_t = \theta_t dt + dW_t, \; \tilde{W}_0 = 0, \tag{D.1}$$

is there a measure $\mathbb{Q}$ equivalent[a] to $\mathbb{P}$ where $\tilde{W}_t$ is a Brownian motion under $\mathbb{Q}$?

For a random variable X on $(\Omega, \mathscr{F})$, its expectation under $\mathbb{Q}$ is given by:

$$\langle X \rangle_{\mathbb{Q}} = \left\langle X \frac{d\mathbb{Q}}{d\mathbb{P}} \right\rangle_{\mathbb{P}}, \tag{D.2}$$

where $d\mathbb{Q}/d\mathbb{P}$ is the Radon-Nikodym derivative (see Sec. 2.2.4). This last equation can be rewritten using the densities of a process under these measures. Let's take, for instance, W_T:

$$\begin{aligned} \langle W_T \rangle_{\mathbb{Q}} &= \int_{-\infty}^{\infty} w f_{\mathbb{Q}}(w) dw \\ &= \int_{-\infty}^{\infty} w \frac{f_{\mathbb{Q}}(w)}{f_{\mathbb{P}}(w)} f_{\mathbb{P}}(w) dw \\ &= \left\langle W_T \frac{f_{\mathbb{Q}}(W_T)}{f_{\mathbb{P}}(W_T)} \right\rangle_{\mathbb{P}}. \end{aligned} \tag{D.3}$$

Therefore,

$$\frac{d\mathbb{Q}}{d\mathbb{P}} = \frac{f_{\mathbb{Q}}(W_T)}{f_{\mathbb{P}}(W_T)}. \tag{D.4}$$

W_T is normally $\mathscr{N}(0,T)$ distributed under $\mathbb{P}$, whereas, according to Eq. D.1, it is $\mathscr{N}(-\theta T, T)$ distributed under $\mathbb{Q}$ if we assume θ_t being constant over time. Therefore, we can write:

$$\begin{aligned} \frac{d\mathbb{Q}}{d\mathbb{P}} &= \frac{\frac{1}{\sqrt{2\pi T}} \exp\left\{-\frac{(W_T + \theta T)^2}{2T}\right\}}{\frac{1}{\sqrt{2\pi T}} \exp\left\{-\frac{W_T^2}{2T}\right\}} \\ &= \exp\left\{-\theta W_T - \frac{1}{2}\theta^2 T\right\} \\ &= \exp\left\{-\int_0^T \theta dW_t - \frac{1}{2}\int_0^T \theta^2 dt\right\}. \end{aligned} \tag{D.5}$$

---

[a] Two measures are equivalent if $\mathbb{P}(A) = 0 \iff \mathbb{Q}(A) = 0 \; \forall A \in \mathscr{F}_t$.

DOI: 10.1201/9781003127956-D

This is known as the *Doléans-Dade exponential*[b] which also appears as the solution of the martingale:

$$dZ_t = -Z_t \theta dW_t,\ Z_0 = 1. \tag{D.6}$$

Applying Ito's lemma to $\ln(Z)$, one gets $\partial f/\partial t = 0$, $\partial f/\partial Z = 1/Z$, and $\partial^2 f/\partial Z^2 = -1/Z^2$. Thus:

$$\begin{aligned} d\ln(Z) &= -\frac{1}{2}\theta^2 dt - \theta dW_t \\ \ln(Z) &= \exp\left\{-\int_0^t \theta dW_s - \frac{1}{2}\int_0^t \theta^2 ds\right\}. \end{aligned} \tag{D.7}$$

The integral inside the exponential is finite if:

$$\left\langle \exp\left\{\frac{1}{2}\int_0^T |\theta_s|^2 ds\right\}\right\rangle_{\mathbb{P}} < \infty. \tag{D.8}$$

This is known as the *Novikov condition*[c] [279, 280].

The Cameron-Martin-Girsanov[d] theorem [281] proves that under these conditions, $\tilde{W}_t$ is a Brownian motion under the measure $\mathbb{Q}$.

For example, for the Ito process:

$$dX_t = \mu dt + \sigma dW_t, \tag{D.9}$$

we can set $\theta = (\mu - \nu)/\sigma$. Therefore, according to Eq. D.1:

$$dW = d\tilde{W} - \frac{\mu - \nu}{\sigma} dt. \tag{D.10}$$

Eq. D.9 can then be written as:

$$\begin{aligned} dX_t &= \mu dt + \sigma\left(d\tilde{W} - \frac{\mu - \nu}{\sigma} dt\right) \\ &= \nu dt + \sigma d\tilde{W}. \end{aligned} \tag{D.11}$$

Since $\tilde{W}$ is a Brownian motion under the measure $\mathbb{Q}$, X has drift ν and the same variance σ^2 under this new measure.

---

[b] Catherine Doléans-Dade (1942-2004) French mathematician.

[c] Alexander Novikov, Russian mathematician.

[d] Robert Horton Cameron (1908-1989) American mathematician, advisee of Monroe David Donkser among others; William Ted Martin (1911-2004) American mathematician; Igor Vladimirovich Girsanov (1934-1967) Russian mathematician.

E Feynman-Kack Formula

Let's consider a partial differential equation:

$$\frac{\partial u(x,t)}{\partial t} + a(x,t)\frac{\partial u(x,t)}{\partial x} + \frac{1}{2}b(x,t)^2\frac{\partial^2 u(x,t)}{\partial x^2} = ru(x,t) \tag{E.1}$$

defined for $x \in \mathbb{R}$ and $t \in [t_0, T]$ subject to the boundary condition $u(x,T) = \varphi(x)$.

Let's define a stochastic differential equation (SDE) in the interval $[t_0, T]$ that represents $u(x,t)$:

$$dX = a(X,t)dt + b(X,t)dW, \tag{E.2}$$

such that $X(t_0) = x$.

Ito's lemma applied to this process produces:

$$\begin{aligned} du(X,t) = & \left(\frac{\partial u(X,t)}{\partial t} + a(X,t)\frac{\partial u(X,t)}{\partial X} + \frac{1}{2}b(X,t)\frac{\partial^2 u(X,t)}{\partial X^2} \right) dt + \\ & + b(X,t)\frac{\partial u(X,t)}{\partial X}dW \\ & = ru(X,t)dt + b(X,t)\frac{\partial u(X,t)}{\partial X}dW. \end{aligned} \tag{E.3}$$

In order to solve this equation, we can create an auxiliary variable:

$$\begin{aligned} f(X,t) &= u(X,t)e^{-rt} \\ df(X,t) &= e^{-rt}du - ru(X,t)e^{-rt}dt \\ &= e^{-rt}\left(ru(X,t)dt + b(X,t)\frac{\partial u(X,t)}{\partial X}dW \right) - ru(X,t)e^{-rt}dt \\ &= b(X,t)e^{-rt}\frac{\partial u(X,t)}{\partial X}dW. \end{aligned} \tag{E.4}$$

Integrating in its temporal domain:

$$\begin{aligned} f(X,T) - f(X,t_0) &= \int_{t_0}^{T} b(X,s)e^{-rs}\frac{\partial u(X,s)}{\partial X}dW_s \\ u(X,t_0)e^{-rt_0} &= u(X,T)e^{-rT} - \int_{t_0}^{T} b(X,s)e^{-rs}\frac{\partial u(X,s)}{\partial X}dW_s. \end{aligned} \tag{E.5}$$

Taking the expectation on both sides of the equation:

$$u(x,t_0) = e^{-r(T-t_0)}\left\langle \varphi(X) \right\rangle. \tag{E.6}$$

DOI: 10.1201/9781003127956-E

This is known as the *Feynman*[a]*-Kac*[b] *formula.* Note that the coefficients $a(X,t)$ and $b(X,t)$ depend only on the current value of $X(t)$ and therefore describe a Markovian process. Therefore, the Feynman-Kac formula does not hold if these coefficients depend on the history of the process.

[a]Richard Phillips Feynman (1918-1988) American physicist, advisee of John Archibald Wheeler, adviser of James Maxwell Bardeen (son of John Bardeen) among others. Feynman was awarded the Nobel prize in physics in 1965.

[b]Mark Kac (1914-1984) Polish mathematician.

F Boltzmann Equation

Let's consider a non-negative function $f : \mathbb{R}^3 \times \mathbb{R}^3 \times \mathbb{R}^+ \rightarrow \mathbb{R}^+_*$ representing the density of particles in a gas. After some short interval Δ, the position and momentum of every particle evolve according to the semiclassical equations of motion:

$$\begin{aligned} \mathbf{r}(t) &= \mathbf{r}(t-\Delta) - \mathbf{v}(t-\Delta)\Delta \\ \mathbf{p}(t) &= \mathbf{p}(t-\Delta) - \mathbf{F}(t-\Delta)\Delta, \end{aligned} \tag{F.1}$$

where $\mathbf{F}$ is an external force. Therefore, in the absence of collisions, the distribution at position $\mathbf{r}$, momentum $\mathbf{p}$, and instant t is given by:

$$f(\mathbf{r}(t),\mathbf{p}(t),t) = f(\mathbf{r}(t-\Delta) - \mathbf{v}(t-\Delta)\Delta, \mathbf{p}(t-\Delta) - \mathbf{F}(t-\Delta)\Delta, t-\Delta). \tag{F.2}$$

Some particles, though, are deflected due to collisions, and some particles that have arrived at $\mathbf{r},\mathbf{p},t$ may also have moved because of past collisions. Therefore, we must correct the previous expression as:

$$\begin{aligned} f(\mathbf{r}(t),\mathbf{p}(t),t) =& f(\mathbf{r}(t-\Delta) - \mathbf{v}(t-\Delta)\Delta, \mathbf{p}(t-\Delta) - \mathbf{F}(t-\Delta)\Delta, t-\Delta) \\ &+ \left(\frac{\partial f(\mathbf{r},\mathbf{p},t)}{\partial t}\right)_{out} \Delta + \left(\frac{\partial f(\mathbf{r},\mathbf{p},t)}{\partial t}\right)_{in} \Delta. \end{aligned} \tag{F.3}$$

After some little algebra and taking the limit $\Delta \rightarrow 0$:

$$\frac{\partial f(\mathbf{r},\mathbf{p},t)}{\partial t} + \mathbf{v}\cdot\nabla_r f(\mathbf{r},\mathbf{p},t) + \mathbf{F}\cdot\nabla_p f(\mathbf{r},\mathbf{p},t) = \left(\frac{\partial f(\mathbf{r},\mathbf{p},t)}{\partial t}\right)_{collision}. \tag{F.4}$$

This is known as the *Boltzmann transport equation*. The second and third terms on the left hand side of the equation are known as the *diffusion* and *drift* terms, respectively.

Given a scattering probability $W_{p,p'}$, the collision term, on the right hand side of the equation, is given by:

$$\left(\frac{\partial f(\mathbf{r},\mathbf{p},t)}{\partial t}\right)_{collision} = -\int d\mathbf{p}'^3 \left\{W_{p,p'} f(\mathbf{p})\left[1-f(\mathbf{p}')\right] - W_{p',p} f(\mathbf{p}')\left[1-f(\mathbf{p})\right]\right\}, \tag{F.5}$$

where the first product term in the integrand indicates the probability of a particle changing its momentum $\mathbf{p}$ to $\mathbf{p}'$, for example.

The Boltzmann equation is often simplified using a *relaxation-time approximation*:

DOI: 10.1201/9781003127956-F

$$\left(\frac{\partial f(\mathbf{r},\mathbf{p},t)}{\partial t}\right)_{collision} = -\frac{f(\mathbf{p}) - f^0(\mathbf{p})}{\tau(\mathbf{p})}, \quad \text{(F.6)}$$

where dt/τ is the probability that a particles suffers a collision in an interval dt and f^0 is an equilibrium distribution.

Typically for rare gases, we are interested in the spatially uniform distribution and we use a simplified version of Boltzmann equation:

$$\frac{\partial f(\mathbf{p},t)}{\partial t} = I_f(\mathbf{p},t), \quad \text{(F.7)}$$

where $I_f(\mathbf{p},t)$ is a *collision operator* that describes the binary interactions among particles. This operator must be positivity preserving[a], and since the number of agents is normally kept the same in a simulation, it has to obey:

$$\frac{\partial}{\partial t}\int f(\mathbf{p},t)d^3\mathbf{p} = \int \frac{\partial f(\mathbf{p},t)}{\partial t}d^3\mathbf{p} = \int I_f(\mathbf{p},t)d^3\mathbf{p} = 0. \quad \text{(F.8)}$$

We can use Eq. F.7 to find the collision operator by making:

$$\frac{\partial}{\partial t}\int f(\mathbf{p},t)\varphi(\mathbf{p})d\mathbf{p} = \int I_f(\mathbf{p},t)\varphi(\mathbf{p})d\mathbf{p}, \quad \text{(F.9)}$$

where $\varphi(\mathbf{p})$ is some observable.

[a]Maps a non-negative density to a non-negative future density.

G Liouville-BBGKY

G.1 LIOUVILLE THEOREM

Consider the phase space volume $d\Omega = \prod_{i=1}^{N} d^3\mathbf{r}_i d^3\mathbf{p}_i$ centered around $(\mathbf{r}, \mathbf{p})$ and occupied by dN pure states. After an interval Δ, these states move to another volume $d\Omega' \prod_{i=1}^{N} d^3\mathbf{r}_i' d^3\mathbf{p}_i'$ centered around:

$$\begin{aligned} \mathbf{r}_i' &= \mathbf{r}_i + \frac{\partial \mathbf{r}_i}{\partial t}\Delta \\ \mathbf{p}_i' &= \mathbf{p}_i + \frac{\partial \mathbf{p}_i}{\partial t}\Delta. \end{aligned} \tag{G.1}$$

The respective differential elements are given by:

$$\begin{aligned} d\mathbf{r}_i' &= d\mathbf{r}_i + \frac{\partial \dot{\mathbf{r}}_i}{\partial \mathbf{r}_i} d\mathbf{r}_i \Delta \\ d\mathbf{p}_i' &= d\mathbf{p}_i + \frac{\partial \dot{\mathbf{p}}_i}{\partial \mathbf{p}_i} d\mathbf{p}_i \Delta. \end{aligned} \tag{G.2}$$

For each differential element pair we get:

$$d\mathbf{r}_i' d\mathbf{p}_i' = d\mathbf{r}_i d\mathbf{p}_i \left[1 + \left(\frac{\partial \dot{\mathbf{r}}_i}{\partial \mathbf{r}_i} + \frac{\partial \dot{\mathbf{p}}_i}{\partial \mathbf{p}_i}\right)\Delta + O(\Delta^2)\right]. \tag{G.3}$$

The N particles obey Hamilton's equations, hence:

$$\begin{aligned} \dot{\mathbf{p}}_i &= -\frac{\partial \mathscr{H}}{\partial \mathbf{r}_i} \\ \dot{\mathbf{r}}_i &= \frac{\partial \mathscr{H}}{\partial \mathbf{p}_i}, \end{aligned} \tag{G.4}$$

where $\mathscr{H}$ is the Hamiltonian of the system, which typically takes the form:

$$\mathscr{H} = \frac{1}{2}\sum_{i=1}^{N} \frac{\mathbf{p}_i^2}{m_i} + \sum_{i=1}^{N} V(\mathbf{r}_i) + \sum_{\{i,j\}} U(\mathbf{r}_i - \mathbf{r}_j), \tag{G.5}$$

where U is a two-body interaction potential and V is a general potential.

Thus, using the results from Eq. G.4 in G.3 we get:

$$d\mathbf{r}_i' d\mathbf{p}_i' = d\mathbf{r}_i d\mathbf{p}_i \left[1 + \left(\frac{\partial^2 \mathscr{H}}{\partial \mathbf{r}_i \partial \mathbf{p}_i} - \frac{\partial^2 \mathscr{H}}{\partial \mathbf{p}_i \partial \mathbf{r}_i}\right)\Delta\right] = d\mathbf{r}_i d\mathbf{p}_i. \tag{G.6}$$

DOI: 10.1201/9781003127956-G

Therefore, although the particles move to another location, they occupy the same volume and the phase space density behaves as an incompressible fluid. This is known as *Liouville*[a] *theorem.*

The incompressibility condition for the phase space density can be written as:

$$f(\mathbf{r}',\mathbf{p}',t+\Delta) = f(\mathbf{r}+\dot{\mathbf{r}}\Delta,\mathbf{q}+\dot{\mathbf{q}}\Delta,t+\Delta) = f(\mathbf{r},\mathbf{p},t). \tag{G.7}$$

After some algebra:

$$\begin{aligned} &\frac{\partial f}{\partial t} + \sum_{i=1}^{N} \frac{\partial f}{\partial \mathbf{r}} \cdot \frac{d\mathbf{r}_i}{dt} + \sum_{i=1}^{N} \frac{\partial f}{\partial \mathbf{q}_i} \cdot \frac{d\mathbf{q}_i}{dt} = 0 \\ &\frac{\partial f}{\partial t} = -\sum_{i=1}^{N} \left(\frac{\partial f}{\partial \mathbf{r}_i} \cdot \frac{\partial \mathscr{H}}{\partial \mathbf{p}_i} - \frac{\partial f}{\partial \mathbf{q}_i} \cdot \frac{\partial \mathscr{H}}{\partial \mathbf{r}_i} \right) = \{f, \mathscr{H}\}. \end{aligned} \tag{G.8}$$

G.2 BBGKY HIERARCHY

The total number of particles lying at some location $(\mathbf{r},\mathbf{p})$, the one-particle distribution function, can be found discarding one particle since they are all identical:

$$f_1(\mathbf{r},\mathbf{p},t) = \int_{(\mathbb{R}^d\times\mathbb{R}^d)^{N-1}} \prod_{i=2}^{N} d^3\mathbf{r}_i d^3\mathbf{p}_i f(\mathbf{R},\mathbf{P}) = \int_{(\mathbb{R}^d\times\mathbb{R}^d)^{N-1}} f d\mathbf{R}^{(1)} d\mathbf{P}^{(1)}, \tag{G.9}$$

where we adopted a simplified notation with $\mathbf{R} = (\mathbf{r}_1,\ldots,\mathbf{r}_N)$, and the subscript "(1)" indicates that the first item is absent.

The evolution of the one-particle distribution function can be found by integrating the Liouville equation (G.8). The first element in the equation gives:

$$\int_{(\mathbb{R}^d\times\mathbb{R}^d)^{N-1}} \frac{\partial f}{\partial t} d\mathbf{R}^{(1)} d\mathbf{P}^{(1)} = \frac{\partial}{\partial t} \int_{(\mathbb{R}^d\times\mathbb{R}^d)^{N-1}} f d\mathbf{R}^{(1)} d\mathbf{P}^{(1)} = \frac{\partial f_1}{\partial t}. \tag{G.10}$$

The first element inside the parenthesis gives:

$$\begin{aligned} &\int_{(\mathbb{R}^d\times\mathbb{R}^d)^{N-1}} \frac{\partial f}{\partial \mathbf{r}_i} \cdot \frac{\partial \mathscr{H}}{\partial \mathbf{p}_i} d\mathbf{R}^{(1)} d\mathbf{P}^{(1)} = \int_{(\mathbb{R}^d\times\mathbb{R}^d)^{N-1}} \frac{\partial f}{\partial \mathbf{r}_i} \cdot \frac{\mathbf{p}_i}{m_i} d\mathbf{R}^{(1)} d\mathbf{P}^{(1)} = \\ &= \int_{(\mathbb{R}^d)^{N-1}} \frac{\mathbf{p}_i}{m_i} \cdot \left(\int_{(\mathbb{R}^d)^{N-1}} \frac{\partial f}{\partial \mathbf{r}_i} d\mathbf{R}^{(1)} \right) d\mathbf{P}^{(1)} = \begin{cases} \frac{\mathbf{p}_i}{m_i} \frac{\partial f_1}{\partial \mathbf{r}_1} & \text{if } i = 1 \\ 0 & \text{otherwise.} \end{cases} \end{aligned} \tag{G.11}$$

The remaining element gives:

---

[a]Joseph Liouville (1809-1882) French mathematician advisee of Siméon Poisson and adviser of Eugéne Charles Catalan, among others.

$$\int_{(\mathbb{R}^d\times\mathbb{R}^d)^{N-1}} \frac{\partial f}{\partial \mathbf{q}_i}\cdot\frac{\partial \mathscr{H}}{\partial \mathbf{r}_i} d\mathbf{R}^{(1)}d\mathbf{P}^{(1)} = \\ = \int_{(\mathbb{R}^d\times\mathbb{R}^d)^{N-1}} \frac{\partial f}{\partial \mathbf{q}_i}\cdot\left(\sum_{j=1}^{N}\frac{\partial V(\mathbf{r}_j)}{\partial \mathbf{r}_i}+\sum_{\{k,j\}}\frac{\partial U(\mathbf{r}_k-\mathbf{r}_j)}{\partial \mathbf{r}_i}\right) d\mathbf{R}^{(1)}d\mathbf{P}^{(1)} \tag{G.12}$$

Let's split the right hand side of the equation into two integrals I_1 and I_2. For the first, we get:

$$I_1 = \int_{(\mathbb{R}^d\times\mathbb{R}^d)^{N-1}} \frac{\partial f}{\partial \mathbf{q}_i}\cdot\frac{\partial V}{\partial \mathbf{r}_i} d\mathbf{R}^{(1)}d\mathbf{P}^{(1)} = \begin{cases} \frac{\partial f_1}{\partial \mathbf{q}_1}\cdot\frac{\partial V}{\partial \mathbf{r}_1} & \text{if } i=1 \\ 0 & \text{otherwise.}\end{cases} \tag{G.13}$$

For the second integral, when $i = 1$, we get:

$$I_2 = \sum_{j=2}^{N}\int_{(\mathbb{R}^d\times\mathbb{R}^d)^{N-1}} \frac{\partial f}{\partial \mathbf{q}_1}\cdot\frac{\partial U(\mathbf{r}_1-\mathbf{r}_j)}{\partial \mathbf{r}_1} d\mathbf{R}^{(1)}d\mathbf{P}^{(1)}. \tag{G.14}$$

Since we are dealing with a binary interaction, we may write:

$$I_2 = (N-1)\int_{(\mathbb{R}^d\times\mathbb{R}^d)^{N-1}} \frac{\partial f}{\partial \mathbf{q}_1}\cdot\frac{\partial U(\mathbf{r}_1-\mathbf{r}_j)}{\partial \mathbf{r}_1} d\mathbf{R}^{(2)}d\mathbf{P}^{(2)}, \tag{G.15}$$

where the second marginal can be identified. Also, when $i > 1$, $I_2 = 0$ as in the previous cases. Hence, putting it all together we end up with:

$$\frac{\partial f_1}{\partial t}+\frac{\mathbf{p}_i}{m_i}\frac{\partial f_1}{\partial \mathbf{r}_1}-\frac{\partial f_1}{\partial \mathbf{q}_1}\cdot\frac{\partial V}{\partial \mathbf{r}_1}-(N-1)\int_{(\mathbb{R}^d\times\mathbb{R}^d)^{N-1}} \frac{\partial f_2}{\partial \mathbf{q}_1}\cdot\frac{\partial U(\mathbf{r}_1-\mathbf{r}_j)}{\partial \mathbf{r}_1} d\mathbf{R}^{(2)}d\mathbf{P}^{(2)} = 0. \tag{G.16}$$

This is known as the BBKGY[b] hierarchy for the first two marginals. The first marginal depends on the second and it is possible to show that the second depends on the third and so on.

[b]Nikolay Bogolyubov (1909-1992) Russian physicist, winner of many awards including the Dirac Prize in 1992; Max Born (1882-1970) German physicist, advisee of Joseph John Thomson and adviser of Victor Weisskopf, Enrico Fermi, and Robert Oppenheimer, among others. Born was awarded the Nobel Prize in Physics in 1954; Herbert Sydney Green (1920-1999) English physicist, advisee of Max Born; John Gamble Kirkwood (1907-1959) American physicist and chemist; Jacques Yvon (1903-1979) French physicist.

References

1. R. Inklaar, H. de Jong, J. Bolt, and J. van Zanden. Rebasing 'maddison': new income comparisons and the shape of long-run economic development. GGDC Research Memorandum GD-174, Groningen Growth and Development Centre, University of Groningen, 2018.
2. P. Duhem. *The aim and structure of physical theory*. Princeton University Press, Princeton, NJ, 1991.
3. R. N. Mantegna and H. E. Stanley. *Introduction to Econophysics: Correlations and Complexity in Finance*. Cambridge University Press, Cambridge, UK, 2007.
4. K. C. Dash. *The story of econophysics*. Cambridge Scholars Publishing, Newcastle, UK, 1st edition, 2019.
5. F. A. von Hayek. The use of knowledge in society. *Am. Econ. Rev.*, 35(4):519–530, 1945.
6. F. A. von Hayek. Scientism and the study of society i. *Economica*, 9:267–291, 1942.
7. F. A. von Hayek. *Law, Legislation, and Liberty*, volume 2, pages 267–290. Routlege Classics, Abingdon, UK, 2013.
8. John Locke. *Some considerations on the consequences of the lowering of interest and the raising of the value of money*. McMaster University Archive for the History of Economic Thought, London, UK, 1691.
9. Adam Smith. *An inquiry into the nature and causes of the wealth of nations*. W. Strahan and T. Cadell, London, UK, 1776.
10. A. Cournot. *Researches into the Mathematical Principles of the Theory of Wealth*. The Macmillan Company, London, UK, 1838.
11. A. Marshall. Principles of economics. 1890.
12. S. H. Strogatz. *Dynamics and chaos: with applications to physics, biology, chemistry, and engineering*. CRC Press, Reading, MA, 1994.
13. N. Kaldor. A classificatory note on the determinateness of equilibrium. *Economic Studies*, 1(2):122–136, 1934.
14. Irving Fisher. *Stabilizing the dollar*. The Macmillan Company, New York, NY, 1920.
15. J. F. Muth. Rational expectations and the theory of price movements. *Econometrica*, 29(3):315–335, 1961.
16. H. A. Simon. A behavioral model of rational choice. *Q. J. Econ*, 69(1):99–118, 1955.
17. A. Tversky and D. Kahneman. Judgment under uncertainty: heuristics and biases. *Science*, 185:1124–1131, 1974.
18. R. H. Thaler and C. R. Sunstein. *Nudge: Improving decisions about health, wealth, and happiness*. Yale University Press, New Haven & London, 1st edition, 2008.
19. R. M. Raafat, N. Chater, and C. Frith. Herding in humans. *Trends Cogn. Sci.*, 13(10):420–428, 2009.
20. J. Regnault. *Calcul des chances et philosophie de la bourse*. Mallet-Bachelier and Castel, Paris, France, 1863.
21. L. J.-B. A. Bachelier. Théorie de la spéculation. *Ann. Sci. Éc. Norm. Supér.*, 3(17):21–86, 1900.
22. B. Mandelbrot. The variation of certain speculative prices. *J. Business*, 36:394–419, 1963.

23. E. Fama. The behaviour of stock market prices. *Journal of Business*, 38:34–105, 1965.
24. P. Samuelson. Proof that properly anticipated prices fluctuate randomly. *Industrial Management Review*, 6:41–59, 1965.
25. S. J. Grossman and J. E. Stiglitz. On the impossibility of informationally efficient markets. *Am. Econ. Rev.*, 70(3):393–408, 1980.
26. M. Musiela and M. Rutkowski. *Martingale methods in financial modelling*. Stochastic modelling and applied probability. Springer-Verlag, Berlin, Heidelberg, 2nd edition, 2005.
27. A. N. Kolmogorov. *Foundations of the theory of probability*. Dover Publications, Inc., Mineloa, NY, 2nd edition, 2018.
28. D. S. Lemons. *An introduction to stochastic processes in physics*. The Johns Hopkins University Press, Blatimore, MD, 2002.
29. P. Devolder, J. Janssen, and R. Manca. *Basic stochastic processes*. Mathematics and Statistics Series. John Wiley & Sons, Inc., Hoboken, NJ, 2015.
30. H. M. Taylor and S. Karlin. *An introduction to stochastic modeling*. Academic Press, Inc., London, UK, 3rd edition, 1998.
31. J. Medhi. *Stochastic processes*. New Academic Science Limited, Kent, UK, 3rd edition, 2012.
32. N. G. van Kampen. *Stochastic processes in physics and chemistry*. North-Holland, Amsterdam, Netherlands, 3rd edition, 1992.
33. P. Lévy. L'addition des variables aléatoires définies sur une circonférence. *Bull. Soc. Math. Fr.*, 67:1–41, 1939.
34. P. Lévy. *Processus Stochastiques et Mouvement Brownien*. Monographies des Probabilities. Gauthier-Villars, Paris, France, 2nd edition, 1965.
35. K. I. Sato. *Lévy processes and infinitely divisible distributions*. Cambdrige studies in advanced mathematics. Cambridge University Press, Cambridge, UK, 1999.
36. M. Capiński, E. Kopp, and J. Traple. *Stohastic calculus for finance*. Cambridge University Press, Cambridge, UK, 2012.
37. B. de Finetti. Sulla funzione a incremento aleatorio. *Atti. Accad. NAz. Lincei Sez.*, 1(6):163–168, 1929.
38. K. I. Sato. *Lévy processes and infinitely divisible distributions*. Cambridge University Press, Cambridge, UK, 1999.
39. P. Lévy. *Calcul des Probabilites*. Gauthier-Villars et Cie, Paris, France, 1925.
40. A. Ya. Khinchin and P. Lévy. Sur les loi stables. *C. R. Acad. Sci. Paris*, 202:374–376, 1936.
41. Jayakrishnan Nair, Adam Wierman, and Bert Zwart. The fundamentals of heavy-tails: Propertis, emergence, and identification. In Mor Harchol-Balter, John R. Douceur, and Jun Xu, editors, *International Conference on Measurement and Modeling of Computer Systems (SIGMETRICS)*, pages 387–388, Pittsburgh, PA, 2013. ACM.
42. B. B. Mandelbrot. *Fractals and scaling in finance. Dicontinuity, concentration, risk*. Selecta Volume E. Springer-Verlag, New York, NY, 1997.
43. B. M. Hill. A simple general approach to inference about the tail of a distribution. *Ann. Stat.*, 3(5):1163–1174, 1975.
44. M. O. Lorenz. Methods of measuring the concentration of wealth. *Am. Stat. Assoc.*, 9(70):209–219, 1905.
45. C. Gini. Concentration and dependecy ratios. *Riv. Pol. Econ.*, 87:769–789, 1997.
46. H. O. Wold. *A Study in the Analysis of Stationary Time Series*. Almqvist & Wiksell, Stockholm, Sweden, 2nd edition, 1938.

47. George Box, Gwilym Jenkins, and G. C. Reinsel. *Time Series Analysis: Forecasting and Control*. Wiley, Hoboken, NJ, 4th edition, 2008.
48. G. E. P. Box and D. A. Pierce. Distribution of residual autocorrelations in autoregressive-integrated moving average time series models. *Journal of the American Statistical Association*, 65:1509–1526, 1970.
49. G. Ljung and G. Box. On a measure of lack of fit in time series models. *Biometrika*, 65:297–303, 1978.
50. J. Durbin. The fitting of time series models. *Rev. Inst. Int. Stat.*, 28:233–243, 1960.
51. N. Levinson. The wiener rms error criterion in filter design and prediction. *J. Math. Phys.*, 25:261–278, 1947.
52. G. U. Yule. On a method of investigating periodicities in disturbed series, with special reference to wolfer's sunspot numbers. *Philosophical Transactions of the Royal Society of London, Ser. A*, 226:267–298, 1927.
53. G. Walker. On periodicity in series of related terms. *Proceedings of the Royol Society of London, Ser. A*, 131:518–532, 1931.
54. R. F. Engle. Autoregressive conditional heteroscedasticity with estimates of the variance of united kingdom inflation. *Econometrica*, 50:987–1007, 1982.
55. J. A. Yan. *Introduction to stochastic finance*. Springer, Singapore, 2018.
56. T. Bollerslev. Generalized autoregressive conditional heteroskedasticity. *J. Econometrics*, 31:307–327, 1986.
57. S. J. Taylor. *Modeling Financial Time Series*. World Scientific Publishing Company, London, UK, 2nd edition, 1986.
58. R. K. Merton. The unanticipated consequences of purposive social action. *American Sociological Review*, 1(6):894–904, 1936.
59. N. Kaldor. A model of economic growth. *The Economic Journal*, 67:591–624, 1957.
60. D. Hischman. Stylized facts in the social sciences. *Sociological Science*, 3:604–626, 2016.
61. J. Tinbergen. *Selected Papers*, pages 37–84. North-Holland Publishing Co., Amsterdam, Holland, 1959.
62. Rama Cont. Empirical properties of asset returns: stylized facts and statistical issues. *Quantitative Finance*, 1:223–236, 2001.
63. C. R. da Cunha and R. da Silva. Relevant stylized facts about bitcoin: Fluctuations, first return probability, and natural phenomena. *Physica A*, 550:124155, 2020.
64. O. Knill. *Probability theory and stochastic processes with applications*. Overseas Press, New Delhi, India, 2009.
65. R. Mansuy. The origins of the word "martingale". *Electron. J. Hist. Prob. Stat.*, 5(1), 2009.
66. L. Mazliak. How paul lévy saw jean ville and martingales. *Electron. J. Hist. Prob. Stat.*, 5(1), 2009.
67. Z. F. Huang and S. Solomon. Finite market size as a source of extreme wealth inequality and market instability. *Physica A*, 294(3-4):503–513, 2001.
68. R. K. Merton. The matthew effect in science. *Science*, 159(3810):56–63, 1968.
69. R. Gibrat. *Les Inégalités économiques. Applications aux inégalités des richesses, à la concentration des entreprises, aux populations des villes, aux statistiques des familles, etc. d'une loi nouvelle. La loi de l'effet proportionnel*. Recueil Sirey, Paris, France, 1931.
70. F. Galton and H. W. Watson. On the probability of the extinction of families. *J. Royal Anthrop. Inst.*, 4:138–144, 1875.

71. K. Frenken and R. Boschma. A theoretical framework for economic geography: industrial dynamics and urban growth as a branching process. *J. Econ. Geogr.*, 7(5):635–649, 2007.
72. A. A. Markov. Extension of the law of large numbers to dependent quantities (in russian). *Izvestiia Fiz.-Matem. Obsch. Kazan Univ. (2nd Ser.)*, 15:135–156, 1906.
73. G. A. Pavliotis. *Stochastic Processes and Applications: Diffusion Processes, the Fokker-Planck and Langevin Equations*. Texts in Applied Mathematics. Springer, New York, NY, 2014.
74. K. Zhu, K. L. Kraemer, V. Gurbaxani, and S. X. Xu. Migration to open-standard interorganizational systems: Network effects, switching costs, and path dependency. *MIS Q.*, 30:515–539, 2006.
75. M. Stack and M. Gartland. Path creation, path dependency, and alternative theories of the firm. *J. Econ. Issues*, 37(2):487, 2003.
76. P. A. David. Clio and the economics of qwerty. *Am. Econ. Rev.*, 75:332–337, 1985.
77. M. Tabor. *Chaos and integrability in nonlinear dynamic system: An introduction*. John Wiley & Sons, Inc., New York, NY, 1989.
78. J. L. McCauley. Integrability and attractors. *Chaos Solitons Fractals*, 4(11):1969–1984, 1994.
79. L. Lam, editor. *Introduction to nonlinear physics*. Springer-Verlag, New York, NY, 2003.
80. A. Kolmogorov. über die analytischen methoden in der wahrscheinlichkeitsrechnung. *Math. Ann.*, 104:415–458, 1931.
81. S. Chapma. On the brownian displacements and thermal diffusion of grains suspended in a non-uniform fluid. *Proc. Roy. Soc. Lond. A*, 119:34–54, 1928.
82. W. H. Furry. On fluctuation phenomena in the passage of high energy electrons through lead. *Phys. Rev.*, 52:569, 1937.
83. G. E. Uhlenbeck and L. S. Ornstein. On the theory of brownian motion. *Phys. Rev.*, 36:823, 1930.
84. A. Einstein. über die von der molekularkinetischen theorie der wärme geforderte bewegung von in ruhenden flüssigkeiten suspendierten teilchen. *Annalen der Physik*, 322(8):549–560, 1905.
85. D. Sondermann. *Introduction to stochastic calculus for finance*. Lecture notes in economics and mathematical systems. Springer-Verlag, Berlin, Heidelberg, 2006.
86. S. M. Iacus. *Simulation and inference for stochastic differential equations with R examples*. Springer series in statistics. Springer, New York, NY, 2008.
87. O. Vasicek. An equilibrium characterization of the term structure. *Journal of Financial Economics*, 5(2):177–188, 1977.
88. L. D. Landau and E. M. Lifshitz. *Statistical physics*, volume 5. Butterworth-Heinemann, Oxford, UK, 3rd edition, 1980.
89. W. Feller. Two singular diffusion problems. *Ann. Math.*, 54(1):173–182, 1951.
90. J. C. Cox, J. E. Ingersoll, and S. A. Ross. A theory of the term structure of interest rates. *Econometrica*, 53:385–407, 1985.
91. D. L. Snyder. *Random point processes*. John Wiley & Sons, Inc., Hoboken, NJ, 1975.
92. A. G. Hawkes. Spectra of some self-exciting and mutually exciting point processes. *Biometrika*, 58:83–90, 1971.
93. T. W. Epps. Comovements in stock prices in the very short run. *Journal of the American Statistical Association*, 74:291–298, 1979.

94. E. Bacry, S. Delattre, M. Hoffmann, and J. F. Muzy. Modelling microstructure noise with mutually exciting point processes. *Quantitative Finance*, 13(1):65–77, 2013.
95. M. D. Donsker. An invariance principle for certain probability limit theorems. *Mem. Am. Math. Soc.*, 6:1–10, 1951.
96. A. Etheridge. *A course in financial calculus*. Cambridge University Press, Cambridge, UK, 2002.
97. J. C. Hull. *Fundamentals of futures and options markets*. Pearson Education Limited, Essex, England, 2017.
98. J. C. Cox, S. A. Ross, and M. Rubinstein. Option pricing: a simplified approach. *J. Financ. Econ.*, 7:229–263, 1979.
99. F. Black and M. Scholes. The pricing of options and corporate liabilities. *J. Pol. Econ.*, 81:654–673, 1973.
100. W. F. Sharpe. Mutual fund performance. *Journal of Business*, 39:119–138, 1966.
101. Moawia Alghalith. Pricing the american options using the black-scholes pricing formula. *Physica A*, 507:443–445, 2018.
102. M. Musiela and M. Rutkowski. Modifications of the black-scholes model. In *Martingale Methods in Financial Modeling*, volume 36 of *Applications of Mathematics (Stochastic Modelling and Applied Probability)*. Springer, Springer, Berlin, Heidelberg, 1997.
103. Svetlana I Boyarchenko and Sergei Z Levendorskii. *Non-Gaussian Merton-Black-Scholes Theory*, volume 9 of *Advanced Series on Statistical Science & Applied Probability*. World Scientific Publishing Co., Singapore, 2002.
104. Ian Steward. *In pursuit of the unknown: 17 Equations that changed the world*. Basic Books, New York, NY, illustrated edition, 2012.
105. C. S. Wehn, C. Hoppe, and G. N. Gregoriou. *Rethinking valuation and pricing models. Lessons learned from the crisis and future challenges*. Academic Press, Inc., Oxford, UK, 1st edition, 2013.
106. G. Galles. *Pathways to policy failure*. American Institute for Economic Research, Great Barrington, MA, 2020.
107. G. Reisman. *The government against the economy*. TJS Books, Laguna Hills, CA, 2012.
108. H. Markovitz. Portfolio selection. *The Journal of Finance*, 7:77–91, 1952.
109. M. W. Brandt, P. Santa-Clara, and R. Valkonov. Parametric portfolio policies: Exploiting characteristics in the cross-section of equity returns. *Review of Financial Studies*, 22:3411–3447, 2009.
110. W. F. Sharpe. Capital asset prices: A theory of market equilibrium under conditions of risk. *Journal of Finance*, 19:425–442, 1964.
111. W. F. Sharpe. The sharpe ratio. *The Journal of Portfolio Management*, 21:49–58, 1994.
112. J. Tobin. Liquidity preference as behavior towards risk. *Review of Economic Studies*, 25:65–86, 1958.
113. F. H. Knight. *Risk, Uncertainty, and Profit*. Cosimo Classics, New York, NY, 2006.
114. J. L. Treynor. *Market Value, Time, and Risk (1961) - Toward a Theory of Market Value of Risky Assets (1962)*, pages 15–22. Risk Books, 1999.
115. J. Litner. The valuation of risk assets and the selection of risky investments in stock portfolios and capital budgets. *Review of Economics and Statistics*, 47:13–37, 1965.
116. J. Mossin. Equilibrium in a capital asset market. *Econometrica*, 34:768–783, 1966.
117. A. Edelman and N. R. Rao. Random matrix theory. *Acta Num.*, 14:1–65, 2005.

118. M. L. Mehta. *Random matrices*. Pure and applied mathematics series. Elsevier Academic Press, San Diego, CA, 3rd edition, 2004.
119. G. W. Anderson, A. Guionnet, and O. Zeitouni. *An introduction to random matrices*. Cambdrige studies in advanced mathematics. Cambridge University Press, Cambridge, UK, 2010.
120. E. J. Heller and S. Tomsovic. Postmodern quantum mechanics. *Phys. Today*, 46:38–46, 2008.
121. V. A. Marčenko and L. A. Pastur. The distribution of eigenvalues in certain sets of random matrices. *Mat. Sb.*, (72):507–536, 1967.
122. E. P. Wigner. Results and theory of resonance absorption. Technical Report ORNL-2309, Oak Ridge National Laboratory, 1957.
123. Y. Choueifaty and Y. Coignard. Toward maximum diversification. *J. Portf. Manag.*, 35(1):40–51, 2008.
124. S. Maillard, T. Roncalli, and J. Teïletche. The properties of equally weighted risk contribution portfolios. *J. Portf. Manag.*, 36(4):60–70, 2010.
125. L.D. Kudryavtsev. *Homogeneous function*. Springer-Verlag, Switzerland, 2011.
126. Florin Spinu. An algorithm for computing risk parity weights. *SSRN*, page 2297383, 2013.
127. M. Kelbert, I. Stuhl, and Y. Suhov. Weighted entropy: basic inequalities. *Mod. Stoch.: Theory Appl.*, 4(3):233–252, 2017.
128. J. R. Wu, W. Y. Lee, and W. J. P. Chiou. Diversified portfolios with different entropy measures. *Appl. Math. Comput.*, 241:47–63, 2014.
129. M. J. Naylor, L. C. Rose, and B. J. Moyle. Topology of foreigh exchange markets using hierarchical structure methods. *Physica A*, 389:199–208, 2007.
130. R. Rammal, G. Toulouse, and M. A. Virasoro. Ultrametricity for physicists. *Rev. Modern. Phys.*, 58:765–788, 1986.
131. M. Mezard, G. Parisi, and M. A. Virasoro. *Spin glass theory and beyond. An introduction to the replica method and its applications*, volume 9 of *World Scientific Lecture Notes in Physics*. World Scientific Publishing, Singapore, 2004.
132. A. Z. Górski, S. Drożdż, and J. Kwapień. Scale free effects in world currency exchange network. *Eur. Phys. J. B*, 66:91–96, 2008.
133. R. N. Mantegna. Hierarchical structure in financial markets. *Eur. Phys. J. B*, 11:193–196, 1999.
134. M. Schulz. Statistical physics and economics: Concept, tools, and applications. 2003.
135. R. C. Prim. Shortest connection networks and some generalizations. *Bell System Techn. J.*, 36:1389–1401, 1957.
136. R. Sibson. Slink: An optimally efficient algorithm for the single-link cluster. *Comp. J.*, 16(1):30–34, 1973.
137. C. Mackay. *Extraordinary Popular Delusions, the Money Mania: The Mississippi Scheme, the South-Sea Bubble, & the Tulipomania*. Cosimo Classics, New York, NY, abridged edition, 2008.
138. F. A. von Hayek. *Hayek's triangles: two essays on the business cycle*. Laissez Faire Books, Maltimore, MD, 2014.
139. R. W. Garrison. *Time and Money, The macroeconomics of capital structure*. Routledge, New York, NY, 1st edition, 2000.
140. L. von Mises. *The Theory of Money and Credit*. Skyhorse, New York, NY, 1st edition, 2013.

141. J. P. Keeler. Empirical evidence on the austrian business cycle theory. *Rev. Austrian Econ.*, 14(4):331–351, 2001.
142. G. Kim and H. M. Markowitz. Investment rules, margin and market volatility. *J. Portf. Manag.*, 16:45–52, 1989.
143. I. Stewart. Bifurcation theory. *Math. Chronicle*, 5:140–165, 1977.
144. R. Thom. *Structural Stability and Morphogenesis*. CRC Press, Boca Raton, FL, 2018.
145. I. Katerelos and N. Varotsis. A cusp catastrophe model of tax behavior. *Nonlinear Dynamics Psychol. Life Sci.*, 21(1):89–112, 2017.
146. C. Diks and J. Wang. Can a stochastic cusp catastrophe model explain housing market crashes? *J. Econ. Dyn. Cont.*, 69:68–88, 2016.
147. J. Barunik and M. Vosvrda. Can a stochastic cusp catastrophe model explain stock market crashes? *J. Econ. Dyn. Cont.*, 33:1824–1836, 2009.
148. I. Cobb. Parameter estimation for the cusp catastrophe model. *Behav. Sci.*, 26(1):75–78, 1981.
149. M. Gardner. Mathematical games - the fantastic combinations of john conway's new solitaier game "life". *Scientific American*, 223:120–123, 1970.
150. R. D. Beer. Autopoiesis and cognition in the game of life. *Artificial Life*, 10:309–326, 2004.
151. N. M. Gotts. Ramifying feedback networks, cross-scale interactions, and emergent quasi-individuals in conway's game of life. *Artificial Life*, 15:351–375, 2009.
152. W. H. Zureck, editor. *Complexity, entropy and the physics of information*, volume VIII of *A proceedings volume in the Santa Fe Institute studies in the sciences of complexity*. CRC Press, Boca Raton, FL, 1990.
153. P. Bak, C. Tang, and K. Wiesenfeld. Self-organized criticality: an explanation of 1/f noise. *Physical Review Letters*, 59(4):381–384, 1987.
154. F. Omori. On the aftershocks of earthquakes. *Journal of the College of Science, Imperial University of Tokyo*, 7:111–200, 1894.
155. T. Utsu. A statistical study of the occurrence of aftershocks. *Geophysical Magazine*, 30:521–605, 1961.
156. B. Gutenberg and C. F. Richter. Magnitude and energy of earthquakes. *Annali di Geofisica*, 9:1–15, 1965.
157. Bogdan Negrea. A statistical measure of financial crises magnitude. *Physica A: Statistical Mechanics and its Applications*, 397:54–75, 03 2014.
158. S. M. Potirakis, P. I. Zitis, and K. Eftaxias. Dynamical analogy between economical crisis and earthquake dynamics within the nonextensive statistical mechanics framework. *Physica A*, 392(13):2940–2954, 2013.
159. J. von Neumann. Zur theorie der gesellschaftsspiele. *Mathematische Annalen*, 100(1):295–320, 1928.
160. J. von Neumann and O. Morgenstern. *Theory of Games and Economic Behavior*. Princeton University Press, Princeton, NJ, 1953.
161. A. W. Brian. Inductive reasoning and bounded rationality. *Am. Econ. Rev.*, 84:406–411, 1994.
162. R. Stalnaker. The problem of logical omniscience, i. *Synthese*, 89(3):425–440.
163. M. Friedman. *The methodology of positive economics*. University of Chicago Press, Chicago, IL, 1953.
164. J. S. Mill. On the definition of political economy; and on the method of investigation proper to it. In *Essays on Some Unsettled Questions of Political Economy*. John W. Parker, West Strand, London, UK, 1844.

165. W. S. Jevons. *The theory of political economy*. Palgrave Classics in Economics. Palgrave Macmillan, London, UK, 2013.
166. D. Kahneman and A. Tversky. Prospect theory: an analysis of decision under risk. *Econometrica*, 47(2):263–291, 1979.
167. D. Challet and Y. C. Zhang. Emergence of cooperation and organization in an evolutionary game. *Physica A*, 246:407–418, 1997.
168. D. Challet, M. Marsili, and Y. C. Zhang. Modeling market mechanism with minority game. *Physica A*, 276:284–315, 2000.
169. M. Levy, H. Levy, and S. Solomon. A microscopy model of the stock market: Cycles, booms and crashes. *Economics Letters*, 45:103–111, 1994.
170. D. Challet and M. Marsili. Phase transition and symmetry breaking in the minority game. *Phys. Rev. E*, 60(6):R6271–R6274, 1999.
171. L. S. Shapley. Notes on the n-person game ii: the value of an n-person game. Technical Report RAND RM670, August 1951.
172. W. Güth, R. Schmittberger, and B. Schwarze. An experimental analysis of ultimatum bargaining. *J. Econ. Behav. Org.*, 3:367–388, 1982.
173. M. Marsili and Y. C. Zhang. Fluctuations around nash equilibria. *Physica A*, 245(1-2):181–188, 1997.
174. M. M. Pillutla and J. K. Murnighan. Unfairness, anger and spite: Emotional rejections of iltimatum offers. *Organ. Behav. Hum. Decis. Process.*, 68(3):208–224, 1996.
175. E. Hoffman, K. A. McCabe, and V. L. Smith. On expectations and the monetary stakes in ultimatum games. *Int. J. Game Theory*, 25:289–301, 1996.
176. A. Gunnthorsdottir, D. Houser, and K. McCabe. Disposition, history and contributions in public goods experiments. *J. Econ. Behav. Org.*, 62(2):304–315, 2007.
177. F. Guala and L. Mittone. Paradigmatic experiments: The dictator game. *J. Socio-Econ.*, 39(5):578–584, 2010.
178. M. M. Flood. Some experimental games. Technical report, June 1952.
179. A. W. Tucker. On jargon: The prisoner's dilemma. *UMAP Journal*, 1:101, 1980. Originalmente: "A Two-Person Dilemma" (1950) mimeo, Stanford University.
180. E. Fehr and J.-R. Tyran. Money illusion and coordination failure. *Games Econ Behav*, 58(2):246–268, 2007.
181. G. Stigler. The theory of economic regulation. *Bell Journal of Economics and Management Science*, 2(1):3–21, 1971.
182. R. H. Coase. The federal communications commision. *J. Law Econ.*, 2:1–40, 1959.
183. I. M. Kirzner. *The meaning of market process: essays in the development of modern Austrian economics*. Routledge, New York, NY, 1st edition, 2002.
184. R. H. Coase. The nature of the firm. *Economica*, 4(16):386–405, 1937.
185. W. B. Arthur. Competing technologies, increasing returns, and lock-in by historical events. *Econ. J.*, 97:642–665, 1989.
186. G. Saint-Paul, D. Ticchi, and A. Vindigni. A theory of political entrenchment. *Econ. J.*, 126(593):1238–1263, 2016.
187. L. Walras. Elements of theoretical economics, or the theory of social wealth. 2014.
188. A. M. Colell, M. D. Whinston, and J. R. Green. *Microeconomic Theory*. Oxford University Press, Oxford, England, 1st edition, 2012.
189. I. Goldstein and A. Pauzner. Demand-deposit contract and the probability of bank runs. *J. Finance*, 60:1293–1327, 2005.
190. S. Morris and H. S. Shin. Unique equilibrium in a model of self-fulfilling currency attacks. *Am. Econ. Rev.*, 88(3):587–597, 1998.

191. S. Morris and H. S. Shin. Coordination risk and the price of debt. *Eur. Econ. Rev.*, 48(1):133–153, 2004.
192. L. E. J. Brouwer. über abbildungen von mannigfaltigkeiten. *Math. Ann.*, 71:97–115, 1911.
193. J. Nash. Equilibrium points in n-person games. *Proceedings of the National Academy of Sciences*, 36(1):48–49, 1950.
194. W. F. Lloyd. *Two lectures on the checks to population.* Oxford University Press, Oxford, UK, 1833.
195. G. Hardin. The tragedy of the commons. *Science*, 162:1243–1248, 1968.
196. E. Ostrom, J. Burger, C. B. Field, R. B. Norgaard, and D. Policansky. Revisiting the commons: Local lessos, global challenges. *Science*, 284(5412):278–282, 1999.
197. R. Axelrod. *The evolution of cooperation.* Basic Books, New York, NY, revised edition, 2006.
198. T. C. Schelling. *The strategy of conflict.* Harvard University Press, Cambridge, MA, 1981.
199. R. Higgs and A. A. Rkirch, Jr. *Crisis and Leviathan. Critical episodes in the growth of American government.* The Independent Institute, Oakland, CA, 2012.
200. G. P. Harmer and D. Abbott. Losing strategies can win by parrondo's paradox. *nature*, 402:864, 1999.
201. G. P. Harmer, D. Abbott, and P. G. Taylor. The paradox of parrondo's games. *Proc. Roy. Soc. Lond. A*, 456:247–259, 2000.
202. B. Cornell and R. Roll. Strategies for pairwise competitions in markets and organizations. *Bell J. Econ.*, 12(1):201–213, 1981.
203. M. Kandori. *Evolutionary game theory in economics.* Cambridge University Press, Cambridge, 1997.
204. D. Friedman. Evolutionary games in economics. *Econometrica*, 59(3):637–666, 1991.
205. D. Friedman and K. C. Fung. International trade and the internal organization of firms: An evolutionary approach. *J. Int. Econ.*, 41:113–137, 1996.
206. D. Friedman. On economic applications of evolutionary game theory. *J. Evol. Econ.*, 8:15–43, 1998.
207. J. Maynard Smith and G. R. Price. The logic of animal conflict. *Nature*, 246:15–18, 1973.
208. A. J. Lotka. Contribution to the theory of periodic reaction. *J. Phys. Chem.*, 14(3):271–274, 1910.
209. A. J. Lotka. Analytical note on certain rhytmic relations in organic systems. *Proc. Natl. Acad. Sci. USA*, 6:410–415, 1920.
210. A. J. Lotka. *Elements of Physical Biology.* Williams & Wilkins, Baltimore, MD, 1925.
211. R. Syms and L. Solymar. A dynamic competition model of regime change. *J. Oper. Res. Soc.*, 66:1939–1947, 2015.
212. R. M. Goodwin. *A Growth Cycle. Socialism, Capitalism and Economic Growth*, pages 165–170. Palgrave Macmillan, London, UK, 1982.
213. B. Powell. *Some implications of capital heterogeneity*, chapter 9. Edward Elgas Publishing Limited, Cheltenham, UK, 2012.
214. A. W. Phillips. The relationship between unemployment and the rate of change of money wages in the united kingdom 1861-1957. *Economica*, 25(100):283–299, 1958.
215. M. Friedman. The role of monetary policy. *Am. Econ. Rev.*, 58(1):1–17, 1968.
216. E. S. Phelps. Money-wage dynamics and labor market equilibrium. *J. Pol. Econ.*, 76(S4):678–711, 1968.

217. E. S. Phelps. Philips curves, expectations of inflation and optimal unemployment over time. *Economica*, 34(135):254–281, 1967.
218. F. A. von Hayek. Full employment, planning and inflation. *Inst. Pub. Affair Rev.*, 4(6):174–184, 1950.
219. P. Hartman. A lemma in the theory of structural stability of differential equations. *Proc. A.M.S.*, 11(4):610–620, 1960.
220. D. M. Grobman. Homeomorphism of systems of differential equations. *Doklady Akad. Nauk SSSR*, 128:880881, 1959.
221. David R. Howell and Arne L. Kalleberg. Declining job quality in the united states: Explanations and evidence. *RSF: The Russell Sage Foundation Journal of the Social Sciences*, 5(4):1–53, 2019.
222. G. Reisman. *Capitalism: A treatise on economics*. TJS Books, Laguna Hills, CA, 2012.
223. W. Weaver. Science and complexity. *Am. Sci.*, 36(4):536–544, 1948.
224. Y. Holovatch, R. Kenna, and S. Thurner. Complex systems: physics beyond physics. *Eur. J. Phys.*, 38-2:023002, 2017.
225. S. Wolfram. *A new kind of science*. Wolfram media, Champaign, IL, 1st edition, 2002.
226. E. W. Dijkstra. A note on two problems in connexion with graphs. *Numer. Math.*, 1:269–271, 1959.
227. R. Bellman. On a routing problem. *Q. Appl. Math.*, 16:87–90, 1958.
228. R. W. Floyd. Algorithm 97: Shortest path. *Commun. ACM*, 5(6):345, 1962.
229. S. Warshall. A theorem on boolean matrices. *J. ACM*, 9(1):11–12, 1959.
230. R. Cont and J. P. Bouchaud. Herd behavior and aggregate fluctuations in financial markets. *Marcoecon. Dyn.*, 4:170–196, 2000.
231. L. C. Freeman. Centrality in social networks conceptual clarification. *Social Networks*, 1:215–239, 1978.
232. A. Bavelas. Communication patterns in task-oriented groups. *J. Acoust. Soc. Am.*, 22(6):725–730, 1950.
233. L. Freeman. A set of measures of centrality based upon betweeness. *Sociometry*, 40(1):35–41, 1977.
234. P. Bonacich. Power and centrality: A family of measures. *Am. J. Sociol.*, 92(5):1170–1182, 1986.
235. P. Erdös and A. Renyi. On random graphs. *Publicationes Mathematicae*, 6:290–297, 1959.
236. M. Kochen, editor. *The small world*. Ablex Publishing Co., Norwood, NJ, 1989.
237. S. Milgram. The small world problem. *Psychology Today*, 2:60–67, 1967.
238. J. Travers and S. Milgram. An experimental study of the small world problem. *Sociometry*, 32:425–443, 1969.
239. G. U. Yule. A mathematical theory of evolution based on the conclusions of dr. j. c. willis, f.r.s. *Philosophical Transactions of the Royal Society*, 213(402-410):8560–8566, 1925.
240. H. A. Simon. On a class of skew distribution functions. *Biometrika*, 42(3-4):425–440, 1955.
241. A. Réka and A. L. Barabási. Statistical mechanics of complex networks. *Rev. Mod. Phys.*, 74(1):47–97, 2002.
242. A.-L. Barabási and R. Albert. Emergence of scaling in random networks. *Science*, 286(5439):509–512, 1999.
243. A. L. Barabási. *Network science*. Cambridge University Press, Cambridge, UK, 2016.

244. Watts D. J and S. H. Strogatz. Collective dynamics of 'small-world' networks. *Nature*, 393(6684):440–442, 1998.
245. W. Lenz. Beiträge zum verständnis der magnetischen eigenschaften in festen körpern. *Physikalische Zeitschrift*, 21:613–615, 1920.
246. E. Ising. Beitrag zur theorie des ferromagnetismus. *Z. Phys.*, 31:253–258, 1925.
247. R. H. Swendsen and J. S. Wang. Nonuniversal critical dynamics in monte carlo simulations. *Phys. Rev. Lett.*, 58(2):86–88, 1987.
248. U. Wolff. Collective monte carlo updating for spin systems. *Phys. rev. Lett.*, 62:361, 1989.
249. T. C. Schelling. Dynamic models of segregation. *J. Math. Sociol.*, 1(2):143–186, 1971.
250. D. Stauffer and S. Solomon. Ising, schelling and self-organising segregation. *Eur. Phys. J. B*, 57:473–479, 2007.
251. C J. Mod. Phys. Self-organazing, two-temperature ising model describing human segregation. *Int. J. Mod. Phys. C*, 19(3):393–398, 2008.
252. A. V. Mantzaris, J. A. Marich, and T. W. Halfman. Examining the schelling model simulation through an estimation of its entropy. *Entropy*, 20(9):623, 2018.
253. L. Gauvin, J. Vannimenus, and J. P. Nadal. Phase diagram of a schelling segregation model. *Eur. Phys. J. B*, 70:293–304, 2009.
254. I. Benenson, E. Hatna, and E. Or. From schelling to spatially explicit modeling of urban ethnic and economic residential dynamics. *Sociol. Methods Res.*, 37(4):463, 2009.
255. R. McCrea. Explaining sociospatial patterns in south east queensland, australia. *Am. Sociol. Rev.*, 41(9):2201–2214, 2009.
256. S. E. Asch. *Effects of group pressure upon the modification and distortion of judgment*. Carnegie Press, Pittsburgh, PA, 1951.
257. A. V. Banerjee. A simple model of herd behavior. *Q. J. Econ*, 107(3):797–817, 1992.
258. R. Hegselmann and U. Krause. Opinion dynamics and bounded confidence models, analysis, and simulation. *JASSS*, 5(3):1–33, 2002.
259. G. Deffuant, D. Neau, F. Amblard, and G. Weisbuch. Mixing beliefs among interacting agents. *Adv. Complex Syst.*, 03(01n04):87–98, 2000.
260. A. Kirman. Ants, rationality, and recruitment. *Q. J. Econ*, 108(1):137–156, 1993.
261. S. Bornholdt. Expectation bubbles in a spin model of markets: Intermittency from frustration across scales. *Int. J. Mod. Phys. C*, 12(5):667–674, 2001.
262. T. Lux. The socio-economic dynamics of speculative markets: interacting agents, chaos, and the fat tails of return distributions. *J. Econ. Behav. Organ.*, 33:143–165, 1998.
263. T. Lux and M. Marchesi. Scaling and criticality in a stochastic multi-agent model of a financial market. *Nature*, 397:398–500, 1999.
264. L. Pareschi and G. Toscani. *Interacting Multiagent Systems: Kinetic equations & Monte Carlo methods*. Oxford University Press, Oxford, England, 2014.
265. A. Chakraborti and B. K. Chakrabarti. Statistical mechanics of money: How saving propensity affects its distributions. *Eur. Phys. J. B*, 17:167–170, 2000.
266. S. Ispolatov, P. L. Krapivsky, and S. Redner. Wealth distributions in asset exchange models. *Eur. Phys. J. B*, 2:267–276, 1998.
267. M. Gallegati, S. Keen, T. Lux, and P. Ormerod. Worrying trends in econophysics. *Physica A*, 370(1):1–6, 2006.
268. J. H. McCulloch. The austrian theory of the marginal use and of orginal marginal utility. *J. Econ*, 37(3-4):249–280, 1977.

269. F. Slanina. Inelastically scattering particles and welath distribution in an open economy. *Phys. Rev. E*, 69:046102, 2004.
270. L. Pareschi and G. Toscani. Self-similarity and power-like tails in non-conservative kinetic models. *J. Stat. Phys.*, 124:747–779, 2006.
271. A. Drăgulescu and V. M. Yakovenko. Exponential and power-law probability distributions of wealth and income in the united kingdom and the united states. *Physica A*, 299:213–221, 2001.
272. H. Tanaka. Professor gisiro maruyama, in memoriam. *Probability Theory and Mathematical Statistics*, 1299:1–6, 1988.
273. N. Metropolis, A. W. Rosenbluth, M. N. Rosenbluth, A. H. Teller, and E. Teller. Equations of state calculations by fast computing machines. *Journal of Chemical Physics*, 21(6):1087–1092, 1953.
274. W. K. Hasting. Monte carlo sampling methods using markov chains and their applications. *Biometrika*, 57(1):97–109, 1970.
275. A. D. Fokker. Die mittlere energie rotierender elektrischer dipole im strahlungsfeld. *Ann. Phys.*, 348:810–820, 1914.
276. M. Planck. über einen satz der statistischen dynamik und seine erweiterung in der quantentheori. *Sitzungsberichte der Preussischen Akademie der Wissenschaften zu Berlin*, 24:324–341, 1917.
277. H. A. Kramers. Brownian motion in a field of force and the diffusion model of chemical reactions. *Physica*, 7:284–304, 1940.
278. J. E. Moyal. Stochastic processes and statistical physics. *Journal of the Royal Statistical Society - B*, 11(2):150–210, 1949.
279. A. Novikov. On an identity for stochastic integrals. *Theory Probab. its Appl.*, 17(4):717–720, 1972.
280. A. Novikov. On conditions for uniform integrability of continuous non-negative martingales. *Theory Probab. its Appl.*, 24(4):820–824, 1979.
281. I. V. Girsanov. On transforming a certain class of stochastic processes by absolutely continuous substitution of measures. *Theory Probab. its Appl.*, 5(3):285–301, 1960.

Index

Printed in the United States
by Baker & Taylor Publisher Services